STUDIES IN GEOPHYSICS

The Earth's Electrical Environment

Geophysics Study Committee
Geophysics Research Forum
Commission on Physical Sciences,
   Mathematics, and Resources
National Research Council

NATIONAL ACADEMY PRESS
Washington, D.C.   1986

NATIONAL ACADEMY PRESS   2101 Constitution Avenue, N.W.   Washington, DC 20418

The Geophysics Study Committee is pleased to acknowledge the support of the National Science Foundation (Grant EAR-8216205), the Defense Advanced Research Projects Agency, the National Aeronautics and Space Administration, the National Oceanic and Atmospheric Administration, the U.S. Geological Survey (Grant 14-08-001-G1111), and the Department of Energy (Grant DE-FGO2-82ER12018) for the conduct of this study.

Library of Congress Cataloging-in-Publication Data

The Earth's electrical environment.

(Studies in geophysics)
Based on papers presented at the American Geophysical Union meetings in June 1983, Baltimore, MD.
Includes bibliographies and index.
1. Atmospheric electricity—Environmental aspects—Congresses.   2. Man—Influence of environment—Congresses.   I.   National Research Council (U.S.). Geophysics Study Committee.   II.   American Geophysical Union.   III.   Series.
QC960.5.E27   1986   551.5'6   86-8782
ISBN 0-309-03680-1

Printed in the United States of America

The National Academy Press was created by the National Academy of Sciences to publish the reports issued by the Academy and by the National Academy of Engineering, the Institute of Medicine, and the National Research Council, all operating under the charter granted to the National Academy of Sciences by the Congress of the United States.

Panel on the Earth's Electrical Environment

E. PHILIP KRIDER, University of Arizona, *Co-chairman*
RAYMOND G. ROBLE, National Center for Atmospheric Research, *Co-chairman*
R. V. ANDERSON, Naval Research Laboratory
KENNETH V. K. BEARD, University of Illinois at Urbana-Champaign
WILLIAM L. CHAMEIDES, Georgia Institute of Technology
ARTHUR A. FEW, JR., Rice University
GIOVANNI P. GREGORI, Istituto di Fisica dell'Atmosfera, Rome
WOLFGANG GRINGEL, Universität Tübingen
DAVID J. HOFMANN, University of Wyoming
WILLIAM A. HOPPEL, Naval Research Laboratory
EDWIN KESSLER, NOAA Severe Storms Laboratory
PAUL R. KREHBIEL, New Mexico Institute of Mining and Technology
LOUIS J. LANZEROTTI, AT&T Bell Laboratories
ZEV LEVIN, Tel Aviv University
HARRY T. OCHS, Illinois State Water Survey
RICHARD E. ORVILLE, State University of New York at Albany
GEORGE C. REID, NOAA Aeronomy Laboratory
ARTHUR D. RICHMOND, National Center for Atmospheric Research
JAMES M. ROSEN, University of Wyoming
W. DAVID RUST, NOAA Severe Storms Laboratory
ISRAEL TZUR, National Center for Atmospheric Research
MARTIN A. UMAN, University of Florida
JOHN C. WILLETT, Naval Research Laboratory

Staff

THOMAS M. USSELMAN

iii

Geophysics Study Committee

Geophysics
Research Forum

vi

Commission on Physical Sciences, Mathematics, and Resources

Studies in Geophysics*

---

*Published to date.

FUNDAMENTAL RESEARCH ON ESTUARIES: THE IMPORTANCE OF AN
INTERDISCIPLINARY APPROACH
Charles B. Officer and L. Eugene Cronin, *panel co-chairmen*, 1983, 79 pp.

EXPLOSIVE VOLCANISM: INCEPTION, EVOLUTION, AND HAZARDS
Francis R. Boyd, Jr., *panel chairman*, 1984, 176 pp.

GROUNDWATER CONTAMINATION
John D. Bredehoeft, *panel chairman*, 1984, 179 pp.

ACTIVE TECTONICS
Robert E. Wallace, *panel chairman*, 1986, 266 pp.

THE EARTH'S ELECTRICAL ENVIRONMENT
E. Philip Krider and Raymond G. Roble, *panel co-chairmen*, 1986, 263 pp.

Preface

This study is part of a series of *Studies in Geophysics* that have been undertaken for the Geophysics Research Forum by the Geophysics Study Committee. One purpose of each study is to provide assessments from the scientific community to aid policymakers in decisions on societal problems that involve geophysics. An important part of such assessments is an evaluation of the adequacy of current geophysical knowledge and the appropriateness of current research programs as a source of information required for those decisions.

The Earth's Electrical Environment was initiated by the Geophysics Study Committee and the Geophysics Research Forum with consultation of the liaison representatives of the agencies that support the Geophysics Study Committee, relevant committees and boards within the National Research Council, and members of the scientific community.

How does atmospheric electricity affect man and his technological systems? Is our electrical environment changing as a result of air pollution, the release of radioactive materials, the construction of high-voltage power lines, and other activities? It is clear that modern technological advances can be seriously affected by various atmospheric electrical processes and that man is also beginning to affect the electrical environment in which he resides.

The study reviews the recent advances that have been made in independent research areas, examines the interrelations between them, and projects how new knowledge could be applied for benefits to mankind. The study also indicates needs for new research and for the types of coordinated efforts that will provide significant new advances in basic understanding and in applications over the next few decades. It emphasizes a need to consider the interactions between various atmospheric, ionospheric, and telluric current systems that will be necessary to achieve an overall understanding of global electrical phenomena.

The preliminary scientific findings of the authored chapters were presented at an

American Geophysical Union symposium in Baltimore in June 1983. In completing their chapters, the authors had the benefit of discussion at this symposium as well as the comments of several scientific referees. Ultimate responsibility for the individual chapters, however, rests with their authors.

The Overview of the study summarizes the highlights of the chapters and formulates conclusions and recommendations. In preparing the Overview, the panel co-chairmen and the Geophysics Study Committee had the benefit of meetings that took place at the symposium and the comments of the panel of authors and other referees. Responsibility for the Overview rests with the Geophysics Study Committee and the co-chairmen of the panel.

Contents

III. GLOBAL AND REGIONAL ELECTRICAL PROCESSES

STUDIES IN GEOPHYSICS

Overview and Recommendations

INTRODUCTION

How does atmospheric electricity affect man and his technological systems? Is our electrical environment changing as a result of air pollution, the release of radioactive materials, the construction of high-voltage power lines, and other activities? It is clear that modern technological advances can be seriously affected by various atmospheric electrical processes and that man is also beginning to affect the electrical environment in which he resides. Our need to assess these technological and environmental impacts requires a better understanding of electrical processes in the Earth's atmosphere than we now possess. Further research is needed to understand better the natural electrical environment and its variability and to predict its future evolution.

We live in an environment that is permanently electrified. Certainly, the most spectacular display of this state occurs during intense electrical storms. Lightning strikes the Earth 50 to 100 times each second and causes the death of hundreds of people each year. Lightning is also a major cause of electric power outages, forest fires, and damage to communications and computer equipment; and new sophisticated aircraft are becoming increasingly vulnerable to possible lightning damage. Lightning contributes to the production of fixed nitrogen in the atmosphere, a gas that is essential for the growth of plants, and other trace gases. It is well known that the intense electric fields that are produced by thunderstorms can cause a person's hair to stand on end and produce corona discharges from antennas, trees, bushes, grasses, and sharp objects; these fields may also affect the development of precipitation in thunderstorms. Even in fair weather, there is an electric field of several hundred volts per meter near the ground that is maintained by worldwide thunderstorm activity.

In the Earth's upper atmosphere near 100-km altitude, a current of a million amperes flows in the high-latitude auroral zones; changes in the upper atmosphere currents, through electromagnetic induction, cause telluric currents to flow within power

1

and communication lines as well as within the Earth and oceans. The upper atmospheric current systems are highly variable and are strongly related to solar-terrestrial disturbances. Power failures and communication disruptions have occurred during intense geomagnetic storms. It also appears that the electromagnetic transients that are produced by lightning and man-made power systems can affect trapped particle populations in the magnetosphere and cause particle precipitation into the upper atmosphere at low geomagnetic latitudes.

The practical needs for understanding many of the basic questions about atmospheric electricity were brought into clear focus on November 14, 1969. Thirty-six seconds after lift-off from the NASA Kennedy Space Center, Apollo 12 was struck by lightning, and 16 seconds later it was struck again. The first discharge disconnected all the fuel cells from the spacecraft power busses, and the second caused the inertial platform in the spacecraft guidance system to tumble. Fortunately, the rocket was still under control of the Saturn V guidance system at the time of the strikes; and, as a result, the astronauts, who had never practiced for such a massive electrical disturbance, were able to reset their circuit breakers, reach Earth orbit, realign their inertial platform, and ultimately land on target on the Moon. Although permanent damage to Apollo 12 was minimal, the potential for disaster of this lightning incident called attention to the important unanswered questions regarding lightning and atmospheric electricity.

Research in atmospheric electricity traditionally has been divided into several broad areas: (1) ion physics and chemistry, (2) cloud electrification, (3) lightning, (4) fair-weather electrical processes, (5) ionospheric and magnetospheric current systems, and (6) telluric current systems. Most of this research has been pursued independently by scientists and engineers in different disciplines such as meteorology, physics, chemistry, and electrical engineering.

This study reviews the recent advances that have been made in these independent research areas, examines the interrelations between them, and projects how new knowledge could be applied for benefits to mankind. The study also indicates needs for new research and for the types of coordinated efforts that will provide significant new advances in basic understanding and in applications over the next few decades. It emphasizes a need to consider the interactions between various atmospheric, ionospheric, and telluric current systems that will be necessary to achieve an overall understanding of global electrical phenomena.

LIGHTNING

Lightning is a large electric discharge that occurs in the atmosphere of the Earth and other planets and can have a total length of tens of kilometers or more. The continental United States receives about 40 million cloud-to-ground (CG) lightning strikes each year; on average, there are probably 50 to 100 discharges each second throughout the world (Chapter 1). Most lightning is produced by thunderclouds, and well over half of all discharges remain within the clouds. Most of our knowledge about the physics of lightning has come from the study of CG discharges. Most CG flashes effectively lower negative charge to ground, however recent evidence shows that positive charge can also be lowered (see Chapter 3 on positive lightning). Cloud-to-ground lightning kills about a hundred people and causes hundreds of millions of dollars in property damage each year in the United States; it is clearly among the nation's most severe weather hazards.

Most CG discharges begin within the cloud where there are large concentrations of positive and negative space charge (see Chapter 8). After several tens of milliseconds, the preliminary cloud breakdown initiates an intermittent, highly branched discharge that propagates horizontally and downward and that is called the stepped-leader (Figure 1). When the tip of any branch of the stepped-leader gets close to the ground, the large electric field that is produced near the surface causes one or more upward propa-

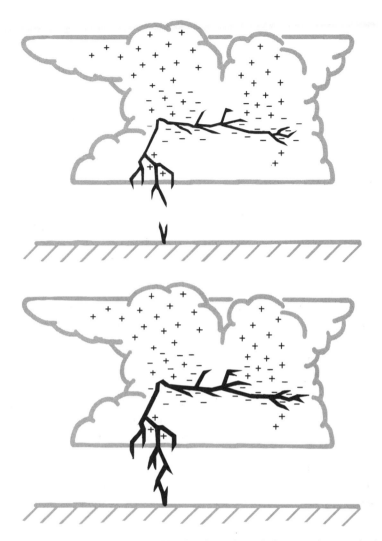

FIGURE 1   The top figure shows the stepped-leader channel just before attachment; the bottom figure shows attachment and the development of the return stroke. The estimated time between the two figures is on the order of 0.001 sec.

gating discharges to form. When an upward discharge makes contact with the stepped-leader, the first return stroke begins. The return stroke is an intense wave of ionization that starts at or just above the ground and that propagates up the leader channel at about one third the speed of light. The return stroke is typically the brightest phase of lightning. The peak currents in these return strokes can reach several hundred thousand amperes; a typical value is about 40,000 A. The peak electric power that is dissipated by a return stroke is on the order of 100 million watts per meter of channel; and the peak channel temperatures approach 30,000 K. A shock wave is produced by the rapid expansion of the hot, high-pressure channel, and this eventually becomes thunder with its own characteristics that depend on the nature of the discharge and the atmospheric environment (see Chapter 4).

   The currents in the return stroke carry the ground potential upward and effectively neutralize most of the leader channel. After a pause of 40 to 80 milliseconds, most CG flashes produce a new leader, the dart leader, that propagates down the previous return-stroke channel and initiates a subsequent return stroke. Most flashes contain two to four return strokes, with each affecting a different volume of cloud charge (see Chapter 8). If a dart leader forges a different path to ground than the previous stroke, then the lightning will actually strike the ground in more than one place and will have a forked appearance. Chapters 2 and 3 discuss the physics of lightning in greater detail.

Among the more important advances that have been made in recent years has been the discovery that both in-cloud and CG discharges produce very fast-rising currents, i.e., rise times of tens to hundreds of nanoseconds and rates of change of current (dI/dt) on the order of 10^{11} A/sec. Most of the standard waveforms that are used to test the performance of lightning protectors and the integrity of lightning protection systems currently have current rise times and dI/dts that are substantially slower than the above values; therefore, these standards may not be an adequate simulation of the true lightning threat to aircraft and other structures (see Chapter 5).

A variety of nonequilibrium trace gases are produced within high-temperature lightning channels and by the shock wave that can affect tropospheric and stratospheric chemistry (see Chapter 6).

Recent spacecraft observations have shown that lightning may be present in the atmospheres of Jupiter, Venus, and Saturn; the upcoming Galileo probe will carry a lightning detector to Jupiter. In the future a study of lightning in atmospheres that are radically different from the Earth's may lead to a better understanding of the formation and characteristics of lightning on Earth.

CLOUD ELECTRIFICATION

Although the vast majority of terrestrial clouds form and dissipate without ever producing precipitation or lightning, they can be weakly electrified. In some clouds, the electrification intensifies as convective activity increases, and strong electrification usually begins when there is rapid vertical and horizontal growth of the cloud and the development of precipitation. Most lightning on Earth is produced by cumulonimbus clouds that are strongly convective (i.e., they contain a vigorous system of updrafts and downdrafts) and that contain both supercooled water and ice. A small fraction of warm clouds are also reported to produce lightning.

The updrafts and downdrafts and the interactions between cloud and precipitation particles act in some still undetermined manner to separate positive and negative charges within the cloud. These processes usually transfer an excess of positive charge to the upper portion of the cloud and leave the lower portion with a net negative charge. Recent research has shown that the negative charge is usually concentrated at altitudes where the atmospheric temperature is between $-10°C$ and $-20°C$ (i.e., 6 to 8 km above sea level in summer thunderstorms and 1 to 3 km in winter storms) and that this altitude remains constant as the storm develops. This finding sets important criteria that must be met by any proposed thunderstorm charging mechanisms. The positive charge that is above the negative may be spread through deeper layers and does not exhibit as clear a relationship with temperature as does the negative charge. Positive charges are found at levels between $-25°C$ and $-60°C$ depending on the size of the storm, and this temperature range usually lies between 8 and 16 km above sea level.

Cloud electrification processes can be viewed as acting over two spatial scales: a microscale separation that ultimately leads to charged ice and water particles and then a larger-scale separation that produces large volumes of net positive and negative charge and eventually lightning. The microscale separation includes the creation of ion pairs, ion attachment, and charge that may be separated by collisions between individual cloud and precipitation particles. The larger cloud-scale separation may be due to precipitation or large-scale convection or some combination of the two.

Numerous mechanisms have been proposed for the electrification of clouds and thunderstorms, and several of these might be acting simultaneously. Feedback can occur through changes in the ion concentrations and electric field, and thus it is difficult to identify or evaluate the primary causes of electrification in a cloudy environment. Currently, there is a great need for more measurements to determine the locations, magnitudes, and movements of space charges within and near the cloud boundary. There is also a need to determine the charge-size relationship that is present on both cloud and precipitation particles, how these charges evolve as a function of time, and how these distributions are affected by lightning. Laboratory experiments have provided valuable information about the physics of selected microscale processes and are expected to continue to provide important data on the relative magnitude of various processes. Theory and numerical models also have played an important role in simulating and evaluating possible charging mechanisms on both the microscale and the cloud scale.

During the early nonprecipitating cloud stage, charging can occur by diffusion, drift, and selective capture of ions. Later, during the rain stage, there can be additional electrification due to drop breakup and other mechanisms based on electrostatic induction. Drift, selective ion capture, breakup, and induction are probably responsible for the charges and fields that are found in stratiform clouds; however, it is difficult to explain with just these mechanisms the stronger electrification that is found in convective clouds more than a few kilometers deep. For clouds in the hail stage, thermoelectric and interface charging mechanisms can provide strong electrification on the microscale.

In thunderclouds, the charges that are generated on a microscale can be subsequently separated on a larger cloud scale by convection and/or gravitational settling. Particles near the boundary of the cloud will become electrified by ion attachment, and the convection of these charges may play an important role in the electrification. Convection also plays a role in the formation and growth of cloud particles by forcing the condensation of water vapor until the particles are large enough to coalesce. Interactions between cloud particles, particularly when there are rebounding collisions, may also produce charge separation. If the larger particles tend to carry charge of predominantly one sign, they will fall faster and farther with respect to the convected air and leave the oppositely charged, smaller particles at higher altitudes.

As the populations of charged particles increase, the mechanisms that discharge these particles become more effective. Two kinds of discharging are possible: (1) discharge by ionic conduction, point discharge, or lightning and (2) discharge by collision and/or coalescence with cloud particles of opposite polarity. The attachment of ions to cloud particles will be a function of the particle charge and the electric field of the cloud, and strong fields may also produce corona discharges from large water drops and the corners of ice crystals. Corona ions and lightning will increase the local electrical conductivity, and this, in turn, may prevent or reduce any further buildup of space charge in this region of the cloud.

Collisional discharge will take place at all stages of cloud particle growth. These mechanisms are enhanced if the interacting particles are highly charged and of opposite polarity; therefore, if a charging mechanism is to be effective, it must separate charge at a rate that is sufficiently high to overcome the discharging processes. It is worth noting that the electric forces on charged elements of precipitation can be several times larger than gravity; therefore, the terminal velocities and frequency of collisions of these particles will be a function of the electric field. More detailed discussions of the various processes involved in cloud electrification are given in Chapters 8, 9, and 10.

Recently, there have been attempts to analyze the patterns of the Maxwell current density that thunderclouds produce at the ground in order to define better the characteristics of the cloud as an electrical generator. The Maxwell, or total current density, contains components due to ohmic and non-ohmic (corona) ion conduction, convection, precipitation, displacement, and the charge-separating currents within the cloud. Under some conditions, there is evidence that the total current density may be coupled directly to the meteorological structure of the storm and/or the storm dynamics, but the lack of simultaneous Maxwell current measurements both on the ground and aloft has not allowed the details of this relationship to be determined. Such measurements will also be complicated by the complexity of the meteorological and electrical environment outside the storm.

ELECTRICAL STRUCTURE OF THE ATMOSPHERE

It has been known for over two centuries that the solid and liquid Earth and its atmosphere are almost permanently electrified. The surface has a net negative charge, and there is an equal and opposite positive charge distributed throughout the atmosphere above the surface. The fair-weather electric field is typically 100 to 300 V/m at the surface; there are diurnal, seasonal, and other time variations in this field that are caused by many factors. The atmosphere has a finite conductivity that increases with altitude; this conductivity is maintained primarily by galactic cosmic-ray ionization. Near the Earth's surface, the conductivity is large enough to dissipate any field in just 5 to 40 minutes (depending on the amount of pollution); therefore, the local electric field must be maintained by some almost continuous current source.

Ever since the 1920s, thunderstorms have been identified to be the dominant generator in the global circuit. Most cloud-to-ground lightning transfers negative charge to the ground, and the point discharge currents under a storm transfer positive space charge to the atmosphere. In addition, there are precipitation and other forms of convection currents and both linear and nonlinear conduction currents that must be considered when attempting to understand the charge transfer to the earth by a thunderstorm. The electrical structure of a thunderstorm is complex (see Chapter 8), but it is often approximated simply as a vertical electric dipole.

The conductivity of the fair-weather atmosphere near the surface is on the order of 10^{-14} mho/m, and it increases nearly exponentially with altitude to 60 km with a scale

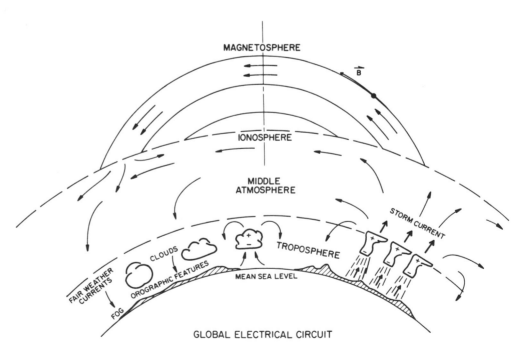

FIGURE 2   Schematic of various electrical processes in the global electrical circuit.

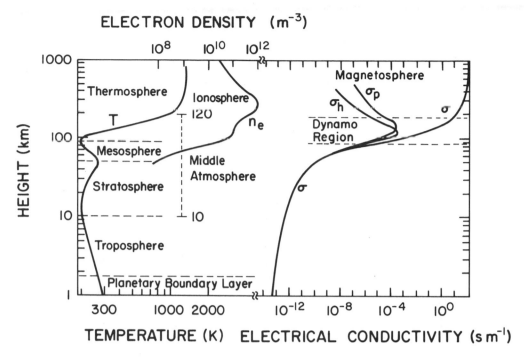

FIGURE 3 Nomenclature of atmospheric regions based on profiles of electrical conductivity (σ), neutral temperature, and electron number density.

height of about 6 km. The main charge carriers below about 60 km are small positive and negative ions that are produced primarily by galactic cosmic rays. Above 60 km, free electrons become more important as charge carriers and their high mobility produce an abrupt increase in conductivity throughout the mesosphere. Above 80 km, the conductivity becomes anisotropic because of the influence of the geomagnetic field, and there are diurnal variations due to solar photoionization processes.

The atmospheric region above about 60 km is known as the equalization layer and is usually assumed to be an isopotential surface and the upper conducting boundary of the global circuit. Currents flow upward from the tops of thunderstorms to this layer where they are rapidly distributed throughout the world. Worldwide thunderstorms maintain a potential difference of 200 to 600 kV between the equalization layer and the surface—the Earth-ionosphere potential. This potential difference, in turn, drives a downward conduction current that is on the order of 2×10^{-12} A/m^2 in fair-weather regions and constant with altitude.

Today, there are still many details that need to be clarified about the role of thunderstorms as the generators in the global circuit (Figure 2). Upward currents have been detected above thunderclouds, but how these currents depend on storm dynamics, stage of development, lightning frequency, precipitation intensity, and cloud height, for example, is still not known. There is a need for further measurements to quantify the relationships between diurnal variations of the ionospheric potential, the electric field or air-earth current, and worldwide estimates of thunderstorm frequency.

Many electrical processes interact within the global circuit, and the following subsections will describe selected processes that occur within certain atmospheric regions (Figure 3). It should be recognized that the global circuit includes mutual electrical interactions between all atmospheric regions.

This report will not present an encyclopedic review of all the electrical phenomena that occur in the atmosphere but will simply give some examples that illustrate a few of the basic processes and some of the important interrelationships. Notably absent, but important to an overall understanding of atmospheric electricity, are discussions dealing with electromagnetic phenomena such as sferics, Schumann resonances, and whistlers. These and many other subjects, however, have been reviewed in the two-volume *Handbook of Atmospherics*, edited by Volland (1982), and recent proceedings of international conferences on atmospheric electricity (Dolezalek and Reiter, 1977; Orville, 1985). Human and biological effects of atmospheric electricity are also important research areas that are not considered in this study.

The Planetary Boundary Layer

The planetary boundary layer (PBL) is the lowest few kilometers of the atmosphere where interactions with the surface, man, and the biosphere are the most pronounced. Galactic cosmic rays are the main source of ionization in the PBL; however, near land surfaces, ionization is also produced by decays of natural radioactive gases emanating from the soil surface and by radiations emitted directly from the surface. Ionization from radioactive sources depends on soil type and surface structure and on the meteorological dispersal rate; this ionization normally decreases rapidly with altitude, and at about 1 km its contribution to the total ionization is less than that from cosmic rays. Other sources of ions in the PBL include lightning; electrification due to waterfalls, ocean surf, and man-made sprays; a variety of combustion processes; point discharge or corona currents that are produced whenever the ambient electric field exceeds breakdown; and frictional processes associated with blowing dust, snow, or volcanic ejecta.

In the troposphere, atmospheric trace gases are numerous and variable, and the ion chemistry is complicated by clustering processes and the relatively long lifetime of the terminal ions. As a further complication, clouds and other aerosols play an important role as sinks for small ions and thereby alter the ion distribution (see Chapter 11). Over continental areas, the loss of ions by attachment to aerosols can be larger than the loss by recombination. Some atmospheric aerosols are hygroscopic, and the particle size increases with relative humidity. At large humidities, fog and cloud droplets form and produce a large decrease in the electrical conductivity of the atmosphere. Since a decrease in conductivity can be a precursor of fog, it might be possible to improve forecasts of the onset of fog by electrical measurements.

Turbulent transport and convection within the PBL are important processes that govern the momentum, heat, and moisture exchanges between the atmosphere, geosphere, and biosphere. These processes influence the mean wind profile, the vertical distribution of temperature, water vapor, trace gases, aerosols, and the ion distribution throughout the troposphere. Turbulent mixing and convection can prevent the buildup of radioactive emanations near the ground and can also disperse aerosols to a greater altitude in the troposphere.

Electrical processes in the PBL are complex, highly variable, and span a tremendous range of space and time scales. The electrical variables respond to many of the lower atmospheric processes but usually have little influence on the phenomena to which they respond. Within the PBL, local turbulent fluctuations of space-charge density impose a time-varying electric field that is comparable in magnitude with or even greater than the electric field maintained by global thunderstorm activity. Since the PBL is the region of the atmosphere with the greatest resistance, it is this layer as well as the generators that control the currents in the global circuit. Electrical processes in the PBL are discussed in greater detail in Chapter 11.

Mid-Troposphere-Stratosphere

The main source of ionization in the mid-troposphere-stratosphere ($\sim$ 2-50-km) region is cosmic radiation; the ionization rate depends on magnetic latitude and on solar activity. At about 50° geomagnetic latitude and at 20-km altitude, the ion-production rate during sunspot maximum is about 30 percent smaller than during sunspot minimum; at 30-km altitude (same latitude) it is about 50 percent smaller. Following solar flares that produce energetic charged particles (solar proton events), the ion-production rate in the stratosphere may increase by orders of magnitude for periods of hours to days, but deeper in the atmosphere the effect is much smaller. Solar flares are usually followed by a reduction (Forbush decreases) in the ion-production rate for periods of hours to weeks that is caused by a temporary reduction of the incoming cosmic-ray flux.

The composition and chemistry of the ions that establish the bulk electrical properties in the mid-troposphere-stratosphere are relatively unknown. The ion concentrations are also affected by aerosols whose distributions are quite variable in both space and time. Aerosols tend to accumulate at temperature inversion boundaries and can cause a general loss of visibility that can be seen by airline passengers as they pass through such layers. Such a buildup of aerosols causes a general decrease in the small-ion concentration, and, thus, the electrical conductivity is also reduced—resulting in an increase in the local electric field.

The concentration of particles with a radius greater than about 0.1 micrometer decreases with altitude above the PBL, and a relative minimum occurs in the upper troposphere. The particle concentrations increase within the lower stratosphere, peak near 20 km, and then decrease again with altitude. This persistent structure is frequently referred to as the 20-km sulfate layer; the character of this layer is controlled largely by gases emitted during volcanic eruptions, as discussed in Chapter 12.

Aerosol particles that have radii on the order of 0.01 micrometer are referred to as condensation nuclei (CN) and are uniformly mixed throughout the troposphere above the PBL. Near the surface, the CN concentration may be large owing to local sources; above the tropopause the concentration decreases with altitude. In recent years, a CN layer has frequently been observed near 30 km. As a result of the El Chichon volcanic eruption in 1982, the normal CN concentration at 30-km altitude increased by at least two orders of magnitude and measurably affected the ion concentration and electrical conductivity.

Under steady-state conditions, the air-earth current density is constant with altitude if there is large-scale horizontal homogeneity and if no thunderstorms or other localized electrical disturbances are in the vicinity. The air-earth current varies with magnetic latitude because of the magnetic variations in cosmic-ray fluxes. The current is generally enhanced over orographic features such as mountain ranges because of the decreased columnar resistance (mountains are closer to the ionosphere than the near-sea-level surface). Estimates have been made that indicate that as much as 20 to 30 percent of the total global current flows into the high mountain peaks.

Mesosphere

In the mesosphere (50-85 km altitude), the major daytime source of ionization is solar Lyman-alpha photoionization of nitric oxide (NO). The major source of NO for this region is the thermosphere, where NO is produced by extreme ultraviolet (EUV) radiation (wavelengths less than 100 nm) and auroral particle precipitation. Meteorological processes in the upper atmosphere transport NO from the thermosphere to the mesosphere, where its distribution is variable. Somewhat smaller sources of ionization in the upper mesosphere include solar x-ray ionization and the photoionization of oxygen in a metastable state. At high latitudes, energetic electrons, protons, and bremsstrahlung ultraviolet radiation associated with auroral particle precipitation are variable sources related to geomagnetic activity.

Solar protons are a sporadic and intense source of ionization at high latitude following intense solar flares. These solar proton events can increase the electrical conductivity of the magnetic polar-cap mesosphere by several orders of magnitude (at altitudes down to about 50 km) during such events. In addition, the current carried by the bombarding solar protons can often exceed the local air-earth conduction current flowing in the circuit.

The principal primary positive ions produced in the mesosphere are N_2^+, O_2^+, and NO^+, but all participate in a wide range of reactions that lead to a rich spectrum of ambient positive ions. An equally rich range of negative ions is generated by reactions initiated by the attachment of electrons to form the main primary species, O_2^- and O^-. Rocketborne mass-spectrometer measurements have shown that below the mesopause the positive ions are proton hydrates with as many as 20 water molecules clustering to individual ions. The positive-ion chemistry of the mesosphere is better understood than is the negative-ion chemistry (see Chapter 13).

The interaction of the terminal ions with aerosol particles is probably a significant sink for ions in the polar aerosol layers near the summer mesopause where noctilucent clouds are commonly observed.

The electrical conductivity of the mesosphere is important because it governs the electrical properties of the equalization layer in the global circuit. Below about 60 km, the terminal small ions are the main charge carriers; but above 60 km, free electrons can exist and their high mobility is responsible for the abrupt increase in electrical conductivity observed in the mesosphere. Furthermore, above 70 km, collisions between electrons and air molecules become infrequent enough so that electrons are confined to spiral about a magnetic field line and the motion perpendicular to the field becomes more difficult than motion along the field. The electrical conductivity becomes anisotropic, and this anisotropy has a dominant influence on the electrical properties of the global circuit above 70 km.

Rocketborne measurements of the upper atmosphere conductivity and electric field indicate some puzzling features. There appear to be regions in the upper stratosphere and mesosphere that have abrupt increases and decreases in vertical conductivity profiles. The decreases are probably associated with aerosol layers, but the increases are difficult to interpret. On occasion, the electric field near 50- to 70-km altitude has been observed to increase enormously from what is expected if the mesosphere is a passive element in the global circuit. The mesosphere may not be electrically passive but may, in fact, contain active electrical generators that are not currently known.

Ionosphere and Magnetosphere

The major sources of ionization above about 85 km are extreme-ultraviolet (EUV) radiation and auroral particle precipitation (see Chapter 14). The ionizing portion of the solar spectrum (i.e., wavelengths below 102.7 nm) is absorbed in the thermosphere and creates an ionosphere that consists of positive molecular and atomic ions (e.g., N_2^+, NO^+, O_2^+, O^+) and negative electrons. The solar EUV radiation and the electron and ion densities throughout the ionosphere are highly dependent on solar activity; there are known variations with the 11-yr sunspot cycle, the 27-day rotation of the Sun, and solar flares.

Auroral particle precipitation is responsible for large variations in ion and electron densities at high latitudes. The bulk of the precipitation occurs within the auroral oval that encircles the geomagnetic pole in magnetic conjugate polar caps. Observations over many years show that there is always auroral activity within the oval. The activity varies considerably over the day and even from hour to hour owing to interactions of the solar-wind plasma with the Earth's magnetic field. The total power dissipated by particles bombarding the upper atmosphere is typically 10^9 W, but during large geomagnetic storms it can approach 10^{12} W.

The sources and composition of the ions that maintain the bulk electrical properties of the upper atmosphere are generally known on the dayside of the Earth, but at night there are still uncertainties with regard to the ionization sources.

In the classical view of the global circuit (see Chapter 15), the ionosphere is assumed to be at a uniform potential with respect to the surface; however, the known upper-atmosphere generators are not included. The two major generators that operate in the ionosphere above about 100 km are the ionospheric wind dynamo and the solar-wind/magnetosphere dynamo (see Chapter 14). Atmospheric winds have the effect of moving the weakly ionized ionospheric plasma through the geomagnetic field. This movement produces an electromotive force and generates electric currents and fields. This process is complicated by the variability of the ionospheric winds and the anisotropic electrical conductivity in the ionosphere.

The magnitude of the horizontal electric field associated with the wind-driven dynamo is on the order of 1 mV/m. A total current of about 100,000 A flows horizontally in the ionosphere because of the combined action of the wind and electric field, mainly on the sunlit side of the Earth. This current flows in two counterrotating vortices on opposite sides of the equator, and these patterns dominate at low latitudes and mid-latitudes. Global-scale horizontal potential differences of about 5 to 10 kV are generated by the ionospheric wind dynamo.

The ionospheric winds that drive the dynamo are mainly caused by upward propagating tides from the lower atmosphere that have large day-to-day fluctuations. During geomagnetic storms, however, thermospheric winds increase in response to high-latitude auroral heating and cause disturbances at low latitudes to the fields and currents of the ionospheric wind dynamo.

The solar-wind/magnetosphere dynamo results from the flow of the solar wind around and perhaps partly into and within the Earth's magnetosphere. The motion of this plasma through the geomagnetic field produces an electromotive force and currents at high latitudes that result in an antisunward flow of plasma over the magnetic polar cap and a sunward flow of ions in the vicinity of the dawn and dusk auroral zones. This motion is described by a two-cell counterrotating ion circulation with one cell on the dawn side and the other on the dusk side of the magnetic polar caps. The polar-cap electric field is typically 20 mV/m, with an ionospheric convection velocity of 300 m/sec. Larger fields of about 50 to 100 mV/m occur in the vicinity of the auroral ovals. The large-scale potential difference that is associated with this horizontal ion flow over the polar caps has a total dawn-to-dusk drop of about 50 kV. This potential drop and the configuration of the two-cell pattern are highly variable. The potential drop has values of 20 to 30 kV during geomagnetic quiet conditions that increase to 100 to 200 kV during geomagnetic storms. These fields are mainly confined to the polar caps because of the shielding from currents within the magnetosphere. During geomagnetic storms, however, the shielding currents can be altered and electric fields have been observed to propagate all the way from the polar caps to the equator.

Currents are an integral part of the complex electrical circuit associated with the solar-wind/magnetosphere dynamo. Currents flowing along the direction of the magnetic field couple the auroral oval and high-latitude ionosphere with outer portions of the magnetosphere. Typically about a million amperes of current flow in the solar-wind/magnetosphere dynamo. The dynamo currents and fields with this high-latitude system are extremely complex and highly variable (see Chapter 14).

The large-scale horizontal fields (scale sizes 100 to 1000 km) within the ionosphere can propagate or map downward in the direction of decreasing electrical conductivity. Horizontal fields of a smaller scale (1 to 10 km) on the other hand are rapidly attenuated. The larger-scale horizontal electric fields that do map to the surface become vertical at the surface because of the high surface conductivity. The solar-wind/magnetosphere field can alter the surface fields at high latitudes by 20 to 50 V/m depending on the level of geomagnetic activity and the magnitude of the dawn-to-dusk potential drop across the magnetic polar cap.

Telluric Currents

Telluric currents consist of both natural and man-induced electric currents flowing in the solid earth and oceans. The fundamental causes of the natural currents are electromagnetic induction resulting from a time-varying, external geomagnetic field or the motion of a conducting body (such as seawater) across the Earth's internal magnetic field. These telluric currents, in turn, produce magnetic fields of their own that add to the external geomagnetic field and that produce a feedback on the ionospheric current system. The complexities associated with telluric currents arise from the complexities in the external current sources and the conductivity structure of the Earth (see Chapter 16).

The external inducing field also has various scale sizes that contribute to the complexities in the telluric current systems. The ionospheric dynamo currents that are associated with the solar diurnal and lunar tides have a planetary-scale size. The ionospheric current variations, however, also have smaller-scale features that are associated with auroral and equatorial electrojets. At low frequencies, the external inducing sources can be approximated by a planetary-scale field that is occasionally altered by strong spatial gradients during geomagnetically disturbed conditions. At higher frequencies (magnetic storms, substorms, or geomagnetic pulsations), the source can often be quite localized and highly time dependent.

Electromagnetic induction caused by ionospheric and magnetospheric current variations has a pronounced effect on telluric currents and on man-made systems. These effects have been detected by a number of investigators, and it is now well recognized that there is a direct electromagnetic coupling from the ionosphere to the telluric currents. The large variation in conductivity of the solid earth can give rise to various channeling effects within the Earth, thereby considerably complicating the flow patterns of the telluric currents. The current patterns are different for different frequencies of external induction. The longer the period of the time-varying field, the deeper into the Earth the induced currents are expected to flow. For example, a signal with a period of about 24 hours is generally believed to have a skin depth of 600 to 800 km. The distribution of sediments, the degree of hydration, differences in porosity, and other properties of the Earth all have an influence on the signal response. Properly interpreted, telluric currents can be a tool to study both shallow and deep structures within the Earth.

TECHNIQUES FOR EVALUATING THE ELECTRICAL PROCESSES AND STRUCTURE

The cornerstone of our understanding of the Earth's electrical environment is an integration of measurements, theory, and modeling. The new instruments and techniques that have been developed in recent years are diverse, and various chapters in this volume contain details on techniques beyond those illustrated in this Overview.

Remote measurements of electric and magnetic fields can now be used to infer many properties of lightning and lightning currents. Also, the amplitude and time characteristics of thunder and various radio-frequency (rf) noise emissions can be used to trace the geometrical development of lightning channels within clouds as a function of both space and time. There are now large networks of ground-based lightning detectors that can discriminate between in-cloud and cloud-to-ground lightning and accurately determine the locations of ground-strike points. With such a detection capability, it should now be possible to determine whether and how the characteristics of individual cloud-to-ground discharges depend on their geographic location, the local terrain, and/or the meteorological structure of the storm.

The rf noise that is generated by lightning in the hf and vhf bands appears in the form of discrete bursts, and within a burst there are hundreds to thousands of separate pulses. If the difference in the time of arrival of each pulse is carefully measured at widely separated stations, the location of the source of each pulse can be computed, and the geometrical development of the rf bursts can be mapped as a function of time.

Satellite observations of lightning have provided rough estimates of the global flashing rates and the geographic distribution of lightning as a function of season. Optical detectors, such as those now in orbit on the Defense Meteorological Satellite Program (DMSP) satellites, are limited in their temporal and spatial coverage, but they have provided data that show a progression of lightning activity toward the summer hemisphere and notable absences of lightning over the ocean during the observing intervals (see Chapter 1). The data to date are only for local midnight, dawn, and dusk; there is a need to obtain data at other times. Measurements of hf radio noise by the Ionosphere Sounding Satellite-B have also been used to estimate a global lightning flash rate.

Global detection of lightning is necessary to determine the global flashing rate and how this rate relates to other parameters in the global circuit. In recent years, the National Aeronautics and Space Administration has developed new optical sensors that could be used to detect and locate lightning in the daytime or at night and with continuous coverage by using satellites in geosynchronous orbits. These sensors are capable of measuring the spatial and temporal distribution of lightning over extended periods with good spatial resolution and offer significant new opportunities for research—without the inherent sampling biases of low-altitude orbiting satellites—and for many applications.

Artificial triggering of lightning now provides the capability of studying both the physics of the discharge process and the interactions of lightning with structures and other objects in a partially controlled environment (see Chapter 2). When a thunderstorm is overhead and the surface electric field is large, a small rocket is launched to carry a grounded wire rapidly upward. When lightning is triggered by the wire, the first stroke is not like natural lightning, but subsequent return strokes appear to be almost identical to their natural counterparts.

Triggered lightning is now being used to investigate the luminous development of lightning channels, the characteristics of lightning currents, the velocities of return strokes, the relationships between currents and electromagnetic fields, the mechanisms of lightning damage, the performance of lightning protection systems, and many other problems. The main benefit of this triggering technique is that it can be used to cause lightning to strike a known place at a known time, thus enabling controlled experiments to be performed. Although lightning cannot be reproduced in full in the laboratory, several lightning simulators have been developed and have provided some quantitative information on the generation of thunder.

Cloud electrification and charge-separation processes are closely coupled to the cloud microphysics and the storm dynamics. The natural storm environment is extremely complicated, and its quantification involves a host of electrical and meteorological parameters. Many of these parameters and their measurements are treated in Chapters 7 and 8 and the three-volume publication, *Thunderstorms: A Social, Scientific, and Technological Documentary*, edited by Kessler (1982). One of the greatest needs is for an in-cloud instrument that can measure in a thunderstorm environment the charge on the smaller cloud particles as a function of particle size and type (see Chapter 8).

In recent years, cooperative field programs, such as the Thunderstorm Research International Program (TRIP), have improved our knowledge about the overall electrical structure of thunderstorms. These programs have also provided a framework wherein a number of different investigators using different techniques can study the same thunderstorms at the same location at the same time. For instance, in situ and remote electrical measurements have been made in New Mexico in conjunction with Doppler radar studies of the cloud precipitation and dynamics and in-cloud sampling of the larger cloud particles. Laboratory experiments also continue to provide information on the effects of electricity on cloud microphysics and charge separation mechanisms that are critical to the interpretation of data collected by such field programs. Finally, numerical models of the electrical development and structure of thunderstorms have provided an important framework in which to interpret the cloud measurements and laboratory experiments.

The techniques that are used to determine the electrical structure of the fair-weather atmosphere are diverse. Usually, vertical profiles of one or more atmospheric-electrical variables—typically electric field, conductivity, and current density—are measured over a relatively short time span. Sensors are carried aloft on aircraft, balloons, or rockets, and the data are presented both as profiles and as numerically integrated results. The vertical profiles represent almost instantaneously measured parameters rather than time averages. Aircraft or constant-pressure balloons, however, do have the capability of measuring temporal variations at a given level. Profiles have been measured over land because of the convenience and to study specific terrain effects and over water in attempts to eliminate distortions caused by land. Profiles have led to the detection of convection currents in the planetary boundary layer that are comparable in magnitude with the total current, the electrode effect over water under stable conditions, the response of columnar resistance to pollution, and the diurnal variation in ionospheric potential.

In addition to the standard electrical parameters, determinations of ion and aerosol contents and compositions, mobility, and chemistry are all critical to an understanding of conductivity. Again, these quantities are usually displayed as vertical profiles and, to a certain extent, are characteristic of the type of platform on which the instrumentation is carried aloft (e.g., some airplane measurements provide horizontal profiles but are limited in their vertical extent). Vertical measurements are probably a good first approximation to the global electrical structure; however, horizontal variations should also be measured as a function of time to complement the profile data. Tethered balloons can provide the time variations of selected electrical properties at a few locations, and this has been attempted on an experimental basis. For understanding the global circuit, it would be valuable to have a number of vertical profiles taken at the same time and, in addition, to repeat certain profiles to obtain the time variations.

Knowledge of the upper-atmosphere current systems is important for understanding the interactions among the ionosphere, magnetosphere, and the solar wind. Some of these current systems were studied during the International Magnetospheric Study, 1976-1980, and during the NASA Dynamics Explorer satellite program. The goals of these programs have been to investigate the coupling of the solar-wind energy through the magnetosphere and into the ionosphere; but little effort was made to couple these current systems into the global electrical circuit. Additional measurements of magnetosphere-ionosphere currents are planned for the International Solar-Terrestrial Physics program.

SOCIETAL IMPACT

Severe weather phenomena that disrupt our lives include tornadoes, hail, high winds, hurricanes, floods, snowstorms, and lightning. Among them, lightning ranks as the number one killer, followed closely by tornadoes. Lightning is much less dramatic than a tornado passing through an area or a severe snowstorm that paralyzes a city, but lightning can strike quickly and kill with little or no warning.

Lightning is a leading cause of outages in electrical power systems and was the initial cause of the massive power blackout in New York City on July 13, 1977. The possible effects of lightning on advanced aircraft, nuclear power stations, and sophisticated military systems are problems of increasing concern.

The detailed physics of how lightning strikes a structure, a power line, or an aircraft and its effects are still not known. The approaching leader is not influenced by the object that is about to be struck until it is perhaps a few tens of meters away. At that time, an upward-moving streamer leaves the object and similar discharges may also leave other objects nearby. When the upward-moving streamer attaches to the downward-moving leader, the return stroke begins. When the details of this attachment are better understood we should be able to predict with higher probability what will and what will not be struck under various conditions and thereby provide better lightning protection. For example, the positioning of overhead ground wires above power transmission lines and the protection of complex structures could be optimized (see Chapter 5).

The current rise time is an important parameter for lightning protection because if the current interacts with an inductive load, the voltage on that load is proportional to the rate of increase of the current. Most of the standard surge waveforms that are used to verify the performance of protectors on power and telecommunications circuits specify that open-circuit voltage should have a rise time of 0.5, 1.2, or 10 microseconds and that the short-circuit current should have a rise time of 8 or 10 microseconds. These values are substantially slower than recently measured lightning current rise times, which are in the range of tens to hundreds of nanoseconds; therefore, it is probable that the degree of protection that is provided by devices tested to present-day standards will not be adequate for protection against direct lightning surges.

The unusually destructive nature of lightning that lowers positive charge to ground is only partially documented and is poorly understood (see Chapter 3). Because of the large and long continuing currents, positive lightning may ignite a disproportionately large number of fires, especially in grasslands and forests. The apparent pattern is for positive lightning to strike preferentially outside areas of rainfall, and this further enhances the likelihood of its starting a fire. Positive lightning may be correlated with storm severity and tornado occurrence, and its detection could enhance our present severe-storm detection and warning systems.

Newly developed lightning-detection equipment now makes it possible to make real-time decisions on the preparations for repairs of utility systems, early warning and detection of lightning-caused forest fires, and a variety of other warning functions in situations that allow protective action to be taken, such as launches at the NASA Kennedy Space Center and outdoor recreational activities. Among the main users of lightning location data at present are the Bureau of Land Management (BLM) in the western United States and Alaska and the Electric Power Research Institute in the eastern United States. The BLM and the Forest Services of most Canadian provinces utilize the time and location of lightning storms to determine when and where to look for forest fires. Early detection of these fires provides considerable savings both in the natural resources and in the cost of fighting the fires. In the eastern United States the lightning data are being used to accumulate statistics on lightning occurrence and for real-time applications by electric power utilities. For warnings of lightning-intensive storms, these data are also disseminated in real time to many National Weather Service offices and to a growing number of television stations.

Although cloud electrification processes are ultimately responsible for producing lightning, these processes can also electrify an aircraft flying in a cloud. It is quite common for the potential of an aircraft to be raised by several million volts, and most planes have discharger wicks to control the interference in radio communications when the aircraft goes into point discharge. Lightning will not usually present a hazard to commercial aircraft as long as the present design practices are continued and the standard practice of avoiding large thunderclouds is maintained. However, since many new aircraft are being developed with composite materials instead of aluminum and with the increased use of computers and microcircuit technology, the decreased electrical shielding on the outside and increased sensitivity inside means that there will be an increasing vulnerability to lightning disturbances (see Chapter 5). Thus, cloud electricity and lightning will have to be considered carefully in the design and operation of future aircraft systems.

Natural telluric currents can significantly disturb man-made systems such as communication cables, power lines, pipelines, railways, and buried metal structures. The largest natural disturbances are associated with the intense auroral current systems that flow at high latitudes during geomagnetic storms. There have been frequent reports of these disturbances, inducing currents on long telephone and telegraph wires that are large enough to generate sparks and even permanent arcs. When this occurs, there can be outages and shutdowns in both land and sea cables and fires can be started by overheating the electrical systems. Currents of up to about 100 A are sometimes induced in power transformers at northern latitudes and cause power blackouts and system failures. During the large geomagnetic storm of February 11, 1958, the Toronto area suffered from an induced power blackout.

Long pipelines are also affected by telluric current disturbances. The Alaskan pipeline has been the subject of careful investigation, principally because of its location across the auroral zone. One of the concerns has been the rate of corrosion of the pipeline, which is enhanced by telluric currents. However, telluric currents appear to affect electronic equipment related to operational monitoring and corrosion control rather than to produce specific serious corrosion problems. A relationship between the expected current flow and geomagnetic activity has been derived and suggests that the pipeline is a large man-made conductor that is capable of significantly affecting the local natural regime of telluric currents.

There is also a concern that the long, power-transmission lines planned for future arctic development will be subject to larger induced currents by auroral activity than was previously considered. This would require new protection equipment development for high-latitude applications.

Telluric currents have also been used in the search for natural resources with two different approaches—magnetotellurics and geomagnetic depth sounding. Telluric currents can also be used to study long-period tidal phenomena and water flows and the Earth's astronomical motion and as possible precursors for earthquakes and volcanic eruptions. It has also been suggested that a natural waveguide for telluric currents in the Earth's crust, consisting of an insulating layer of dry rocks sandwiched between an upper hydrated conducting layer and an underlying conducting hot layer, could be used for communications. There are also investigations to determine the feasibility of using the natural resonances in the earth-ionosphere waveguide, Schumann resonances, as a means for long-distance communications.

RECOMMENDATIONS

An increased interest in understanding the Earth's electrical environment has resulted from recent advances in different disciplines, along with the recognition that many of man's modern technological systems can be adversely affected by this environment. This understanding appears to be on the threshold of rapid progress.

There should be a concerted effort of coordinated measurement campaigns, supported by critical laboratory experiments, theory, and numerical modeling of processes, to improve our understanding of the Earth's electrical environment.

Because the study of the electrical environment is commonly divided into three major components—lightning, cloud electricity, and the global circuit (including ion chemistry and physics and ionospheric, magnetospheric, and telluric currents)—the specific needs in these areas are detailed below. However, it should not be forgotten that there are interactions among these components and that the understanding of these interactions may be fundamental to an understanding of an individual component.

1. More needs to be known about the basic physics of the lightning discharge and its effects on structures in order to design proper protection systems.

Most lightning begins within a thundercloud, but the initiation and subsequent development of a flash within the cloud are poorly understood. The physics of electrical breakdown over distances of 10-10,000 m is not understood, nor is the relationship between the channel geometry and the fields and charges that existed before the discharge. The fields and currents that are produced by most of the important lightning processes have large submicrosecond variations, but how the discharge currents develop as a function of space and time and what the ranges of variability of the maximum I and dI/dt parameters are need to be determined. The power and energy balances within the lightning channel and many other important lightning parameters also need to be determined.

There should be a comprehensive and carefully coordinated effort to understand the basic physics of intracloud and cloud-to-ground lightning discharges and their effects on our geophysical environment. This new knowledge should be applied to the development of improved lightning-protection methods.

Several new techniques are now capable of providing much insight into the complex and varied physical processes that occur during a lightning flash. For example, radio interferometry and time-of-arrival methods can be used to trace the three-dimensional development of lightning channels with microsecond resolution. Rockets can be used to trigger lightning under a thunderstorm, so that many of the physical properties of the discharge and its interactions with structures can be studied in a partially controlled environment.

2. The question of how thunderclouds generate electricity has been a fascinating scientific problem for over two centuries, but only in the past decade have cooperative experiments using new experimental techniques provided valuable insights into the complex and varied electrical processes that occur within clouds. Unfortunately, there are no sensors that can determine the charge-size relationship on the smaller cloud particles inside a thunderstorm; thus the data are not adequate to determine which of the many possible mechanisms dominate the generation and separation of charge. In addition to not knowing the charge-size relationship for various cloud particles, it is not known how this relationship evolves with time when there is lightning. The electrical forces on individual elements of precipitation can be several times larger than gravity, but further research is needed to determine whether (and how) these and other electrical effects play a significant role in the formation of precipitation.

In view of the successes of recent research, *significant new understanding of cloud electricity and lightning can be made by continuing to develop new instruments and by making coordinated in situ and remote measurements of selected thunderclouds. These studies should be complemented by measurements of cloud microphysics and dynamics, by comprehensive laboratory studies, and by theory and numerical modeling.*

The complexity of the processes that produce both precipitation and lightning makes it impossible to construct or validate theories of cloud electrification from simple field experiments. It is only through the complementary efforts of comprehensive field observations, laboratory experimentation, and numerical modeling that we can hope to understand the physical processes that are important in thunderstorms. An improved understanding of the major processes that create strong electric fields and their interactions with cloud particles and precipitation might lead to better forecasting of electrical hazards to aviation, forestry, and other outdoor activities.

The first goal of the in-cloud measurements should be to determine the charge-size relationship for various cloud and precipitation particles and the role of screening layers in the upper and lower regions of the storm. The electric current densities that flow above and below the cloud should also be monitored as a function of time. Since the natural storm environment is complicated, laboratory experiments should focus on the detailed physics of mechanisms that appear to be important on the basis of both the in-cloud measurements and the numerical models. Laboratory experiments should also determine the effects of electric fields on drop coalescence efficiencies and the ability of electrified drops to scavenge charged constituents of atmospheric aerosols. Analyses of the in-cloud and laboratory data could be accelerated through the establishment of a common data base, particularly for theory and numerical modeling efforts.

3. Even in fair weather the solid earth and atmosphere are electrified. Thunderstorms have been identified as the dominant generator in the global electric circuit, but many details remain concerning storms as electrical generators and their electrical interactions with their neighboring environment. Lightning and the steady currents above and below thunderclouds play an important role in maintaining an electrical potential between the upper atmosphere and the surface, but the amount and type of lightning and the values of cloud currents that flow to the surface and the upper atmosphere are not well known. The lightning phenomenology and cloud currents may depend on many factors, such as the geographical location of the storm, the season, and the meteorological environment; these dependencies have yet to be determined. The charge transports to the surface under a storm are due to linear and nonlinear field-dependent currents, precipitation and other forms of convection currents, and lightning. Unfortunately, the values of each of these current components and their dependence on the stage of the storm, the lightning-flash frequency, or the local terrain are poorly known. The charge transports to land and ocean surfaces that occur in fair weather, and also to mountainous terrain, need to be determined.

With recent progress in the development of satellite lightning sensors and the technology for measuring the electrical effects of storms with rocket-, balloon-, aircraft-, and ground-based sensors, a new attack on this fascinating problem of atmospheric electricity is needed.

There should be an effort made to quantify further the electrical variables that are acting in the global electric circuit and to determine their relationship to the various current components that flow within and near thunderstorms. There is also an important need for theoretical and numerical studies to quantify further the role of thunderstorms as generators in the global circuit.

The establishment of the ionospheric potential, or some other globally representative parameter, as a geoelectrical index that gives an indication of the state of the global circuit would be extremely useful. This index would be the electrical equivalent of the geomagnetic index that has been used for many years to characterize geomagnetic phenomena.

The effects of stratiform clouds and large-scale cyclones on the global circuit also need to be quantified. Once a globally representative parameter that describes the state of the global circuit has been obtained, it can then be related to other remotely observed quantities such as the global lightning flashing rate or directly observed quantities such as the air-earth current or surface electric field.

4. Electromagnetic and optical sensors, both on the ground and on satellites, can be used to (1) detect and map lightning on a regional, national, and global scale and (2) determine, for the first time, how much lightning actually occurs and its geographic distribution as a function of time. With ground-based sensors, it should be possible to determine whether and how the characteristics of individual lightning flashes depend on their geographical location and the storm structure. If a global detection capability were implemented, it would be possible to map and monitor the intensity of lightning storms and to examine the effects of lightning on the global circuit, the ionosphere, and the magnetosphere. When combined with simultaneous spectroscopic measurements, the satellite data could also be used to determine when and where lightning produces significant concentrations of trace gases in the atmosphere.

A lightning sensor, capable of measuring lightning flashes during both day and night, should be flown on a geosynchronous satellite at the earliest possible date. The resulting data when combined with those from other sensors and data from ground-based detection networks will provide information that could be used to relate lightning to storm size, intensity, location, rainfall, and other important meteorological parameters.

5. Electrical processes in the lower atmosphere and, in particular, within the planetary boundary layer, are important because these, together with global variations, determine the electrical environment of man and the biosphere. Galactic cosmic rays and various radioactive decays produce atmospheric ions that undergo a complex and still only partially understood series of ion-chemical reactions. The composition of the ions is poorly known between the surface and about 50 km, and profile measurements are needed. How the ion characteristics relate to atmospheric aerosols and various trace gases needs to be determined before the bulk electrical properties of the atmosphere can be understood. A significant fraction of the ions attach to atmospheric particles; therefore, smoke and other forms of particulates can significantly affect the electrical properties of the lower atmosphere.

Turbulence and convection in the planetary boundary layer play an important role in establishing the vertical distributions of ions, trace gases, and particles. These processes also transport space charge and drive convection currents that alter the electrical properties of the planetary boundary layer.

The clarification of the chemistry of atmospheric ions, their mobilities, and the physics of electrical processes in the troposphere and stratosphere will require further measurements, particularly in determining how these processes are affected by man's activities and natural events. There is also a need for further laboratory measurements and modeling to determine the important chemical reactions and ion composition in the atmosphere.

A number of meteorological research stations in a variety of geographic locations should begin to measure electrical parameters routinely to determine the relationships between electrical and meteorological processes. Vertical profile measurements of electrical properties should be continued in an attempt to determine their relationships to aerosols and trace-gas chemistry. Provisions should be made for the expansion of such synoptic measurements during planned international programs (e.g., the Global Change Programme of the International Council of Scientific Unions, which is currently in the planning stages).

6. Recent research has indicated that the mesosphere may not be electrically passive but may, in fact, contain active electrical generators that are not understood. In addi-

tion, ground-based and balloonborne measurements have indicated that there is a global electrical response to cosmic-ray and solar variations that is also not understood. The bulk electrical properties of the middle atmosphere are poorly defined, and there is a need to determine the ion composition and chemistry both for quiet conditions and during solar-terrestrial events. There is also new evidence that the electric fields produced by thunderstorms and lightning can produce significant disturbances in the electrical structure of the upper atmosphere and magnetosphere.

The horizontal electric fields that are generated by the ionospheric-wind dynamo and the solar-wind/magnetospheric dynamo propagate downward to the Earth's surface where they can locally perturb the fair-weather electric field by about 1-2 percent and 20-50 percent, respectively. Horizontal currents in the middle atmosphere and the characteristics of the equalization layer need to be determined in order to understand better the electrical interactions that occur between the upper and lower atmosphere.

To determine (1) the electrical properties of the middle atmosphere, (2) the effects of thunderstorms on ionospheric and magnetospheric processes, and (3) the effects of time variations in the cosmic-ray and energetic solar-particle fluxes on the properties of the global circuit, additional measurements are required. Theoretical investigations and modeling are also important components of such investigations.

Lightning has long been known to be a source of whistlers in the Earth's magnetosphere, and recent spacecraft observations suggest that lightning also generates whistler-mode signals on Jupiter. The questions of just how lightning fields couple to a whistler duct and whether these fields have effects on the ionosphere or magnetosphere are important and need further investigation.

REFERENCES

Dolezalek, H., and R. Reiter, eds. (1977). *Electrical Processes in Atmospheres*, Steinkopff, Darmstadt, Germany.

Kessler, E., ed. (1982). *Thunderstorms: A Social, Scientific, and Technological Documentary*, Univ. of Oklahoma Press, Norman, Okla.

Orville, R. E., ed. (1985). *Proceedings of the VIIth International Atmospheric Electricity Conference*, special issue of *J. Geophys. Res. 90* (June 30, 1985).

Volland, H., ed. (1982). *Handbook of Atmospherics*, Vol. I and II, CRC Press, Inc., Boca Raton, Fla.

I
LIGHTNING

Lightning Phenomenology

1

RICHARD E. ORVILLE
State University of New York at Albany

INTRODUCTION

Severe weather phenomena that disrupt our lives include tornadoes, hail, high winds, hurricanes, snowstorms, and lightning. It is not well known that in most years, lightning ranks as the number one killer, followed closely by tornadoes. Much less dramatic than a tornado passing through an area or a severe snowstorm that paralyzes a city, a lightning ground strike can quickly kill one or two people in less than a second with little or no warning. Annually in the United States about 100 people are killed by lightning strikes, and reliable estimates for the world would be in the thousands. Lightning on a global and regional scale is an area of science that brings together the interests of the atmospheric physicist, chemist, and meteorologist in an effort to learn its characteristics.

The phenomenology of lightning involves the frequency of lightning observed over large spatial and time scales. It involves the maximum and average flashing rate per unit area and the variation of flash characteristics with location and storm type. Studies of lightning phenomenology can now be discussed in terms of both satellite and ground-based observations. With the use of satellites, we obtain data on the global lightning flash rates and the distribution of lightning with respect to the continents and oceans. With the extensive use of ground-based observations, we can determine the flashing rates and flash characteristics of individual storms. In addition, we can monitor the variations of the ground flashes as a function of location and storm type.

SATELLITE OBSERVATIONS OF LIGHTNING

Optical Detectors

Significant advances in obtaining a better estimate of global flash rates and distribution have occurred as the result of satellite lightning observations in the last decade. Turman (1978, 1979), Turman *et al.* (1978), and Turman and Edgar (1982), using optical detectors on the Defense Meteorological Satellite Program (DMSP) satellites, showed the distribution of lightning at dawn and dusk for a period of 1 year. One example of this recent result is shown in Figure 1.1, where the dusk lightning distribution for November-December 1977 demonstrates the spatial distribution and the rate. Note that the lightning is found mostly in the southern hemisphere, but significant activity still occurs in the northern hemisphere.

The latitudinal and seasonal variation of the lightning activity is best shown by examining Figure 1.2 (Kowalczyk, 1981; Turman and Edgar, 1982). In this histogram, the lightning rate has been summed over

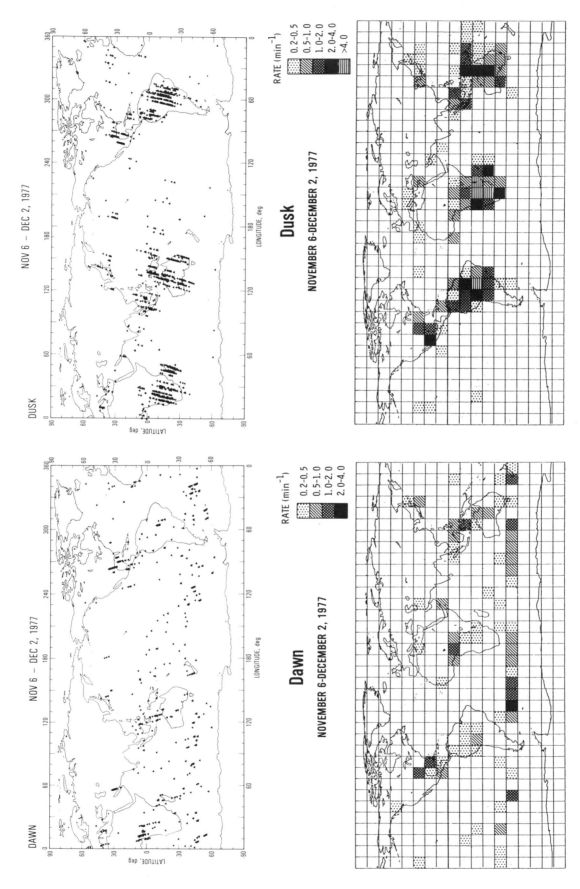

FIGURE 1.1 Lightning activity recorded by a DMSP satellite for the period November 6–December 2, 1977. The dawn and dusk distribution as well as the lightning rates can be compared. From Turman and Edgar (1982) with permission of the American Geophysical Union.

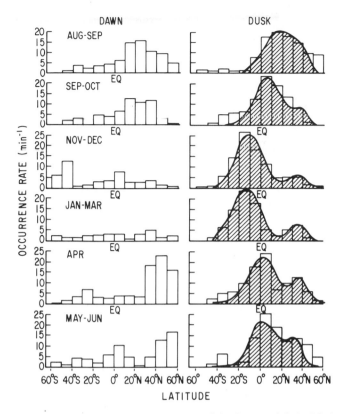

FIGURE 1.2 The latitudinal variation of the dawn and dusk global lightning activity as a function of season. Adapted from Kowalczyk and Bauer (1981) and Turman and Edgar (1982).

longitude at both dawn and dusk and presented as a function of latitude. The dusk distributions show a smooth change as the seasons change. But note the secondary peak at 30-40° that persists through all the dusk distributions except August-September. The dawn distributions do not change as smoothly, but the seasonal shift is apparent. The enhanced polar-front activity at dawn is quite evident in the November-December southern hemisphere and the April-June northern hemisphere. At dawn there appears to be an overall minimum for the period January-March. Analysis of the lightning rates for an entire year show a variation of 10 percent from a global value that is estimated to range from 40 to 120 flashes per second.

Other studies using the DMSP data from a different sensor provide a glimpse at the global midnight lightning activity with a spatial resolution of approximately 100 km (Orville and Spencer, 1979; Orville 1981). A study of the global midnight lightning activity yields a lightning rate of 96 flashes per second, but this could be in error by a factor of 2 (Orville and Spencer, 1979). Orville plotted a series of monthly maps reproduced in Figure 1.3 for the months of September, October, and

November. The progression of the lightning activity toward the southern hemisphere as summer approaches in that hemisphere is evident. The striking absence of lightning over the ocean is apparent in all three months and clearly shows the importance of land in the production of thunderstorms.

Radio Detectors

Recent measurements of high-frequency radio noise by the Ionosphere Sounding Satellite-B have been used by Kotaki et al. (1981) to estimate a global lightning flash rate of approximately 300 per second, in contrast to the optical measurements discussed previously. The radio measurements may overestimate the lightning frequency since it is assumed that all the emissions are produced by lightning. But the satellite optical measurements are uncertain in estimating the lightning rate since the fraction of lightning that is actually detected depends on a calibration factor that represents a best estimate.

Despite the availability of satellites to estimate the global lightning activity, we have made little progress in obtaining a flash rate with small error bars owing to the present experimental limitations of sensor sensitivity, area coverage, and the number of satellite platforms. Nevertheless, the satellite observation provides us with the first reliable estimate of the distribution of global lightning. The resolution of the varying global flash rate estimates may depend on the close coordination of satellite-based and ground-based observations of lightning and the availability of larger-coverage-area platforms, such as geosynchronous satellites.

GROUND OBSERVATIONS OF LIGHTNING

Most of our information on the characteristics of lightning has come, and will continue to come, from ground-based observations of the lightning flash. Many of these studies have focused on the ratio of intracloud to cloud-to-ground flashes, the lightning ground-flash density, and the flashing rate of different types of thunderstorms.

Intracloud Versus Cloud-to-Ground Lightning

The ratio of intracloud to cloud-to-ground lightning is of fundamental importance. How does this ratio vary with latitude and longitude, and how does this ratio vary in the lifetime of a storm? Are there storms that have nearly all intracloud flashes and consequently are less damaging, and are there storms that have almost all ground flashes and consequently are of greater concern?

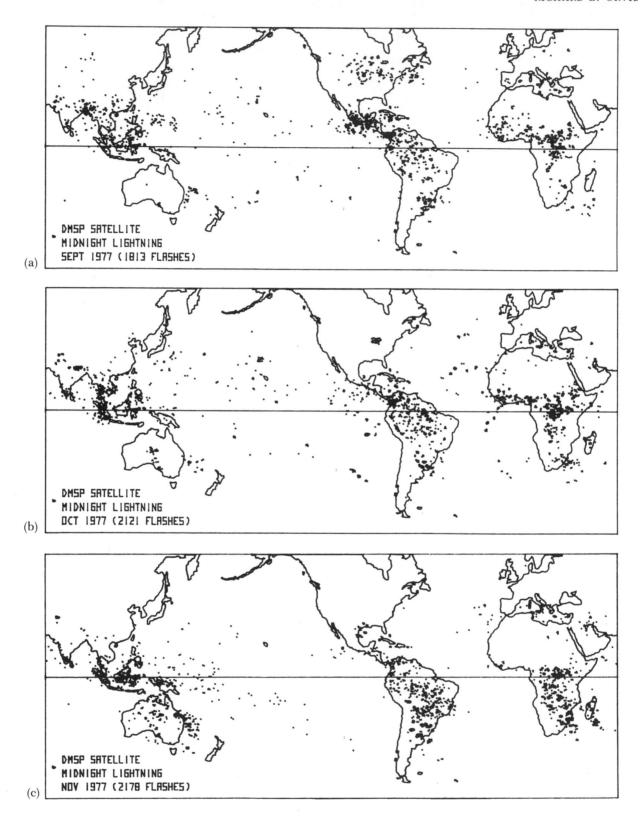

FIGURE 1.3   Three maps showing the progression of monthly lightning for (a) September, (b) October, and (c) November. From Orville (1981), reproduced with permission of the American Meteorological Society.

Studies by Prentice and Mackerras (1977) have summarized much of the available data on the ratio of intracloud to cloud-to-ground flashes (N_c/N_g). From an analysis of 29 data sets from 13 countries, they obtain the following relationship for an average thunderstorm:

$$N_c/N_g = (4.16 + 2.16 \cos 3\lambda)$$
$$\left(0.6 + \frac{0.4T}{72 - 0.98\lambda}\right), \quad (1.1)$$

where T, the number of thunder days per year, is less than or equal to 84 and λ, the latitude in the northern and southern hemispheres, is less than 60°. If the number of thunder days is unknown, then the ratio can be estimated from the relation

$$N_c/N_g = (4.16 + 2.16 \cos 3\lambda). \quad (1.2)$$

This result is plotted in Figure 1.4. Note that the ratio has the highest value in the tropics where most of the lightning was shown to occur by the satellite data. Recall, however, that the satellite data were composed of both intracloud and cloud-to-ground lightning flashes. There is at the moment no way to distinguish between a ground flash and an intracloud flash from a satellite.

Lightning Flash Density

The number of lightning strikes per unit time per unit area, or the flash density, is a fundamental quantity of interest. Most of the available information has been obtained with lightning flash counters.

Prentice (1977) summarized the values for several geographical areas and reported 5 flashes per km^2 per year in Queensland, Australia; 0.2 to 3 flashes per km^2 per year in Norway, Sweden, and Finland; and 0.05 to 15 flashes per km^2 per year in South Africa depending on the location.

Piepgrass *et al.* (1982) reported the results of studying 79 summer storms at the Kennedy Space Center, Florida, which produced 10 or more discharges, during the years 1974-1980. Using field mill sites covering an effective area of 625 km^2, they observed an area flash density for all discharges during June, July, and August to range from 4 to 27 discharges per km^2 per month, with a systematic uncertainty of perhaps a factor of 2 in the sample area. The mean and the standard deviation of the monthly area density over the above years was 12 ± 8 discharges per km^2. Approximately 38 percent of the discharges were ground flashes. Therefore, they were able to estimate the ground flash density to be 4.6 ± 3.1 flashes per km^2 per month.

The most recent estimate of the ground flash density in the United States has been made by Maier and Piotrowicz (1983) using thunderstorm hour statistics and is reproduced as Figure 1.5. They used thunderstorm duration data from approximately 450 aviation weather reporting stations, each with an uninterrupted 30 years of records. The station density available is twice that of any previous thunderstorm frequency analysis of the United States. The maximum annual ground flash densities of 18 per km^2 are found in the western interior of Florida. High flash densities greater than 12 per km^2 are found over much of Florida and westward to eastern Texas. Flash densities greater than 8 per km^2 are found in most of Oklahoma, Kansas, Missouri, Arkansas, Louisiana, Mississippi, and Tennessee. Most western and northeastern states have flash densities that are less than 4 per km^2.

Lightning Flash and Related Characteristics

Data from two summers at the Kennedy Space Center, Florida, have been used to estimate the flashing rates in thunderstorms (Livingston and Krider, 1978). It was observed that large storms evolve through an initial, an active, and a final phase of activity. Most of the lightning activity was observed to occur in the active phase with 71 percent of the lightning, although this phase of the storm occupied only 27 percent of the total storm duration. During the active phase, 42-52 percent of all lightning was to ground, while during the final storm period, only about 20 percent of the lightning was to ground. The discharge rate for all storms observed in 1975 was approximately 4 flashes per minute with a maximum flashing rate of 26 discharges per minute during any 5-minute period. The highest flashing rate averaged over an entire storm was about 9 discharges per minute for over 200 minutes. More recent data from a 4-year interval indicates that the mean rate of flashes is about 2.4 discharges per minute per storm (Piepgrass *et al.*, 1982).

The relationship of rainfall to lightning flash rates has been investigated by Piepgrass *et al.* (1982). They reported that when the meteorological conditions favor the production of lightning, there is almost a direct proportionality between the total rain volume and the total number of flashes. Maier *et al.* (1978) noted in an earlier paper that the lightning counts were proportional to the total storm rainfall and that the proportionality increased with the rain volume until the rainfall reached about 1.2 to 2.7×10^4 m^3 per flash. Beyond these volumes, storms that produced more rainfall tended to produce proportionally less lightning. Piepgrass *et al.* (1982) point out that, "Clearly, these problems warrant further study."

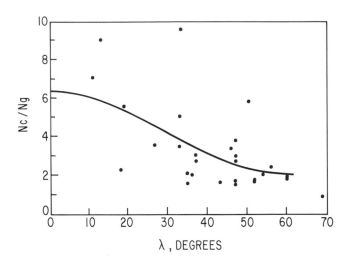

FIGURE 1.4 The ratio of intracloud to cloud-to-ground lightning as a function of latitude. From Prentice and Mackerras (1977).

Lightning Location Networks

The study of lightning phenomenology has made a major advance in the last decade with the introduction of new magnetic direction-finding techniques (Krider *et al.*, 1980) that provide the means to monitor ground strikes over areas exceeding 10^6 km². Extensive networks of lightning direction finders have been established for forest-fire detection in the western United States, Canada, and Alaska. Figure 1.6 shows the coverage as of the summer of 1984, and it can be predicted that within the next few years the entire United States will be covered.

One expanding lightning detection network covers the East Coast and is approaching the Mississippi River to provide coverage of the eastern part of the United States (Orville *et al.*, 1983). This network is operated by the State University of New York at Albany in a multi-drop communication network that links all the direction finders to one computer. Data are now retrieved on the time, location, number of strokes in the flash, polarity of the charge lowered to ground, and amplitude of the peak magnetic radiation field that can be related to the maximum current in the first stroke. These data, in turn, can be analyzed and related to the meteorological patterns producing the observed phenomena.

To report all initial results would far exceed the space available in this brief paper; nevertheless, it is interesting to note a few observations. The highest ground flash rate recorded by the East Coast Network occurred on June 13, 1984, when 50,836 flashes were detected in a 24-hour period over an area of approximately 250,000 km². The highest hourly summary was 7800 flashes with the highest 5-minute rate exceeding 10,000 ground flashes per hour. These results are remarkable when it is realized that these flash rates were from storms in only three states—Pennsylvania, New Jersey, and part of New York—and at the time were producing approximately 3 percent of the entire global lightning activity.

Other results indicate that lightning is recorded in every week of the year along the East Coast and that the polarity of the lightning ground strikes shows a change from negative to positive in the fall and a shift back to negative in the spring. A discussion of positive lightning and its characteristics is presented by Rust (Chapter 3, this volume).

FIGURE 1.5 Lightning flash density estimates on an annual basis. Adapted from Maier and Piotrowicz (1983) and MacGorman *et al.* (1984).

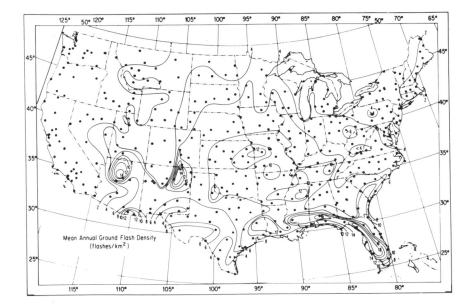

FIGURE 1.6 Coverage of North America by time of arrival (dashed lines) and wide-band magnetic direction finders (solid lines) as of the summer of 1985.

CONCLUSIONS

The past decade has been a period of significant advances in lightning knowledge. Satellite studies have provided the first confirmation of the early estimates of the global lightning flash rates and added new information on the distribution of lightning over the land and over the ocean. The development of widely distributed ground-based lightning networks provides for the first time the ability to monitor and calculate lightning characteristics in near real time. Relating these parameters to the meteorological observations of visible and infrared images from space and to radar observations from the ground poses a major challenge in the near future.

REFERENCES

Kotaki, M., I. Kuriki, C. Katoh, and H. Sugiuchij (1981). Global distribution of thunderstorm activity, *J. Radio Res. Labs. Japan 66.*

Kowalczyk, M., and E. Bauer (1981). Lightning as a source of NO$_x$ in the troposphere, final report FAA-EE-82-4.

Krider, E. P., A. E. Pifer, and D. L. Vance (1980). Lightning direction-finding systems for forest fire detection, *Bull. Am. Meteorol. Soc. 61,* 980-986.

Livingston, J. M., and E. P. Krider (1978). Electric fields produced by Florida thunderstorm, *J. Geophys. Res. 83,* 385-401.

MacGorman, D. R., M. W. Maier, and W. D. Rust (1984). Lightning strike density for the contiguous United States from thunderstorm duration records, prepared for Division of Health, Siting and Waste Management, Office of Nuclear Regulatory Research, U.S. Nuclear Regulatory Commission, Washington, D.C., 44 pp.

Maier, M. W., and J. M. Piotrowicz (1983). Improved estimates of the area density of cloud-to-ground lightning over the United States, presented at 8th International Aerospace and Ground Conference on Lightning and Static Electricity, June 21-23, 1983, Forth Worth, Texas.

Maier, M. W., A. G. Boulanger, and J. Sarlet (1978). Cloud-to-ground lightning frequency over south Florida, preprint, Conference on Cloud Physics and Atmospheric Electricity (Issaquah, Wash.), American Meteorological Society, Boston, Mass., pp. 605-610.

Orville, R. E. (1981). Global distribution of midnight lightning September to November 1977, *Mon. Weather Rev. 109,* 391-395.

Orville, R. E., and D. W. Spencer (1979). Global lightning flash frequency *Mon. Weather Rev. 107,* 934-943.

Orville, R. E., R. W. Henderson, and L. F. Bosart (1983). An East Coast lightning detection network, *Bull. Am. Meteorol. Soc. 64,* 1029-1037.

Piepgrass, M. V., E. P. Krider, and C. B. Moore (1982). Lightning and surface rainfall during Florida thunderstorms, *J. Geophys. Res. 87,* 11193-11201.

Prentice, S. A. (1977). Frequencies of lightning discharges, in *Physics of Lightning,* R. H. Golde, ed., Academic Press, New York, pp. 465-496.

Prentice, S. A., and D. Mackerras (1977). The ratio of cloud to cloud-ground lightning flashes in thunderstorms, *J. Appl. Meteorol. 16,* 545-549.

Turman, B. N. (1978). Analysis of lightning data from the DMSP satellite, *J. Geophys. Res. 83,* 5019-5024.

Turman, B. N. (1979). Lightning detection from space, *Am. Scientist 67,* 321-329.

Turman, B. N., and B. C. Edgar (1982). Global lightning distributions at dawn and dusk, *J. Geophys. Res. 87,* 1191-1206.

Turman, B. N., B. C. Edgar, and L. N. Friesen (1978). Global lightning distribution at dawn and dusk for August-September 1977, *EOS 59,* 285.

Physics of Lightning

2

E. PHILIP KRIDER
University of Arizona

INTRODUCTION

Lightning is a transient, high-current electric discharge that occurs in the atmospheres of the Earth and other planets and that has a total path length on the order of kilometers. Most lightning is produced by thunderclouds, and well over half of all discharges occur within the cloud. Cloud-to-ground flashes (Figure 2.1), although not so frequent as intracloud flashes, are, of course, the primary lightning hazard to people or structures on the ground. The continental United States receives an estimated 40 million cloud-to-ground strikes each year, and lightning is among the nation's most damaging weather hazards (see Chapter 1, this volume). The peak power and total energy in lightning are very large. Thus far, it has not been possible to simulate in the laboratory either the geometrical development of a lightning channel or the full extent of lightning damage. Lightning is a leading cause of outages in electric power and telecommunications systems, and it also is a major source of interference in many types of radio communications. The possible effects of lightning on advanced aircraft, nuclear power plants, and sophisticated military systems are problems of increasing concern.

Besides its many deleterious effects, lightning also has some unique benefits. The chemical effects of lightning may have played an important role in the prebiotic synthesis of amino acids, and today lightning is still an important source of fixed nitrogen, a natural fertilizer, and other nonequilibrium trace gases in the atmosphere (see Chapter 6, this volume). Also, lightning-caused fires have long dominated the dynamics of forest ecosystems throughout the world. The electromagnetic fields that are radiated by lightning can be used to study the physics of radio propagation and have been used for many years in geophysical prospecting. Also, lightning-caused "whistlers" are still being employed to study the characteristics of the ionosphere and magnetosphere. Lightning plays an important role in maintaining an electric charge on the earth and is therefore an important component of the global electric circuit (see Chapter 15, this volume). It is clear, therefore, that an understanding of the physics of lightning is important to further insight into our geophysical environment as well as for the development of optimum protection from the lightning's hazards.

In recent years, new experimental techniques have enabled researchers to obtain a better understanding of the physics of lightning. Among these techniques have been applications of optical, acoustic, and electromagnetic sensors to measure the properties of various discharge processes on time scales ranging from tens of nanoseconds to several seconds. These measurements

30

FIGURE 2.1 Cloud-to-ground lightning over Tucson, Arizona.

have also been used to infer properties of the thunder-cloud charge distribution that is affected by lightning (see Chapter 8, this volume).

Within the category of cloud-to-ground lightning, there are flashes that effectively lower negative charge to ground and those that lower positive charge. Most ground flashes are negative, but there is recent evidence that positive discharges are often unusually deleterious, and, therefore, this type is discussed separately in Chapter 3 (this volume).

Before describing some of the recent advances in lightning research in more detail, the processes that occur during a typical negative lightning flash to ground will be reviewed briefly. There are still many open questions about basic lightning phenomena, such as whether or how the characteristics of individual flashes depend on the type of thunderstorm, the season, and the type of terrain that is struck.

CLOUD-TO-GROUND LIGHTNING

Simplified sketches of the luminous processes that occur during a typical cloud-to-ground discharge are given in Figures 2.2, 2.3, and 2.4. For more detailed discussions of these phenomena, the reader is referred to Schonland (1964), Uman (1969), and Salanave (1980). Figure 2.2a shows an assumed cloud charge-distribution just before the lightning begins. Concentrations of negative charge are shown at altitudes where the ambient air temperature is -10 to $-20°C$, typically 6 to 8 km above mean sea level. The positive charge is more diffuse than the negative and most of it is at higher altitudes (see Chapter 8, this volume).

Cloud-to-ground lightning almost always starts within the cloud with a process that is called the *preliminary breakdown*. The location of the preliminary breakdown is not well understood, but it may begin in the high-field region between the positive and negative charge regions, as shown in Figure 2.2b. After several tens of milliseconds, the preliminary breakdown initiates an intermittent, highly branched discharge that propagates horizontally and downward and that is called the *stepped-leader*. The stepped-leader is sketched in Figures 2.2c and 2.2d, and this process effectively lowers negative charge toward ground. The individual steps in the stepped-leader have lengths of 30 to 90 m and occur at intervals of 20 to 100 μsec. The direction of the branches in a photograph indicates the direction of stepped-leader propagation; for example, in Figure 2.1 each stepped-leader propagated downward.

When the tip of any branch of the stepped-leader gets close to the ground, the electric field just above the surface becomes very large, and this causes one or more upward discharges to begin at the ground and initiate the *attachment process* (see Figure 2.3). The upward propagating discharges rise until one or more attach to the leader channel at a junction point that is usually a few tens of meters above the surface. When contact occurs, the first *return stroke* begins. The return stroke is basically an intense, positive wave of ionization that starts at or just above the ground and propagates up the leader channel at about one third the speed of light (Figure 2.2e). The peak currents in return strokes range from several to hundreds of kiloamperes, with a typical value being about 40 kA. These currents carry the ground potential upward and effectively neutralize most of the leader channel and a portion of the cloud charge. The peak power dissipated by the return stroke

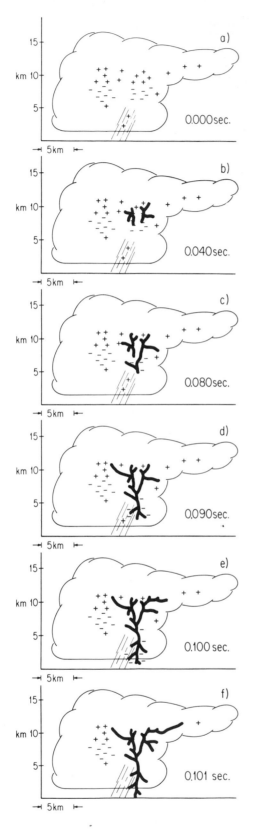

FIGURE 2.2 Sketch of the luminous processes that form the stepped-leader and the first return stroke in a cloud-to-ground lightning flash.

is probably on the order of 10^8 watts *per meter* of channel (Guo and Krider, 1982), and the peak channel temperature is at least 30,000 K (Orville, 1968).

The last few steps of the stepped-leader, the onset of a connecting discharge, and the beginning of a return stroke are illustrated in Figure 2.3. Here, the distance between the object that is about to be struck and the tip of the leader when the connecting discharge begins is called the *striking distance* (SD) and is an important concept in lightning protection. The distance to the actual junction (J) between the leader and the connecting discharge is often assumed to be about half the striking distance.

After a pause of 40 to 80 milliseconds, most cloud-to-ground flashes produce a new leader, the *dart leader*, which propagates without stepping down the previous return-stroke channel and initiates a *subsequent return stroke*. Most flashes contain two to four return strokes, and each of these affects a different volume of cloud charge (see also Chapter 8, this volume). Figure 2.4 shows a sketch of a dart leader and the subsequent return stroke. Visually, lightning often appears to flicker because the human eye can just resolve the time intervals between different strokes. In 20 to 40 percent of all cloud-to-ground flashes, the dart leader propagates down just a portion of the previous return-stroke channel and then forges a different path to ground. In these cases, the flash actually strikes the ground in two places, and the channel has a characteristic forked appearance that can be seen in many photographs. (See the left two flashes in Figure 2.1.)

IMPORTANT RESULTS OF RECENT RESEARCH

Three types of research have recently provided new information about the physics of lightning. These are described briefly, and their importance is indicated. First, we discuss how remote measurements of electric and magnetic fields can be used to infer properties of lightning currents, including some implications of recent measurements. Next, we describe how the sources of radio-frequency (rf) noise can be used to trace the geometrical development of lightning channels as a function of time and to determine other properties of lightning. Finally, we discuss how small rockets can be used to trigger lightning artificially and give some applications of this technique.

Time-Domain Fields and Lightning Currents

Recently, it has become clear that the electric and magnetic fields that are radiated by different lightning

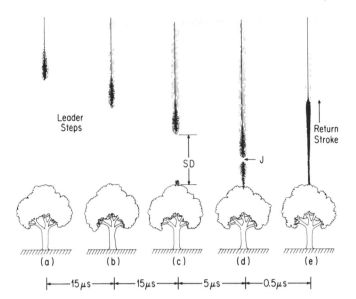

FIGURE 2.3 Sketch of the luminous processes that occur during attachment of a lightning stepped-leader to an object on the ground.

processes have different but characteristic signatures that are reproduced from flash to flash. For example, Figure 2.5 shows three of the many impulses that were radiated by a typical cloud-to-ground flash at a distance of about 50 km. These particular signatures were recorded using a broadband antenna system and an oscilloscope that covered all frequencies from about 1 kHz to 2 MHz. Trace (a) shows a cloud impulse that was radiated during the preliminary breakdown; trace (b) shows the waveform that was radiated by the first return stroke; and trace (c) shows a subsequent return stroke. The small pulses that precede the first return stroke in trace (b) were radiated by individual steps of the stepped-leader just before the attachment occurred (see Figure 2.3). The characteristics of these newly measured signatures have been put to use in the detection and location of cloud-to-ground lightning. For example, there are now large networks of magnetic direction-finders that can discriminate between the shapes of the return-stroke fields and other processes and that can provide accurate locations of the ground-strike points (see Chapters 1 and 5, this volume; Krider *et al.*, 1980).

In a series of recent papers, Uman and co-workers have developed a theoretical model that describes the shapes of the electric and magnetic fields that are produced by return strokes at various distances (Uman *et al.*, 1975; Master *et al.*, 1981; Uman and Krider, 1982). One particularly important result of this work is the prediction that during the first few microseconds of the stroke, i.e., just after the attachment process has been

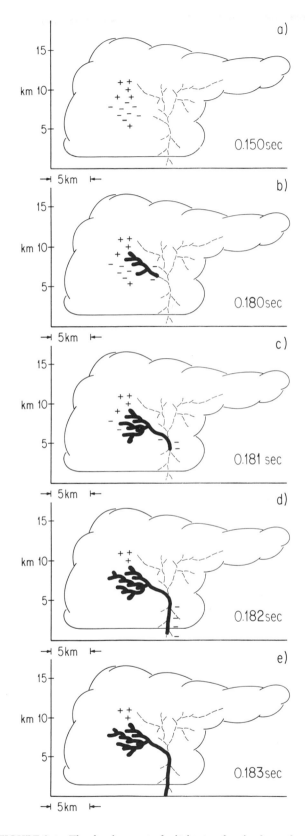

FIGURE 2.4 The development of a lightning dart-leader and a return stroke subsequent to the first in a cloud-to-ground lightning flash.

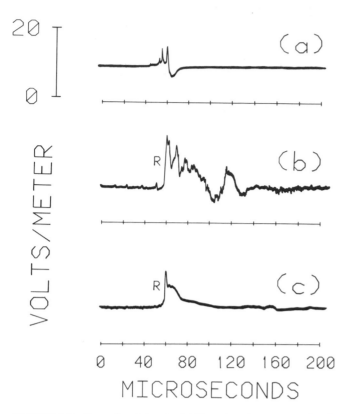

FIGURE 2.5 Examples of electric-field impulses that were produced by a cloud-to-ground flash at a distance of about 50 km. Trace (a) was radiated during the preliminary breakdown, trace (b) is due to the first return stroke, and trace (c) is due to a subsequent return stroke.

completed, the waveform of the distant or radiation field is proportional to the channel current,

$$E_{RAD}(t) = -\frac{\mu_0 \, v}{2 \, \pi \, D} I(t - D/c), \qquad (2.1)$$

where E is the vertical electric field that is measured at the ground at time t, μ_0 the permeability of free space, v the return stroke velocity, c the speed of light, and D the horizontal distance to the flash. A typical first return stroke will produce a peak field of about 8 V/m at a distance of 100 km (Lin et al., 1979). The stroke velocity near the ground is typically on the order of 10^8 m/sec (Idone and Orville, 1982). With these values, Eq. (2.1) predicts a peak current of about 40 kA, a value that is in good agreement with currents that have been measured in direct strikes to instrumented towers (Berger et al., 1975; Garbagnati et al., 1981).

Recently, Weidman and Krider (1978, 1980) examined the microsecond and submicrosecond structure of return-stroke E-field and field derivative, dE/dt, signatures. These investigators found that a typical first stroke produces an electric field "front" that rises in 2-8 μsec to about half of the peak-field amplitude. This

front is followed by a fast transition to peak whose mean 10-90 percent rise time is about 90 nsec (see Figure 2.6). Subsequent stroke fields have fast transitions similar to first strokes, but fronts that last only 0.5 to 1 μsec and that rise to only about 20 percent of the peak field. This fine structure in the initial return-stroke field is illustrated by the waveform shown in Figure 2.7.

Unfortunately, the origin of fronts in return-stroke fields is still not well understood, particularly for first strokes (Weidman and Krider, 1978). If fronts are produced by upward connecting discharges, then these discharges must have lengths in excess of 100 m and peak currents of 10 kA or more. A front may be produced by a slow surge of current in the leader channel prior to the fast transition, but then this surge must contain currents on the order of 10 kA or more, and the associated channel length must be at least 1 km. To date, the available optical data are not adequate to determine whether either of these processes (or both) does actually occur. Clearly, more research will be needed before we shall understand the physics of the important striking process.

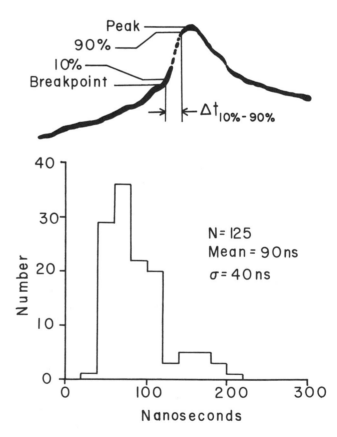

FIGURE 2.6 Histogram of the 10 to 90 percent rise times of the fast portions of return stroke fields over seawater.

FIRST RETURN STROKE
RANGE 19km

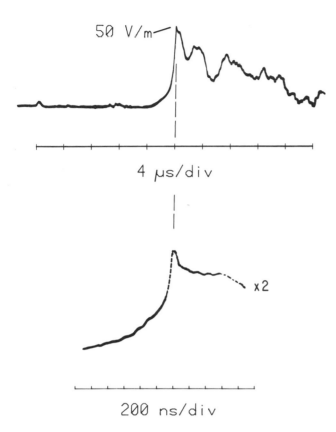

50 V/m

4 µs/div

×2

200 ns/div

FIGURE 2.7 The initial portion of a return-stroke field recorded with fast time resolution. The same signal is shown on both traces, and the peaks coincide in time.

The rise times of the fast field components that are radiated by return strokes are summarized in Figure 2.6. Note that the mean 10 to 90 percent value is about 90 nsec. These submicrosecond components in the field must be caused by submicrosecond components in the current, but few of the currents that have been measured during direct strikes to instrumented towers show components as fast as 90 nsec (Berger *et al.*, 1975; Garbagnati *et al.*, 1981). It is possible that an upward discharge from a tall tower or the electrical characteristics of the tower itself will reduce the rise time that is measured from the value that would actually be present in a strike to normal terrain. Therefore, new measurements of lightning currents and fields with fast time-resolution will be necessary before we can understand the true current rise times.

The actual current rise time is important for the design of lightning protection systems (see Chapter 5, this volume). For example, if a 100-nsec current interacts with a resistive load, the voltage rise time on that load will be 100 nsec. Most of the standard surge waveforms that are used to verify the performance of protectors on power and telecommunications circuits specify that open-circuit voltage should have a rise time of 0.5, 1.2, or 10 µsec and that the short-circuit current should have a rise time of 8 or 10 µsec (see IEEE Standard 587-80 and FCC Docket 19528, Part 68). These values are substantially slower than those shown in Figure 2.6; therefore, it is probable that the degree of protection that is provided by devices that have been tested to the above standards will not be adequate for direct lightning surges.

Measurements of the maximum dE/dts that are radiated by return strokes striking seawater are summarized in Figure 2.8. Here, the average maximum dE/dt is about 33 V/m/µsec when the values are range-normalized to 100 km using an inverse distance relation. To

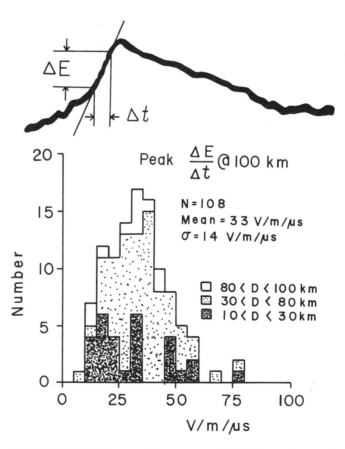

FIGURE 2.8 Histogram of the maximum dE/dt during the initial portion of return-stroke fields. All data have been range-normalized to 100 km using an inverse distance relation.

measure a dE/dt signature on a 10-nsec time scale, it is essential that the field propagation from the lightning source to the measuring station be entirely over saltwater; otherwise there can be a significant degradation in the high-frequency content of the signal due to propagation over the relatively poorly conducting earth. It is possible that strikes to saltwater contain inherently faster rise times than lightning strikes to ground, but Weidman and Krider (1978) argued that this is probably not the case.

If Eq. (2.1) is valid, then the dE/dt values in Figure 2.8 can be used to infer the maximum dI/dt in the lightning channel, i.e.,

$$dI/dt = -\frac{2\pi D}{\mu_0 v} dE/dt. \qquad (2.2)$$

If a typical dE/dt is 33 V/m/μsec at 100 km and v is 10^8 m/sec, then Eq. (2.2) implies that the maximum dI/dt is typically 1.5×10^{11} A/sec during a return stroke, a value that is about a factor of 20 higher than the tower measurements. At this point, it should be noted that, if return strokes do contain large current components with dI/dt on the order of 1.5×10^{11} A/sec, then, if these currents interact with an inductive load, the overvoltage will be substantially larger and faster than has previously been assumed and the lightning hazard will be greater. (See Chapter 5, this volume, for further discussions of lightning protection.)

As a final point, we note that the electromagnetic fields that are produced by lightning are now known to be large and to change rapidly; hence, these fields themselves can be deleterious. For example, Uman *et al.* (1982) showed that, if an object is struck directly by lightning, then the associated electromagnetic disturbance may be substantially more severe than the electromagnetic pulse (EMP) produced by an exoatmospheric nuclear burst at all frequencies below about 10 MHz. Also, Krider and Guo (1983) showed that the typical peak field from a return-stroke at 100 km corresponds to a peak electromagnetic power at the source of at least 20,000 megawatts.

Locations of Lightning Radio Sources

The radio-frequency noise that is generated by lightning in the HF and VHF bands appears in the form of discrete bursts, and within each of these bursts there are hundreds to thousands of separate pulses. If the difference in the time of arrival of each pulse is carefully measured at four widely separated stations, the location of the source of each pulse can be computed, and the geometrical development of the rf bursts can be mapped as

a function of time (Proctor, 1971). Unfortunately, the physical processes that produce HF and VHF radiation in lightning are not well understood. Proctor (1981) reported that the pulses in most bursts are produced by a regular progression of source points, and, therefore, he suggests that bursts are produced by new ionization processes and extensions of old channels.

If the source location of each rf pulse within a burst is plotted, the width of the associated "radio image" of the channel ranges from about 100 m to more than 1 km (Proctor, 1981, 1983; Rustan *et al.*, 1980). Figure 2.9 shows the paths of the central cores of six successive lightning discharges that were reconstructed by Proctor (1983). By combining reconstructions such as these with measurements of the associated changes in the electric field at the ground, Proctor inferred that in-cloud channels usually have a net negative charge and that the average line charge density is about 0.9 C/km. Also, by dividing the length of a channel segment by the time required for that portion to develop, Proctor determined that the average velocity of streamer formation ranges from 4×10^4 m/sec to 8×10^5 m/sec with a mean of $(1.4 \pm 1.2) \times 10^5$ m/sec. If Proctor's average charge density is multiplied by the velocity of channel formation, then the average current in developing channels would appear to be on the order of 100 A, a value that is in reasonable agreement with other estimates (Brook and Ogawa, 1977).

In an analysis of the locations of the first rf sources in 26 lightning flashes, Proctor (1983) found that most discharges begin within or near precipitation, i.e., those regions of the cloud that produce a radar reflectivity greater than 25 dBZ. He also reports that all stepped-leaders begin in a narrow range of altitudes where the ambient air temperature is -5 to $-16°C$. The average altitude of the initial rf sources in the flashes studied by Proctor was about 4 km above ground level ($-10°C$), and the standard deviation was only 440 m.

The geometrical forms of intracloud discharges range from concentrated "knots" or "stars" a few kilometers in diameter to extensive branched patterns up to 90 km in length. Proctor (1983) reports that successive discharges in a storm often form an interconnected system and that some flashes seem to extend the paths of earlier discharges. In one case, two flashes that were separated by just 1.6 sec produced tortuous channels that ran parallel to each other for almost 2 km, but the channels remained about 300 m apart.

Although more data will be required before Proctor's results can be generalized, it is clear that time-of-arrival methods offer great promise for future research, particularly for those phases of lightning that occur within a cloud (see also Taylor, 1978; Rustan *et al.*, 1980). Radio

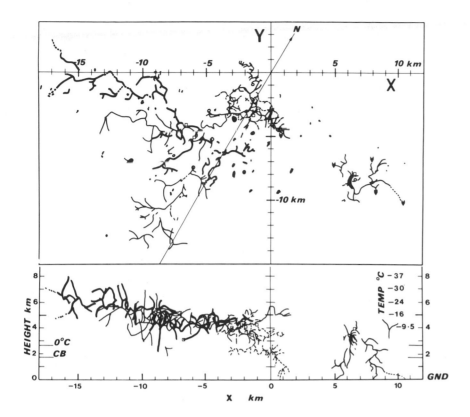

FIGURE 2.9 Geometrical reconstructions of six successive lightning discharges in South Africa (Proctor, 1983). The top panel shows a plane view of the channels from above, and the bottom panel shows an elevation view that is a projection of the same channels on a vertical plane parallel to the x axis.

interferometer observations of lightning have also provided interesting results (Warwick *et al.*, 1979; Hayenga and Warwick, 1981), and perhaps in the future interferometric methods will be developed to the point that they can provide unambiguous three-dimensional reconstructions of the discharge processes.

Artificial Triggering of Lightning

The last development that we describe is the artificial triggering of lightning by small rockets. This technique is particularly important because it provides, for the first time, the capability of studying both the physics of the discharge process and the interactions of lightning with structures and other objects in a partially controlled environment. Although rockets were first used to study atmospheric electricity in the eighteenth century, the first artificial initiation of lightning was clearly demonstrated by Newman *et al.* (1967). The technique has subsequently been improved by researchers in France (Fieux *et al.*, 1975; Fieux and Hubert, 1976; St. Privat d'Allier Research Group, 1982) and is now being used to investigate a variety of lightning problems in France, Japan, and the United States.

When a thunderstorm is overhead and the electrical conditions are favorable, a small rocket is launched and carries a grounded wire aloft. If the rocket is fired when the surface electric field is 3 to 5 kV/m, then about two thirds of all launches will trigger a lightning discharge (Fieux *et al.*, 1978). Most triggers occur when the rocket is at an altitude of only 100 to 300 m, and the first stroke in the flash usually propagates upward into the cloud. The majority of the subsequent strokes follow the first stroke and the wire to ground; but in about one third of the cases, the subsequent strokes actually forge a different path to ground. These latter events are called "anomalous triggers" (Fieux *et al.*, 1978). The first stroke in a triggered discharge is not like natural lightning, but subsequent strokes appear to be almost identical to their natural counterparts.

An example of lightning that was triggered by Hubert and co-workers is shown in Figure 2.10. The upward branching in this photograph was produced by a leader that propagated upward from the wire, and the bright, straight section of channel near the ground shows the path of the wire just before it exploded as a result of the lightning current.

Triggered lightning is now being used to investigate the luminous development of lightning channels, the characteristics of lightning currents, the velocities of return strokes, the relation between currents and fields, the mechanisms of lightning damage, the performance of lightning protection systems, and many other problems (Fieux *et al.*, 1978; Hubert and Fieux, 1981; Horii,

FIGURE 2.10   An example of a rocket-triggered lightning flash in New Mexico.

1982; Miyachi and Horii, 1982; St. Privat d'Allier Research Group, 1982; Hubert *et al.*, 1984).

Among the more important results to date have been a direct experimental verification of the existence of submicrosecond fields and currents during return strokes and the general validity of Eq. (2.1) (Fieux *et al.*, 1978; Djibari *et al.*, 1981; Hubert and Fieux, 1981). Waldteufel *et al.* (1980) also reported a curious case in which a triggered discharge originated everywhere in clear air.

The main benefit of the rocket triggering technique is that it can be used to cause lightning to strike a known place at a known time, thus enabling controlled experiments to be performed. The triggering wire guides the lightning current to a point where a variety of sensors can measure the physical properties of the discharge and its deleterious effects directly. All cameras and data-recording equipment can be turned on and be fully operational just before the rocket is fired. In most locations, the total number of triggers is limited to a few tens of events per year by the frequency of overhead storms, but the number and quality of the measurements can be made quite high to compensate for the relatively few events.

CONCLUSION

We have seen that new experimental techniques now provide an opportunity to investigate many of the important questions that remain unanswered about the physics of lightning. Among the more important unknowns are the following:

How is lightning initiated within a cloud?

Can the initiation of lightning be suppressed or controlled?

What are the mechanisms of stepped- and dart-leader propagation?

What factors control the geometrical development of lightning?

What is the physics of the attachment process?

What are the currents in return strokes?

How rapidly do return-stroke currents change with time?

What physical processes control the propagation of return strokes?

What is the energy balance of the various lightning processes?

What physical phenomena occur during a cloud discharge?

What are the characteristics of the currents in cloud discharge processes?

What processes generate HF and VHF radio noise in lightning?

Recent spacecraft observations have shown that lightning may be present in the atmospheres of Jupiter, Venus, and Saturn, and the upcoming Galileo probe will carry a lightning detector to Jupiter (Lanzerotti *et al.*, 1983). Perhaps a study of lightning in atmospheres that are radically different from that of Earth will help us to better understand lightning on Earth and offer even more challenging questions for future work.

REFERENCES

Berger, K., R. B. Anderson, and H. Kroninger (1975). Parameters of lightning flashes, *Electra 80*, 23-37.

Brook, M., and T. Ogawa (1977). The cloud discharge, in *Lightning*, Vol. 1, R. H. Golde, ed., Academic, New York, pp. 191-230.

Djibari, B., J. Hamelin, C. Leteinturier, and J. Fontain (1981). Comparison between experimental measurements of the electromagnetic field emitted by lightning and different theoretical models—Influence of the upward velocity of the return stroke, *Proc. Int. Conf. on EMC*, March 1981, Zurich.

Fieux, R., and P. Hubert (1976). Triggered lightning hazards, *Nature 260*, 188.

Fieux, R., C. Gary, and P. Hubert (1975). Artificially triggered lightning above land, *Nature 257*, 212-214.

Fieux, R., C. H. Gary, B. P. Hutzler, A. R. Eybert-Berard, P. L. Hubert, A. C. Meesters, P. H. Perroud, J. H. Hamelin, and J. M. Person (1978). Research on artificially triggered lightning in France, *IEEE Trans. Power Appar. Syst. PAS-97*, 725-733.

Garbagnati, E., F. Marinoni, and G. P. Lo Pipero (1981). Parameters of lightning currents—Interpretation of the results obtained in Italy, *Proc. 16th Int. Conf. on Lightning Protection, July 1981*, Szeged, Hungary.

Guo, C., and E. P. Krider (1982). The optical and radiation field signatures produced by lightning return strokes, *J. Geophys. Res. 87*, 8913-8922.

Havenga, C. O., and J. A. Warwick (1981). Two-dimensional interferometric positions of VHF lightning sources, *J. Geophys. Res. 86*, 7451-7462.

Horii, K. (1982). Experiment of artificial lightning triggered with rockets, *Mem. Fac. Eng. (Nagoya U.) 34*, 77-112.

Hubert, P., and R. Fieux (1981). 2—Mesure des courants de foudre à la station d'étude de la foudre de St.-Privat-d'Allier, *Rev. Gen. Electr. 5*, 344-349.

Hubert, P., P. Laroche, A. Eybert-Berard, and L. Barret (1984). Triggered lightning in New Mexico, *J. Geophys. Res. 89*, 2511-2521.

Idone, V. P., and R. E. Orville (1982). Lightning return stroke velocities in the Thunderstorm Research International Program (TRIP), *J. Geophys. Res. 87*, 4903-4915.

Krider, E. P., and C. Guo (1983). The peak electromagnetic power radiated by lightning return strokes, *J. Geophys. Res. 88*, 8471-8474.

Krider, E. P., R. C. Noggle, A. E. Pifer, and D. L. Vance (1980). Lightning direction-finding systems for forest-fire detection, *Bull. Am. Meteorol. Soc. 61*, 980-986.

Lanzerotti, L. J., K. Rinnert, E. P. Krider, M. A. Uman, G. Dehmel, F. O. Gliem, and W. I. Axford (1983). Planetary lightning and lightning measurements on the Galileo Probe to Jupiter's atmosphere, in *Proceeding in Atmospheric Electricity*, J. Latham and L. Ruhnke, eds., Deepak Publ. Co., Hampton, Va., pp. 408-410.

Lin, Y. T., M. A. Uman, J. A. Tiller, R. D. Brantley, E. P. Krider, and C. D. Weidman (1979). Characterization of lightning return stroke electric and magnetic fields from simultaneous two-station measurements, *J. Geophys. Res. 84*, 6307-6314.

Master, M. J., M. A. Uman, Y. T. Lin, and R. B. Standler (1981). Calculations of lightning return stroke electric and magnetic fields above ground, *J. Geophys. Res. 86*, 12127-12132.

Miyachi, I., and K. Horii (1982). Five years' experiences on artificially triggered lightning in Japan, *7th International Conference on Gas Discharges and Applications*, London, pp. 468-471.

Newman, M. M., J. R. Stahman, J. D. Robb, E. A. Lewis, S. G. Martin, and S. V. Zinn (1967). Triggered lightning strokes at very close range, *J. Geophys. Res. 72*, 4761-4764.

Orville, R. E. (1968). A high-speed time-resolved spectroscopic study of the lightning return stroke, Pts. 1, 2, 3, *J. Atmos. Sci. 25*, 827-856.

Proctor, D. E. (1971). A hyperbolic system for obtaining VHF radio pictures of lightning, *J. Geophys. Res. 76*, 1478-1489.

Proctor, D. E. (1981). VHF radio pictures of cloud flashes, *J. Geophys. Res. 86*, 4041-4071.

Proctor, D. E. (1983). Lightning and precipitation in a small multicellular thunderstorm, *J. Geophys. Res. 88*, 5421-5440.

Rustan, P. L., M. A. Uman, D. G. Childers, W. H. Beasley, and C. L. Lennon, (1980). Lightning source locations from VHF radiation data for a flash at Kennedy Space Center, *J. Geophys. Res. 85*, 4893-4903.

Salanave, L. E. (1980). *Lightning and Its Spectrum*, Univ. of Arizona Press, Tucson.

Schonland, B. F. J. (1964). *The Flight of Thunderbolts*, Clarendon Press, Oxford.

St. Privat d'Allier Research Group (1982). Eight years of lightning experiments at St. Privat d'Allier, *Rev. Gen. Electr. 9*, 561-582.

Taylor, W. L. (1978). A VHF technique for space-time mapping of lightning discharge processes, *J. Geophys. Res. 83*, 3575-3583.

Uman, M. A. (1969). *Lightning*, McGraw-Hill, New York.

Uman, M. A, and E. P. Krider (1982). A review of natural lightning: Experimental data and modeling, *IEEE Trans. Electromagn. Compat.*, *EMC-24*, 79-112.

Uman, M. A., D. K. McLain, and E. P. Krider (1975). The electromagnetic radiation from a finite antenna, *Am. J. Phys. 43*, 33-38.

Uman, M. A., M. J. Master, and E. P. Krider (1982). A comparison of lightning electromagnetic fields with the nuclear electromagnetic pulse in the frequency range 10^4-10^7 Hz, *IEEE Trans. Electromagn. Compat.*, *EMC-24*, 410-416.

Waldteufel, P., P. Metzger, J. L. Aouley, P. Laroche, and P. Hubert (1980). Triggered lightning strokes originating in clear air, *J. Geophys. Res. 83*, 2861-2868.

Warwick, J. W., C. O. Hayenga, and J. W. Brosnahan (1979). Interferometric directions of lightning sources at 34 MHz, *J. Geophys. Res. 84*, 2457-2468.

Weidman, C. D., and E. P. Krider (1978). The fine structure of lightning return stroke wave forms, *J. Geophys. Res. 83*, 6239-6247.

Weidman, C. D., and E. P. Krider (1980). Submicrosecond risetimes in lightning return-stroke fields, *Geophys. Res. Lett. 7*, 955-958. [See also C. D. Weidman and E. P. Krider (1982), Correction, *J. Geophys. Res. 87*, 7351.]

Positive Cloud-to-Ground Lightning

3

W. DAVID RUST
NOAA National Severe Storms Laboratory

INTRODUCTION

Of the two common types of lightning flash, cloud-to-ground (CG) and intracloud (IC), the CG flashes have historically received more attention and study. This is undoubtedly because they not only have more visible channels that lend themselves to quantitative observations but also because they are responsible for most death and damage caused by lightning. There are various common names for several visually different features of CG lightning, e.g., streak, forked, and ribbon. However, the parameters often used in the scientific literature to categorize CG lightning are the polarity of charge lowered, the magnitude of the current, and the direction of propagation of the initiating leader and/or ensuing return stroke. While most CG flashes transfer negative charge from the cloud to the ground, early documentation of flashes that lower positive charge from cloud to ground (+CG) is found in Berger's classical study (1967) of lightning at Mount San Salvatore above Lake Lugano in Switzerland. Berger used instruments and photography to document strikes to towers on the mountain and found a minority of +CG flashes. All but one of the +CG flashes had an upward-propagating leader followed by a downward-moving return stroke, altogether opposite to the common negative CG flashes, consisting of downward-propagating leaders followed

by upward return strokes. Because these +CG flashes originated from tall towers, and apparently not even from the mountain peaks without tall structures, they have been termed "triggered" lightning, in contrast to those occurring naturally. They nearly always had only one return stroke. Berger's documentation of +CG flashes to Mount San Salvatore established their devastating nature; they had larger peak currents, charge transfer, action integral, and duration than most negative flashes.

In general the instrumentation used to observe +CG flashes is the same as that for other lightning. Typical measurements include the electrostatic-field change for the entire flash, fast electric- and magnetic-radiation-field wave forms for the return stroke, optical transients, and thunder. Photographic and television recordings are also often made. In addition, one type of modern automatic lightning-strike locating system is being used experimentally in several parts of the world to locate +CG flashes. However, its ability to distinguish between intracloud and +CG flashes and its resulting detection efficiency are still unknown.

RECENT AND ONGOING STUDIES

The recent and renewed interest in +CG flashes has been stimulated by observations of winter storms in Japan, where there is a large percentage (although low total number) of CG flashes that are positive but not triggered by tall structures. These storms have low cloud bases and cloud tops at about 5 km and are much colder in their lower region than typical summer thunderstorms. Although they produce fewer flashes, there have been several reports of strikes to aircraft. Whether the strikes to aircraft were from + CG flashes is unknown, but the prevalence of + CG flashes in these storms has led to questions concerning their correlation to storm structure and their relevance to aviation safety.

The existence of + CG flashes in Japan was first reported by Japanese scientists, who have since conducted collaborative research programs with scientists from the United States and Sweden. The recently published results by Brook et al. (1982) include the best quantitative information on charge transfer for naturally occurring + CG flashes. From their multistation network of electrostatic field-change sensors, they calculated charge transfer and the magnitude of continuing current. As in Berger's earlier work on + CG flashes triggered by tall towers, the naturally occurring + CG flashes in Japan have large currents. While the largest negative flashes are comparable with + CG flashes, it appears that + CG flashes as a group tend to have significantly greater charge transfer and currents.

The combined data from seven Japanese winter storms show a remarkably good correlation between the percentage of + CG flashes per storm and the vertical shear in the horizontal wind in the cloud layer. This suggests that if a storm has its upper positive charge displaced horizontally from its lower negative charge, the production of + CG flashes is facilitated. This may well hold true for such storms that are relatively shallow and only mildly convective; preliminary results from other investigations indicate that the correlation may not be universal. Studies of this are in progress elsewhere.

The finding of naturally occurring + CG flashes in Japan inspired the search for these flashes in the highly sheared, large, and often severe springtime storms over the Great Plains of the United States. The occurrence of + CG flashes during the mature and later stages of severe storms has been verified in the observational program at the National Severe Storms Laboratory (NSSL) in Oklahoma (Rust et al., 1981a). Shown in Figure 3.1 is a sketch of an isolated supercell thunderstorm, which contains an intense updraft, wind shear, turbulence, a large anvil, and a mesocyclone (rotation that can produce a tornado). Flashes to ground in the regions of heavy precipitation have always been observed to lower negative charge, while those from the upshear and downshear anvil near the main storm tower can lower either polarity of charge. Only a relatively few CG

flashes have been observed to emerge from beneath the mesocyclone wall cloud (i.e., the visible manifestation of a mesocyclone; see Figure 3.1), and two have been documented as positive. Rust et al. (1981b) also observed and obtained electric-field change records for 16 + CG flashes in 30 minutes from a downshear anvil, well away from the main tower of a severe storm shortly after it produced a wall cloud. No other flashes to ground were seen from the anvil during that time. In most cases, + CG flashes appear visually to emanate from high in the storm. The + CG flashes can cluster in time and dominate CG activity for certain periods.

Positive CG flashes are observed both in isolated supercells and in squall-line storms. Preliminary analysis of two squall lines indicates that + CG flashes often occur on the back side (relative to squall-line movement) after the squall line has been in existence for several hours. Acoustic recordings of thunder have been analyzed for two of six + CG flashes that were detected within an 8-minute period in the back side of such a squall line. They show a significant number of acoustic sources, and thus channels, throughout a depth of 15 km (the freezing level was about 4 km above ground). Doppler radar data show that these + CG flashes were imbedded in the low radar reflectivities (less than 17 dBZ) associated with very light precipitation behind the squall line. Visual observations indicate that + CG flashes may propagate horizontally through tens of kilometers along the back side of the squall line before coming to ground. Relatively large horizontal extent is also apparent in observations of a few confirmed + CG flashes during the final stage in some of the smaller thunderstorms over the Rocky Mountains where usually there are no or only a few + CG flashes (Fuquay, 1982).

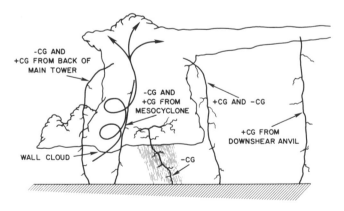

FIGURE 3.1 Sketch of observed locations and polarities of CG flashes from severe thunderstorms. The spiral denotes the updraft region and rotation (mesocyclone). Only negative CG flashes have been observed in the intense precipitation. The horizontal scale of this storm sketch is greatly compressed. After Rust et al. (1981a).

The apparent correlation between + CG flashes and storm development and severity suggested above are quite tentative. This is due to the paucity of + CG flash data, caused in part by the relative infrequency of + CG flashes and in part by the need for corrobative verification of the occurrence of each + CG flash until simpler detection techniques are proven.

Nowhere is this more obvious than in the large data bases recently acquired with a few of the automatic strike-locating systems that have been modified but as yet unproved (as stated by the manufacturer, Lightning Location and Protection, Inc.) for detection of positive as well as negative ground flashes. If we had confidence in the data bases obtained with these systems, much could readily be learned, owing to the large numbers of supposed + CG flashes that have and continue to be recorded. Samples of results from these as yet unproven systems include the following:

1. Orville *et al.* (1983) reported a case study of a cyclone that produced several convective cells that moved through their East Coast strike-locating network. They found that while only 4 percent of all flashes to ground in the storm are identified as + CG flashes, the percentage increased to 37 percent in the last hour of significant CG flash activity. The also reported observations of a higher percentage of + CG flashes in the later stages of other storms.

2. The NSSL strike-locating system has been used to study the diurnal variation of + CG flashes for summertime storms in Oklahoma, which tend to be less severe than storms in the spring. The fraction of CG flashes that were positive, averaged for 1 month, peaks about 2 hours later than the total CG flash activity.

3. Attempts have been made to ascertain if the + CG flashes observed in severe storms are related to storm severity or tornado occurrence. Two tornadic storms have been analyzed using the strike-locating system in Oklahoma. The ratio of + CG to all CG flashes appears to be greater before and during tornadoes than afterward. This result is preliminary, not only because of uncertainties in the performance of the strike locating system but also because only two tornadic storms have been analyzed. It also remains to be shown just what severe storm parameters are related to the production of + CG flashes. Some possibilities that are being examined are mesocyclone strength, updraft speed, shear, and precipitation structure.

PHYSICAL CHARACTERISTICS

Although a few of the characteristics of + CG flashes have been described above, it is worth considering what

we know in total about their characteristics, irrespective of the storm conditions in which they occur. There are two characteristics that appear consistently in all the reported observations of + CG flashes: (1) the vast majority have only a single return stroke, and (2) the return stroke is often followed by continuing current. A representative electrostatic field change for a + CG flash is shown in Figure 3.2. Before the return stroke, there is usually lengthy preliminary activity, which averages about 0.25 sec but can be as long as about 0.8 sec. If in-cloud channels for + CG flashes are primarily horizontal, as suggested by observations in several locations, and if progression speeds are 10^5 m/sec as typically observed, then there may be large horizontal extent to many + CG flashes. Indeed, horizontal movement of + CG flashes before they come to ground has been determined from analysis of multistation field-change data in Florida (Brook *et al.*, 1983) and is indicated also by visual observations of squall lines in Oklahoma.

The field change for the leader to ground has not yet been extensively studied; however, both multistation analysis for several flashes in Japan and photographic evidence for a few flashes in the Rocky Mountains and Oklahoma show that the leader propagates down from the cloud to the ground, in contrast to the initially upward-moving, triggered flashes to Mount San Salvatore. Recent studies in Japan indicate that + CG flashes can be preceded by either a stepped or a nonstepped leader.

The return-stroke wave form (Figure 3.3) is similar in shape to that for negative flashes (Rust *et al.*, 1981b), with a relatively slow initial ramp followed by a faster

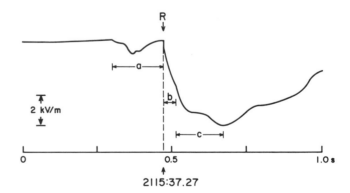

FIGURE 3.2   Typical electric-field change for + CG flash, recorded at 2115:37.270 CST on May 30, 1982. The distance to the flash was about 4 km. Time at the center of the scale (at 0.5 sec) is 2115:37.300 CST. Interval a is the preliminary breakdown and leader; b is the time from the return stroke through the end of the continuing current seen on the streak photograph in Figure 3.4; and c is a larger interval of possible continuing current or additional intracloud breakdown. The return stroke (see Figure 3.3) is labeled R.

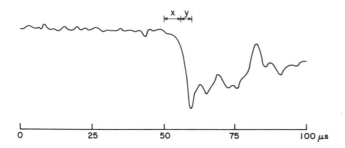

FIGURE 3.3 Electric-field wave form for return stroke of the + CG flash in Figure 3.2. The slow ramp, interval x, is followed in interval y by a faster transition to peak, which is typical of both negative and positive return strokes.

ported to be about twice that for negatives. The peak amplitude of the electrostatic-field change due to the return stroke itself averages about one tenth of the total change for the entire flash. This appears due to the large preliminary breakdown and continuing current, which dominate the field change.

Analysis of multistation electric-field change measurements by Brook *et al.* (1982) reveal + CG flashes with continuing currents up to 10^5 A and positive charge transfer to ground of up to several hundred coulombs. Nakahori *et al.* (1982) made direct measurements of currents in + CG flashes and found peak stroke currents of 31 kA and total charge transfer of 164 coulombs in one flash. The largest magnitudes of charge transfer are often more than 10 times greater than these for negative flashes to ground in summer storms.

The duration of continuing currents in + CG flashes has been reported to vary from a few milliseconds to about 250 msec. However, the longer durations were obtained from single-station field-change measurements. The few streak-film and TV recordings of continuing current obtained thus far indicate that later portions of the slow field change may not always be from continuing current in the channel to ground but may be additional intracloud activity. For example, the streak photograph in Figure 3.4 (for the field change in Figure 3.2) indicates a continuing current duration of 60 msec, but Figure 3.2 alone could be interpreted as indicating at least 200 msec of current flow. Either the luminosity decreased below the threshold for the film, or the cur-

transition to peak. Thus far, the wave forms obtained in several widely separated locations are essentially the same. For 15 visually confirmed + CG flashes in Oklahoma, the average zero-to-peak rise time is 6.9 μsec. In Florida, the average for three visually observed + CG flashes at distances of 20-40 km is about 4 μsec, a value comparable with negative flashes in the same storm. In both locations, some + CG flashes have been observed with fast transition portions of the wave form having rise times of less than 1 μsec. While there was no visual or photographic documentation, apparent + CG flashes in Sweden have yielded zero-to-peak times of 5-25 μsec for flashes at ranges of approximately 100 km; the mean zero-to-peak times for + CG flashes was re-

FIGURE 3.4 Streak-film photograph of + CG flash recorded on May 30, 1982 (see Figures 3.2 and 3.3). Continuing current is evident from the smearing of luminosity. It is visible in the photograph for about 60 msec and occurs during interval b in Figure 3.2.

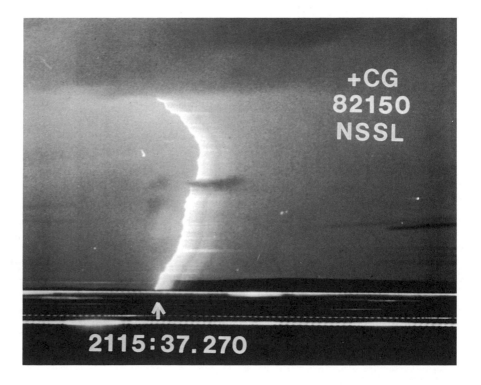

rent in the channel ceased and the remaining activity was intracloud.

PRACTICAL IMPLICATIONS

The destructive nature and practical importance of + CG flashes to the electrical power industry are at least partially documented. The multiline grounding and wire strand fusing in high-voltage transmission lines cannot be explained by normal negative flashes to ground in Japan.

Because of the usual occurrence of continuing current, + CG flashes may ignite a disproportionate number of fires, especially in grasslands and forests. Of the 75 + CG flashes reported in the Rocky Mountain study, all had field changes indicative of continuing current. The apparent pattern is for + CG flashes to strike outside the rainfall, further enhancing the likelihood of their starting a fire. One third of all storm days in a 3-year period had + CG flashes within a 30-km radius of the U.S. Forest Service observing site at Missoula, Montana. Thus on any given day the fire-starting probability from + CG flashes in mountain thunderstorms appears significant.

The results to date of the observations in severe storms suggest that + CG flashes may be correlated with storm severity and tornado occurrence. If this hypothesis were shown to be true, it would enhance existing severe-storm detection and warning capabilities. Additional observations to test this hypothesis are in progress.

An area of increasing importance and study is the effect of lightning on new-generation aircraft (see Chapter 5, this volume). The relevance of + CG flashes to this problem is currently unknown; however, the reports of strikes to aircraft in the winter storms in Japan suggest that + CG flashes were involved. There are several aspects of + CG flashes needing additional clarification to ascertain whether these flashes pose an unusual threat to aircraft. They include the presence of fast return-stroke rise times and large peak currents; the frequent occurrence of continuing current; and the apparent tendency, especially in squall lines, for + CG flashes to propagate through large horizontal and vertical extents and to be in low radar reflectivity regions, which can appear "innocent" on radar.

Because + CG flashes are a small percentage of the total flashes that most storms produce, appear to have large spatial extent, and do not seem to have a unique electric-field change, it is difficult to verify + CG flashes. However, because of their potentially devastating nature and their possible link to storm severity, + CG flashes have become an important research topic in several parts of the world.

AREAS NEEDING ADDITIONAL RESEARCH

It is worth noting those general areas of research that are needed to increase our understanding of + CG flashes and our ability to determine and cope with their effects on important technologies. These research areas are listed below but not in order of priority.

1. Measure electric field changes and wave forms for a large number of confirmed + CG flashes to establish their typical characteristics with greater certainty.

2. Determine what storm types and environmental conditions are conducive to + CG flashes for storms throughout the year.

3. Relate the production of + CG flashes to storm evolution and structure, including also flash initiation and propagation characteristics.

4. Determine typical and extreme peak currents.

5. Evaluate the capabilities of automatic strike-locating systems in identifying + CG flashes, especially their detection efficiency and false identification rate.

6. Determine the significance of + CG flashes to aviation, especially to new-generation aircraft (typified by composite structures and computer-controlled flight).

7. Determine the importance of the threat posed by + CG flashes to power distribution systems, including whether they are the cause of the unexpected large number of faults on power lines in various parts of the United States.

REFERENCES

Berger, K. (1967). Novel observations of lightning discharges: Results of research on Mount San Salvatore, *J. Franklin Inst.* 283, 478-525.

Brook, M., M. Nikano, P. Krehbiel, and T. Takeuti (1982). The electrical structure of the Hokuriku winter thunderstorms, *J. Geophys. Res.* 87, 1207-1215.

Brook, M., P. Krehbiel, D. MacLaughlan, T. Takeuti, and M. Nakano (1983). Positive ground stroke observations in Japanese and Florida storms, in *Proceedings in Atmospheric Electricity*, L. H. Ruhnke and J. Latham, eds., A. Deepak Publishing, Hampton, Va., pp. 365-369.

Fuquay, D. M. (1982). Positive cloud-to-ground lightning in summer thunderstorms, *J. Geophys. Res.* 87, 7131-7140.

Nakahori, K., T. Egawa, and H. Mitani (1982). Characteristics of winter lightning currents in Hokuriku district, manuscript 82WM205-3, IEEE Power Engineering Society 1982 Winter Meeting (Jan. 31-Feb. 5, 1982, New York).

Orville, R. E., R. W. Henderson, and L. F. Bosart (1983). An East Coast lightning detection network, *Bull. Am. Meteorol. Soc. 64*, 1029-1037.

Rust, W. D., W. L. Taylor, D. R. MacGorman, and R. T. Arnold (1981a). Research on electrical properties of severe thunderstorms in the Great Plains, *Bull. Am. Meteorol. Soc. 62*, 1286-1293.

Rust, W. D., D. R. MacGorman, and R. T. Arnold (1981b). Positive cloud-to-ground lightning flashes in severe storms, *Geophys. Res. Lett. 8*, 791-794.

Acoustic Radiations from Lightning

4

ARTHUR A. FEW, JR.
Rice University

ACOUSTIC SOURCES IN THUNDERSTORMS

Electric storms produce a variety of acoustic emissions. The acoustic emissions can be broadly divided into two categories—those that are related to electric processes (i.e., they correlate with lightning) and those that either do not depend on cloud electricity or for which no correlations with electric changes have been observed. Only the first group will be discussed here (see Few, 1982; Georges, 1982, for reviews of nonelectrical acoustics).

Two types of acoustic emissions correlated with electric processes are thunder, which is produced by the rapidly heated lightning-discharge channel, and infrasonic emissions produced by electrostatic fields throughout the charged regions of the cloud. Thunder is probably the most common of all loud natural sounds, while other acoustic emissions are not ordinarily observed without special devices.

THUNDER—THE RADIATION FROM HOT CHANNELS

Spectrographic studies of lightning return strokes (Orville, 1968) show that this electric-discharge process heats the channel gases to a temperature in the 24,000 K range. At high temperatures the expansion speed of the shock wave is roughly 3×10^3 m/sec and decreases rapidly as the shock wave expands; in comparison the measured speeds for various lightning-breakdown processes range from 10^4 to 10^8 m/sec (Uman, 1969; Weber *et al.*, 1982). Therefore, the electric breakdown process in a discharge event is completed in a given length of the channel before the hydrodynamic responses are fully organized. Other electric processes occur over longer periods (e.g., continuing currents), but the energy input to the hot channel is strongly weighted toward the early breakdown processes when channel resistance is higher (Hill, 1971).

Shock-Wave Formation and Expansion

The starting point for developing a theory for a shock-wave expansion to form thunder is the hot ($\sim 24,000$ K), high-pressure ($> 10^6$ Pa) channel left by the electric discharge. Hill's (1971) computer simulation indicated that approximately 95 percent of the total channel energy is deposited within the first 20 μsec with the peak electric power dissipation occurring at 2 μsec; during the 20-μsec period of electric energy input, the shock wave can only move approximately 5 cm. This simulation may actually be slower than real lightning because Hill used a slower current rise time than indicated by

46

more recent measurements (Weidman and Krider, 1978).

The time-resolved spectra of return strokes (Orville, 1968) show the effective temperature dropping from ~30,000 to ~10,000 K in a period of 40 μsec and the pressure of the luminous channel dropping to atmospheric in this same time frame. During this period the shock wave can expand roughly 0.1 m. Even though channel luminosities and currents can continue for periods exceeding 100 μsec, the processes that are important to the generation of thunder occur very quickly (<10 μsec) and in a very confined volume (radius ~5 cm). The strong shock wave propagates outward beyond the luminous channel, which returns to atmosphere pressure within 40 μsec. The channel remnant cools slowly by conduction and radiation and becomes nonconducting at temperatures between 2000 and 4000 K perhaps 100 msec later (Uman and Voshall, 1968).

Turning our attention now to the shock wave itself we can divide its history into three periods—strong shock, weak shock, and acoustic. The division between strong and weak can be related physically to the energy input to the channel; the weak-shock transition to acoustic is somewhat arbitrary. Calculations and measurements have shown that the radiated energy is on the order of 1 percent of the total channel energy (e.g., Uman, 1969; Krider and Guo, 1983), hence most of the available energy is in the form of internal heat energy behind the shock wave.

As the strong shock wave expands it must do thermodynamic work (PdV) on the surrounding fluid. The expected distance though which the strong shock wave can expand will be the distance at which all the internal thermal energy has been expended in doing the work of expansion. Few (1969) proposed that this distance, which he called the "relaxation radius," would be the appropriate scaling parameter for comparing different sources and different geometries. The expressions for the spherical, R_s, and cylindrical, R_c, relaxation radii are

$$R_s = (3E_t/4\pi P_0)^{1/3} \tag{4.1}$$

and

$$R_c = (E_l/\pi P_0)^{1/2}, \tag{4.2}$$

where E_t is the total energy for the spherical shock wave, E_l is the energy per unit length for the cylindrical shock wave, and P_0 is the environmental atmospheric pressure. Table 4.1 gives R_c over a range of values that have been suggested in the literature for E_l. Nondimensional distances denoted by X may be defined for spherical problems by dividing by R_s and for cylindrical problems by dividing by R_c.

Figure 4.1 shows the propagation of the strong shock

TABLE 4.1 Relaxation Radii (R_c) (in meters) for Different Energies per Unit Length (E_l) of Cylindrical Shock Waves

E_l	$P_0 = 100$ kPa (~surface)	$P_0 = 60$ kPa (~4-km height)	$P_0 = 30$ kPa (~9-km height)
10^4	0.18	0.23	0.33
2×10^4	0.25	0.32	0.46
5×10^4	0.40	0.52	0.73
10^5	0.56	0.72	1.02
2×10^5	0.80	1.03	1.46
5×10^5	1.26	1.63	2.30
10^6	1.78	2.30	3.25

into the transition region ($X \sim 1$) and beyond into the weak-shock region. As the shock front passes the relaxation radius ($X = 1$) the central pressure falls below ambient pressure as postulated in the definition of the relaxation radius. The momentum gained by the gas during the expansion carries it beyond $X = 1$ and forces the central pressure to go momentarily below atmospheric. At this point the now weak-shock pulse decouples from the hot-channel remnant and propagates outward. Figure 4.2 shows on a linear coordinate system the final output from Brode's (1955) numerical solution, the weak-shock pulse at a radius of $X = 10.5$.

Figure 4.3 shows Plooster's (1968) cylindrical shock wave near $X = 1$ with Brode's (1955) spherical shock wave. The effects of channel tortuosity will be discussed

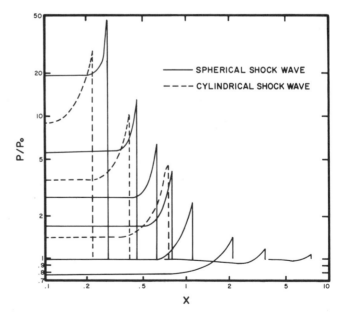

FIGURE 4.1 The expansion of spherical and cylindrical shock waves from the strong-shock region into the weak-shock region. The radii of both spherical and cylindrical geometries have been nondimensionalized using the relaxation radii defined in Eqs. (4.1) and (4.2). The spherical shock wave is that of Brode (1956), and the cylindrical shock wave is from a similarity solution by Sakurai (1954).

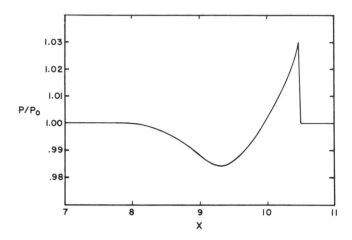

FIGURE 4.2   The weak shock wave formed from the spherical strong shock wave. This is the final pressure profile computed by Brode (1956). For an energy input of 10^5 J/m (R_c = 0.56 m for P_0 = 10^5 Pa) this weak shock wave would be approximately 6 m from the lightning channel.

in greater detail later; for now we note that owing to tortuosity we cannot expect the shock wave to continue to perform as a cylindrical wave once it has propagated beyond a distance equal to the effective straight section of the channel that generated it. If the transition from cylindrical to spherical occurs near X = 1 as suggested by Few (1969), then the spherical weak-shock solutions of Brode provide a good means of estimating the wave shapes of lightning-caused acoustic pulses.

Figure 4.4 presents a graphical summary of the various transitions that are thought to take place. The initial strong shock will behave cylindrically following the dashed line based on Plooster's (1968) computations; this must be the case for the line source regardless of the tortuosity because the high-speed internal waves (3 × 10^3 m/sec) will hydrodynamically adjust the shape of the channel during this phase. The transition from strong shock to weak shock occurs near X = 1, and the transitions from cylindrical divergence to spherical divergence will occur somewhere beyond X = 0.3 and probably beyond X = 1 depending on the particular geometry of the channel at this point. The family of lines labeled χ in Figure 4.4 represent transitions occurring at different points. χ is the effective length, L, of the cylindrical source divided by R_c (χ = L/R_c); it is approximately equal to the value of X at which the transition to spherical divergence takes place.

Comparisons with Numerical Simulations and Experiments

In the numerical solutions of Plooster (1971a, b) and Hill (1971) the energy inputs to the cylindrical problem

were computed as a function of time for specified current wave shapes and channel resistance obtained from the computations in the numerical model. These model results predicted that the energy input to the lightning channel was an order of magnitude or more below the values obtained from electrostatic estimates or from other indirect measurements of lightning energy (Few, 1982). The major differences might be due to the assumed current wave forms used in the models. The recent data obtained with fast-response-time equipment yields current rise times for natural cloud-to-ground lightning in the 35-50 kA/μsec range (Weidman and Krider, 1978). These values are considered as representative of normal strokes; extraordinary strokes have been measured with current rise times in the 100-200 kA/μsec range. By way of comparison, Hill's (1971) current rise time was 2.5 kA/μsec.

Laboratory simulations of lightning have been successfully performed in a series of experiments conducted at Westinghouse Research Laboratories; these results provide us with our best quantitative information on thunder generation. In these tests a 6.4 × 10^6 V impulse generator was used to produce 4-m spark discharges in air (Uman et al., 1970). Circuit instrumentation allowed the measurement of the spark-gap voltage and current from which the power deposition can be computed. Calibrated microphones were used to measure the shock wave from the spark as a function of distance. The results of the research (Uman et al., 1970) have been compared with the theory of Few (1969) and with other

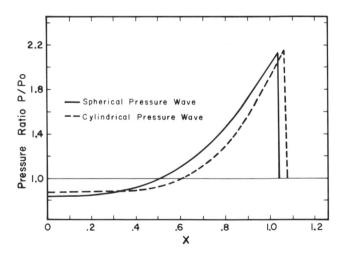

FIGURE 4.3   Comparison of spherical and cylindrical shock-wave shapes near X = 1. These profiles are for the point-source, ideal-gas solutions of Brode (1955) and Plooster (1968). In the transition region of strong shock to weak shock, these wave shapes are nearly identical. From Few (1969) with permission of the American Geophysical Union.

possible interpretations (Plooster, 1971a). The data were found to be consistent with the theory developed by Few.

Figure 4.5 compares a measured spark-pressure pulse with the profile that is predicted from the theory; both represent conditions in the plane perpendicular to the spark channel. Figures 4.6 and 4.7 summarize the extensive series of spark measurements. Figure 4.6 is in the same format as Figure 4.4. The center line passing through the scattered points and labeled $L = 0.5$ m corresponds (using the measured energy input of 5×10^3 J/m, which gives $R_c = 0.126$ m) to $\chi = 4$ in Figure 4.4. The two boundary lines $L = 6.25$ cm and $L = 4.0$ m would correspond to χ values 0.5 and 32. The lower bound is very close to the lower limit value of one third indicated in Figure 4.4. The upper bound of Figure 4.6 ($\chi \simeq 32$) is too large to be depicted in Figure 4.4, where $\chi = 4$ is the last line shown.

The data points of Figure 4.6 corresponding to the larger χ or L values could represent situations where the shock-wave expansion was following the cylindrical behavior over a long distance, hence large χ. However, if the expansions were truly cylindrical to that extent, then

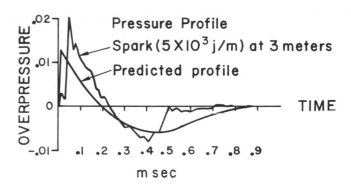

FIGURE 4.5 Comparison of theory with a pressure wave from a long spark. The measured pressure wave from a long spark (Uman *et al.*, 1970) is compared with the predicted pressure from a section of a mesotortuous channel having the same energy per unit length. χ is assumed to be 4/3. From Few (1969) with permission of the American Geophysical Union.

the length of the pulse would be longer, as required by the cylindrical-wave predictions. The data of Figure 4.7 indicate that this cannot be the case. The lengths of the positive-pressure pulses shown in Figure 4.7 are clearly not in the cylindrical regime; if anything, they tend to be even shorter than predicted by the spherical expansion. (See also Figure 4.5.)

It is obvious from both the spark photographs and wave forms in Uman *et al.* (1970) that the spark is tortuous and produces multiple pulses. They found that the wave shapes, more distant from the spark where pulse-transit times were most similar, showed evidence of an in-phase superposition of pulses; at closer range the pulses exhibited greater relative phase shifts and more multiplicity aspects. The in-phase superposition of spherical waves would reproduce the distributions shown in Figures 4.6 and 4.7. The pressure amplitude would be increased relative to a single pulse, but the wavelength would not be substantially affected.

The measured spark wave forms (Uman *et al.*, 1970) were systematically shorter than predicted by the theory. As shown in Figure 4.5, the tail of the wave was compressed, and the data of Figure 4.6 indicate that the positive pulse was similarly shortened. This shortening could be due simply to an inadequacy in the numerical shock-wave model; we think, instead, that the difference results from the energy input being instantaneous in the one case (Brode, 1956) and of longer duration for the spark case. If energy, even in small quantities, continues to be input into the low-density channel core after the shock front has moved outward then the core will be kept at temperatures much higher than predicted by the theories, having an instantaneous energy input followed by expansion. Owing to the elevated sound speed associated with the higher core temperature the part of the

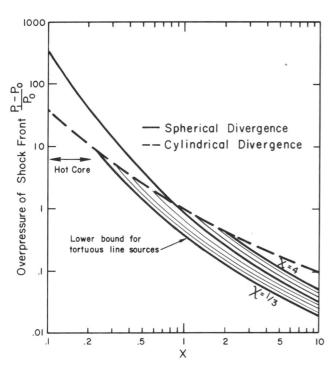

FIGURE 4.4 Line-source shock-wave expansion. The overpressure of the shock front is given for spherical (Brode, 1956) and cylindrical (Plooster, 1968) shock waves. Line sources must initially follow cylindrical behavior, but on expanding to distances of the same size as line irregularities they change to spherical expansion following curves similar to the depicted curves. From Few (1969) with permission of the American Geophysical Union.

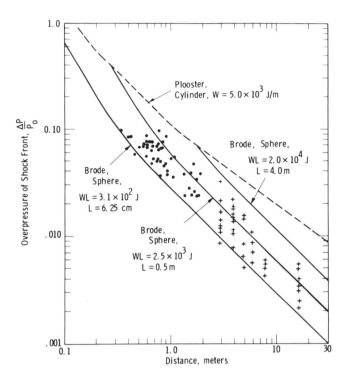

FIGURE 4.6   Shock-front overpressure as a function of distance from the spark. The dots represent data obtained with a piezoelectric microphone; the crosses data obtained with a capacitor microphone. The total electric energy per unit length computed from measurement of the spark voltage and current is 5×10^3 J/m. Also shown are theoretical values for cylindrical and spherical shock waves. From Uman *et al.* (1970).

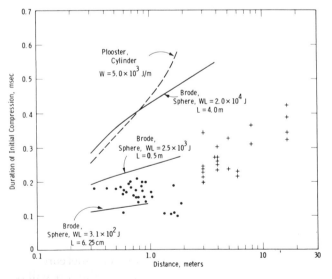

FIGURE 4.7   Duration of positive part of the shock wave from the long spark. For the same data of Figure 4.6, we see here the length of the positive pressure pulse for the 5×10^3 J/m sparks at various distances. From Uman *et al.* (1970).

wave following the shock front will form and propagate outward faster than predicted by theory. We expect, therefore, that the elevated core temperature associated with sparks and lightning can reasonably produce the shortened wave forms.

The wave shape produced by the shock wave is related to the energy per unit length of the lightning flash; thunder is superposition of many such pulses from the lightning channel; hence, the power spectrum of the thunder, with simplifying assumptions, can be related to the energy per unit length of the channel (Few, 1969). Other properties (tortuosity and attenuation) that influence the spectrum of thunder are discussed later.

The assumptions in this theory all affect the thunder spectrum in the same sense; the peak of the theoretical spectrum will occur at higher frequencies than the peak of the real thunder spectrum (Few, 1982). The lightning-channel energy that one estimates from the peak will therefore be an overestimate of the actual lightning-channel energy. Holmes *et al.* (1971) provided the most complete published thunder spectra to date; these spectra show a lot of variation. Most of the spectra are consistent with the qualitative expectations of thunder produced by multiple-stroke lightning, but a few of them exhibit very-low-frequency (<1 Hz) components that are dominant during portions of the record and appear to be totally inconsistent with the thunder-generation theory from the hot explosive channel. Dessler (1973), Bohannon *et al.* (1978), and Balachandran (1979) suggested that these lower-frequency components might be electrostatic in origin; Holmes *et al.* (1971) also considered that this was a possible explanation.

Tortuosity and the Thunder Signature

With respect to the effects of lightning-channel tortuosity on the thunder signal there is almost unanimous agreement among researchers. Lightning channels are undeniably tortuous and are tortuous apparently on all scales (Few *et al.*, 1970). For convenience in discussing channel tortuosity Few (1969) employed the terms microtortuosity, mesotortuosity, and macrotortuosity relative to the relaxtion radius of the lightning shock wave.

For a lightning channel having an internal energy of 10^5 J/m (see Table 4.1), $R_c \simeq 1/2$ m. The microtortuous features smaller than R_c, although optically resolvable, are probably not important to the shock wave as measured at a distance because the high-speed internal waves (3×10^3 m/sec) are capable of rearranging the distribution of internal energy along the channel while the shock remains in the strong-shock regime. At the mesotortuous scale ($\sim R_c$) the outward propagating shock wave decouples from the irregular line source because the acoustic waves from the extended line source can no longer catch up with the shock wave. Somewhere in this mesotortuous range the divergence of the shock waves makes the transition from cylindrical to spherical.

Whereas the mesotortuous channel segments are important in the formation and shaping of the individual pulses being emitted by the channel the macrotortuous segments are fundamental to the overall organization of the pulses and the amplitude modulation of the resulting thunder signature. Few (1974a) computed that 80 percent of the acoustic energy from a short spark was confined to within ± 30° of the plane perpendicular to the short line source. A macrotortuous segment of a lightning channel will direct the acoustic radiations from its constituent mesotortuous, pulse-emitting segments into a limited annular zone. An observer located in this zone (near the perpendicular plane bisecting the macrotortuous segment) will perceive the group of pulses as a loud clap of thunder, whereas another observer outside the zone will perceive this same source as a lower-amplitude rumbling thunder. This relationship between claps, rumbles, and channel macrotortuosity has been confirmed by experiment (Few, 1970) and in computer simulations (Ribner and Roy, 1982).

Loud claps of thunder are produced, as mentioned above, near the perpendicular plane of macrotortuous channel segments; there are three contributary effects (Few, 1974a, 1975) to the formation of the thunder claps. The directed acoustic radiation pattern described above is one of the contributing factors, and this effect is distributed roughly between ± 30° of the plane.

A second effect, which occurs only very close to the plane, is the juxtaposition of several pulses in phase, which increases the pulse amplitude to a greater extent than would a random arrival of the same pulses.

The third effect contributing to thunder clap formation is simply the bunching in time of the pulses. In a given period of time more pulses will be received from a nearly perpendicular macrotortuous segment of channel than from an equally long segment that is perceived at a greater angle owing to the overall difference in the travel times of the composite pulses.

In this section we have examined the complex nature of the formation of individual pulses from hot lightning channels and how a tortuous line source arranges and directs the pulses to form a thunder signature. The resulting thunder signature depends on (1) the number and energy of each rapid channel heating event (leaders and return strokes); (2) the tortuous and branched configuration of the individual lightning channel; and (3) the relative position of the observer with respect to the lightning channel.

Perhaps the most convincing discussion of thunder generation as described above comes not from analytical evidence but from research using sophisticated computer models of thunder. Ribner and Roy (1982) synthesized thunderlike acoustic signals utilizing computer-generated waves formed by the superposition of N wavelets from tortuous geometric sources. The resulting "thunder" is highly similar to natural thunder (see Figures 4.8 and 4.9). Where the computer models are used to simulate laboratory experiments, there is also close agreement.

PROPAGATION EFFECTS

Once generated, the acoustic pulses from the lightning channel must propagate for long distances through the atmosphere, which is a nonhomogeneous, anisotropic, turbulent medium. Some of the propagation effects can be estimated by modeling the propagation using appropriate simplifying assumptions; however, other effects are too unpredictable to be reasonably modeled and must be considered in individual situations.

Three of the largest propagation effects—finite-amplitude propagation, attenuation by air, and thermal refraction—can be treated with appropriate models to account for average atmosphere effects. Reflections from the flat ground can also be easily treated. Once the horizontal wind structure between the source and the receiver are measured, the refractive effects of wind shear and improved transient times may also be calculated. Beyond these effects, elements such as vertical

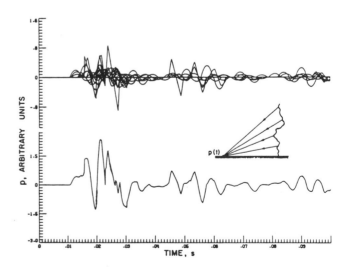

FIGURE 4.8 Schematic depiction of the synthetic generation of thunder by computer by the superimposition (upper trace) of N wavelets from a tortuous line source (Ribner and Roy, 1982); the summed signal is shown on the lower trace.

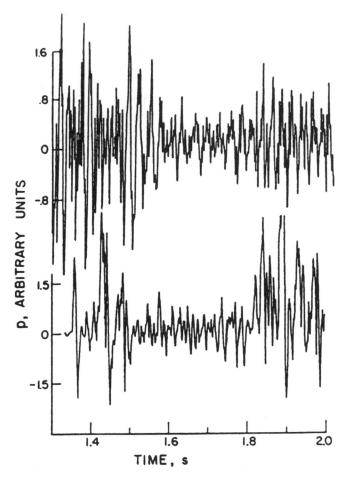

FIGURE 4.9 Comparison of synthetic (upper trace) and real (lower trace) thunder signals (Ribner and Roy, 1982).

winds, nonsteady storm-related horizontal winds, turbulence, aerosol effects, and reflections from irregular terrain produce complications that must be either ignored or examined on a case-by-case basis.

Finite-Amplitude Propagation

As large-amplitude acoustic waves propagate through air, theory predicts that the shape of the wave must evolve with time. A single pulse will evolve to the shape of an N wave (see, for example, the spark wave in Figure 4.5); further propagation of the wave produces a lengthening of this N wave. The best theoretical treatment of this process for application to the thunder problem is the one developed by Otterman (1959). His formulation addressed the lengthening of a Brode-type pulse, such as Figure 4.2, from an initial length (L_0) at an initial altitude (H_0) down to the surface; his treatment differs from many others that do not include the change of ambient pressure (P_0) with altitude. Few

(1982) used the Otterman theory to develop an expression for the lengthening of acoustic pulses generated by mesotortuous lightning-channel elements. The result for the length of the positive-pressure pulse at the ground, L_g, is given by

$$\frac{2}{3}\left(L_g^{3/2} - L_0^{3/2}\right) = \frac{\gamma + 1}{4\gamma} R_0 L_0^{1/2} \Pi_0$$
$$\left[\ln\left(\frac{H_0}{R_0 \cos\theta}\right) - \frac{H_0}{2H_g}\right]. \quad (4.3)$$

R_0 is the distance from the channel to the front of the pulse at the initial state where the fractional overpressure at the pulse front is $\Pi_0 = \delta P_0/P_0$. The angle θ is measured between the acoustic ray path and vertical; γ is the ratio of specific heats; and H_g is the atmosphere scale height.

Equation (4.3) provides the finite-amplitude stretching that should be applied to the waves predicted by strong-shock theory. Uman *et al.* (1970) demonstrated that pulse stretching occurred beyond Brode's final pressure profile shown in Figure 4.2; we see this clearly in Figure 4.7. Few (1969) used linear propagation beyond the profile of Figure 4.2 to estimate the power spectrum of thunder but commented that nonlinear effects may be important. The need for application of nonlinear or finite-amplitude theory to the thunder signal has been voiced in a number of papers in addition to these mentioned above (e.g., Holmes *et al.*, 1971; Few, 1975, 1982; Hill, 1977; Bass, 1980).

If the Brode pressure pulse (shown in Figure 4.4) is used as the initial condition for the finite-amplitude propagation effect, the following values for input to Eq. (4.3) are $R_0 = 10.46R_c$, $L_0 = 0.53R_c$, and $\Pi_0 = 0.03$. In addition, if $\gamma = 1.4$ and $H_g = 8 \times 10^3$ m are used in Eq. (4.3), the following equation is obtained:

$$L_g = R_c\left\{0.386 + 0.147\left[\ln\left(\frac{R_0}{10.46R_c\cos\theta}\right)\right.\right.$$
$$\left.\left. - \frac{H_0}{16 \times 10^3}\right]\right\}^{2/3} \quad (4.4)$$

Equation (4.4) has been used to generate the values in Table 4.2. The relaxation radii (R_c) cover the entire range of values for R_c in Table 4.1. Three values for θ are represented, as are three heights for the source. In general, the finite-amplitude propagation causes a doubling in the length of the positive pulse within the first kilometer, but beyond this range the wavelength remains approximately constant. The theory developed by Otterman did not include attenuation of the signal; because attenuation reduces wave energy, which in turn

TABLE 4.2 Finite-Amplitude Stretching of a Positive Pulse (Length, L_0) for a Range of Cylindrical Relaxation Radii (R_c), Source Heights (H_0), and Angles (θ), See Eq. (4.4)

R_c (m)	0.20	0.40	0.60	0.80	1.00	1.50	2.00	2.50	3.00	3.50
L_0 (m)	0.11	0.21	0.32	0.42	0.53	0.80	1.06	1.33	1.59	1.86
$\theta = 0°$										
L_g (m), H_0 = 1 km	0.24	0.45	0.65	0.84	1.03	1.49	1.93	2.35	2.77	3.17
L_g (m), H_0 = 4 km	0.26	0.49	0.71	0.93	1.14	1.66	2.16	2.65	3.12	3.59
L_g (m), H_0 = 8 km	0.26	0.51	0.74	0.96	1.18	1.72	2.24	2.75	3.25	3.74
$\theta = 45°$										
L_g (m), H_0 = 1 km	0.24	0.46	0.67	0.87	1.06	1.54	1.99	2.44	2.87	3.30
L_g (m), H_0 = 4 km	0.26	0.50	0.73	0.96	1.18	1.71	2.22	2.73	3.22	3.71
L_g (m), H_0 = 8 km	0.27	0.52	0.76	0.99	1.22	1.77	2.31	2.83	3.35	3.85
$\theta = 60°$										
L_g (m), H_0 = 1 km	0.25	0.47	0.69	0.89	1.10	1.59	2.06	2.52	2.98	3.42
L_g (m), H_0 = 4 km	0.27	0.51	0.75	0.98	1.21	1.76	2.29	2.81	3.32	3.82
L_g (m), H_0 = 8 km	0.28	0.53	0.77	1.01	1.25	1.81	2.37	2.91	3.44	3.97

reduces the wave stretching, this theory should be viewed as a maximum estimator of the pulse length.

The finite-amplitude propagation effect does, however, help to resolve the overestimate of lightning-channel energy made by acoustic power-spectra measurements. Few (1969) noted that the thunder-spectrum method yielded a value for E_l that was an order of magnitude greater than an optical measurement by Krider *et al.* (1968). By assuming a doubling in wavelength by the finite-amplitude propagation, the energy estimate is reduced by a factor of 4, bringing the two measurements into a range of natural variations and measurement precision.

Attenuation

There are three processes on the molecular scale that attenuate the signal by actual energy dissipation; the wave energy is transferred to heat. Viscosity and heat conduction, called classical attenuation, represent the molecular diffusion of wave momentum and wave internal energy from the condensation to the rarifaction parts of the wave. The so-called molecular attenuation results from the transfer of part of the wave energy from the translational motion of molecules to their internal molecular rotational and vibrational energy during the condensation part of the wave and back out during the rarifaction part of the wave. The phase lag of the energy transfer relative to the wave causes some of the internal energy being retrieved from the molecules to appear at an inappropriate phase; thus it goes into heat rather than the wave. These three processes can be treated theoretically within a common framework (Kinsler and Frey, 1962; Pierce, 1981). The amplitude of a plane wave, δP, as a function of the distance, x, from the coordinate origin is given by

$$\delta P = \delta P_0 e^{-\alpha x}, \tag{4.5}$$

where δP_0 is the wave amplitude at the origin. The coefficient of attenuation, α, can be shown in the low-frequency regime to be

$$\alpha = \frac{\omega^2 \tau}{2e}. \tag{4.6}$$

In Eq. (4.6), ω is the angular frequency and τ is the relaxation time (or e-folding time) for the molecular process being considered; c is the speed of sound. The low-frequency condition above assumes that $\omega\tau < 1$. The expressions for depend on the particular molecular processes under consideration; it is important to note, however, that α is proportional to ω^2 for the assumed conditions; hence, attenuation alters the spectral shape of the propagating signals.

For thunder at frequencies below 100 Hz it can be shown (Few, 1982) that the total attenuation is insignificant. However, for the many small branches having much lower energy than the main channel, the frequencies will be much higher and attenuation is important. Because of lower initial acoustic energies, spherical divergence, and attenuation it is unlikely that acoustics emitted by the smaller branches and channels can be easily detected over longer distances (see also Bass, 1980; Arnold, 1982).

Scattering and Aerosol Effects

The scattering of acoustic waves from the cloud particles is similar to the scattering of radar waves from the particles; both are strongly dependent on wavelength. The intensity of the scattered sound waves from a plane acoustic wave of wavelength, λ, incident on a hard stationary sphere of radius a is proportional to $(\pi a^2)(a/\lambda)^4$; this is the same relationship that appears in the radar cross-section expression for these parameters. For thunder wavelengths (~ 1 m) and cloud particles ($\sim 10^{-3}$ m) the ratio $(a/\lambda)^4$ is 10^{-12}. The cloud is, therefore, transparent to low-frequency thunder just as it is to meter-wavelength electromagnetic radiation, although insignificant fractions of the radiation do get scattered.

There are, however, eddies in the same size range as low-frequency thunder wavelengths, and these features, owing to small thermal changes and flow shears, produce a distortion of wave fronts and scattering-type effects. For the part of the turbulent spectrum having wavelengths smaller than the acoustic wavelengths of interest, the turbulence can be treated statistically by scattering theory. Larger-scale turbulence must be described with geometric acoustics. For the low-frequency thunder, turbulent scattering will attenuate the high-amplitude beamed parts of the thunder signal; this increases the rumbles at the expense of claps.

In the first part of this subsection we discussed the cloud particles as sources of acoustic scattering; there are other and probably more important ways in which these aerosol components interact with the acoustic waves. First, the surface area of the cloud particles within a volume provides preferred sites for enhanced viscosity and heat conduction; hence, the presence of particles increases the classical attenuation coefficient. Another totally different process produces attenuation by changing the thermodynamic parameters associated with the acoustic wave over the surfaces of cloud particles; this changes the local vapor-to-liquid or vapor-to-solid conversion rates. For example, during the compressional part of the wave the air temperature is increased and the relative humidity is decreased relative to equilibrium; the droplets partially evaporate in response and withdraw some energy from the wave to accomplish it. The opposite situation occurs during the expansion part of the wave. Because the phase-change energy is ideally 180 out of phase with the acoustic-wave energy this process produces attenuation. Landau and Lifshitz (1959) included this effect in their "second viscosity" term. This attenuation process differs from the other microscopic processes in that it can be effective at the lower frequencies. The magnitude of this effect plus the enhanced attenuation by viscous and heat conduction at the surface exceed that of particle-free air by a factor of 10 or greater depending on the type, size, and concentration of the cloud particles (Kinsler and Frey, 1962).

Finally, there is a mass-loading effect with respect to the cloud particles that must be considered. The amplitude of the fluid displacement, ζ, produced by an acoustic wave of pressure amplitude δp and angular frequency ω is (Kinsler and Frey, 1962)

$$\zeta = \frac{\delta p}{\rho_0 c \omega}. \tag{4.7}$$

Using 50 Pa as a representative value of δp for thunder inside a cloud we find for a 100-Hz frequency that $\zeta = 100$ μm. The part of the cloud particle population whose diameter is much smaller than this, say 10 μm, should, owing to viscous drag, come into dynamic equilibrium with the wave flow. [Dessler (1973) computed the response time for a ~ 10-μm droplet to re-establish dynamic equilibrium with drag forces; only 10^{-3} sec is required.] These cloud particles, which participate in the wave motion, add their mass to the effective mass of the air; this effects both the speed of sound and the impedence of the medium. For higher-frequency waves, fewer cloud particles participate, so the effect is reduced; whereas lower-frequency waves include greater percentages of the population and are more strongly affected. Clouds are, therefore, dispersive with respect to low-frequency waves. Also, the cloud boundary acts as a partial reflector of the low-frequency acoustic signals because of the impedence change at the boundary. Assuming a total water content of order 5 g/m^3, we estimate that the order of magnitude of the effect on sound speed and impedence is 10^{-3}; this is not large, but it may be detectable.

The cloud aerosols interact with the acoustic waves in three different ways depending on their size relative to the amplitude of air motion of the sound. The smallest fraction "ride with the wave"—altering the wave-propagation parameters. The largest particles are stationary and act as scatterers of the acoustic waves. The particles in the middle range provide a transition scale for the above effects but are primarily responsible for enhanced viscous attenuation.

In summary, there are several processes that can effectively attenuate higher-frequency components of thunder; this is in support of the conclusions of the previous section. We have, in addition, found three processes that affect the low-frequency components. Low frequencies can be attenuated by turbulent scattering and, in the cloud, by coupling wave energy to phase changes. We have also found that low frequencies interact with the cloud population dynamically; as a result, cloud boundaries may act as partial reflectors and in-cloud propagation may be dispersive.

Refraction

There is a wide range of refractive effects in the environment of thunderstorms. In the preceding section we found that turbulence on the scale of the acoustic wavelength and smaller could be treated with scattering theory. Turbulence larger than acoustic wavelengths, up to and including storm-scale motions, should be describable by geometric acoustics or ray theory. To actually do this is impractical because it requires detailed information (down to the turbulent scale) of temperature and velocity of the air everywhere along the path between the source and the observer. Since the thunder sources are widely distributed we would require complete knowledge of the storm environment down to the meter scale to trace accurately the path of an individual acoustic ray. These requirements can be relieved if we relax somewhat our expectations regarding the accuracy of our ray path. The three fluid properties that cause an acoustic ray to change its direction of propagation are the components of thermal gradient, velocity gradient, and velocity that are perpendicular to the direction of propagation. Beyond the overall thermal structure of the environment, which will be approximately adiabatic, we do not expect that the thermal perturbations due to turbulence will be systematic. In fact, the turbulent thermal perturbations should be random with a zero average value; hence, an acoustic ray propagating through turbulence should not deviate markedly owing to thermal gradients associated with the turbulence from the path predicted by the overall thermal structure of the environment. Similarly, velocity and velocity gradients should produce a zero net effect on the acoustic ray propagating through the turbulence.

This argument of compensating effects is not valid for large eddies whose dimensions are equal to or greater than the path length of the ray because the ray path is over a region containing a systematic component of the gradients associated with the large eddy. We can obtain a worst-case estimate of these effects by examining a horizontal ray propagating from a source at the center of an updraft of 30 m/sec through 2 km to the cloud boundary where the vertical velocity is assumed to be zero; we also assume a linear decrease in vertical velocity between the center and boundary. The ray will be "advected" by 90 m upward during this transit, which requires approximately 6 sec, while the direction of propagation of the ray will be rotated through 5 downward (maximum angle $\simeq \tan^{-1} \Delta V/C$). Owing to this rotation, which is a maximum computation, the "apparent" source by straight-ray path would be 180 m above the real source. These two effects have been estimated independently when, in fact, they are coupled and are to some extent compensatory; when we merely add them the result is an overestimate of the apparent source shift, which in this example is 270 m. If this worst case is the total error in propagation to the receiver at 5

km then this error represents 5 percent of the range; over the length in which it occurs, 2 km, it represents 13 percent error.

Now we turn our attention to the large-scale refraction effects that can be incorporated in an atmospheric model that employs horizontal stratification. The two strongest refractive effects of the atmosphere—the vertical thermal gradient and boundary-layer wind shears—fall into this category along with other winds and wind shears of less importance.

The nearly adiabatic thermal structure of the atmosphere during thunderstorm conditions has been recognized for a long time as a strong influence on thunder propagation (Fleagle, 1949). This thermal gradient is effective because it is spatially persistent and unavoidable. Even though the temperature in updrafts and downdrafts—inside and outside the cloud—may differ (sometimes significantly), the thermal gradients in all parts of the system will be near the adiabatic limit (or pseudoadiabatic in some cases) because of the vertical motion. Hence, the acoustic rays propagate in this strong thermal gradient throughout its existence.

We can employ a simplified version of ray theory to illustrate some of the consequences of this thermal structure. If we assume no wind, a constant lapse rate ($\Gamma = -\partial T/\partial z$), and $\Delta T/T_0 \ll 1$ (ΔT is the change in temperature and T_0 is the maximum temperature along the path), then the ray path may be described as a segment of a parabola

$$l^2 = \frac{4 T_0}{\Gamma} h. \qquad (4.8)$$

In Eq. (4.8), T_0 also corresponds to the vertex of the parabola where the ray slope passes through zero and starts climbing. h and l are, respectively, the height above the vertex and the horizontal displacement from the vertex. To apply Eq. (4.8) to all rays it is necessary to ignore (mathematically) the presence of the ground because the vertices of rays reflecting from the ground are mathematically below ground. In addition, we must in other cases visualize rays extending backward beyond the source to locate their mathematical vertices.

If T_0 is set equal to the surface temperature, a special acoustic ray that is tangent to the surface when it reaches the surface is defined; this is depicted in Figure 4.10. This same ray is applicable to any source, such as S_1, S_2, or S_3, that lies on this ray path. For the conditions assumed in this approximation it is not possible for rays from a point source to cross one another (except those that reflect from the surface). The other acoustic rays emanating from S_2 must pass over the point on the ground where the tangent ray makes contact; this is also true for rays reflecting from the surface inside the tan-

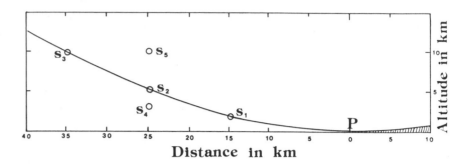

FIGURE 4.10 Parabolic acoustic ray from sources S_1, S_2, or S_3 tangent to the surface at P. This ray was generated utilizing Eq. (4.8) with T_0 = 30°C and Γ = 9.8 K/km. Observers on the surface to the right of P cannot detect sound from sources S_1, S_2, S_3, or S_4; an observer at P can only detect sound originating on or above the parabolic ray shown.

gent point. The shaded zone in Figure 4.10 corresponds to a shadow zone that receives no sound from any point source on the tangent ray beyond the tangent point. Point sources below the tangent ray, such as source S_4 in Figure 4.10, have their tangent ray shifted to the left in this representation and similarly cannot be detected in the shadow zone. However, sources above the tangent ray, S_5 for example, can be detected in some parts of the shadow zone.

For each observation point on the ground one can define a paraboloid of revolution about the vertical generated by the tangent ray through the observation point; the observer can only detect sounds originating above this parabolic surface. For this reason we usually hear only the thunder from the higher parts of the lightning channel unless we are close to the point of a ground strike. For evening storms, which can often be seen at long distances, it is common to observe copious lightning activity but hear no thunder at all; thermal refraction is the probable cause of this phenomenon. For T_0 = 30°C, Γ = 9.8 K/km, and h = 5 km we find that l = 25 km; as noted by Fleagle (1949) thunder is seldom heard beyond 25 km. (See also the discussion in Ribner and Roy, 1982.)

Winds and wind shears also produce curved-ray paths but are more difficult to describe because they affect the rays in a vectorial manner, whereas the temperature was a scalar effect. If you are downwind of a source and the wind has positive vertical shear ($\partial u / \partial z > 0$), the rays will be curved downward by the shear; on the upwind side, the rays are curved upward. Wind shears are very strong close to the surface and can effectively bend the acoustic rays that propagate nearly parallel to the surface. The combined effects of temperature gradients, winds, and wind shears can best be handled with a ray-tracing program on a computer. With such a program one can accurately trace ray paths through a multilevel atmosphere with many variations in the parameters; it is usually necessary in these programs to assume horizontal stratification of the atmosphere. The accuracy of the ray tracing by these techniques can be very high, usually exceeding the accuracy with which temperature and wind profiles can be determined.

MEASUREMENTS AND APPLICATIONS

A number of the experimental and theoretical research papers dealing with thunder generation have been discussed in earlier sections and will not be repeated here. In this section we describe additional results, techniques, and papers that deal with thunder measurements.

Propagation Effects Evaluation

The reader should have, at this point, an appreciation for the difficulty in quantitatively dealing with the propagation effects on both the spectral distribution of thunder and the amplitude of the signal. If, however, we are willing to forfeit the information content in the higher-frequency (> 100 Hz) portion of the thunder signal, which is most strongly affected by propagation, we can recover some of the original acoustic properties from the low-frequency thunder signal.

If the peak in the original power spectrum of thunder is assumed to be below 100 Hz, then the ω^2-attenuation effects deplete the higher frequencies without shifting the position of the peak. Most spectral peaks of thunder tend to be around or below 50 Hz; therefore, this assumption appears to be safe even with finite-amplitude stretching effects considered. Further assume that the spectra are not substantially altered by turbulent scattering and cloud aerosols. To the extent that these assumptions are valid, the finite-amplitude stretching can be removed from the thunder signal and its peak frequency at the source can be estimated. This technique enables a rough estimate of the energy per unit length of the stroke to be made; the result is corrected for first-order propagation effects. Holmes *et al.* (1971) found that the spectral peak overestimated the channel energy using Few's (1969) method; if corrected for stretching these measurements are in closer agreement, with the exception of those events containing other lower-frequency acoustic sources.

There are a number of experiments that could and should be done to evaluate the propagation effects. Using thunder as the acoustic source, several widely separated arrays of microphones could compare signals from the same source at several distances. If carefully executed this experiment could quantify some of the propagation effects. Another approach would be to employ a combination of active and passive experiments such as point-source explosions inside clouds from either balloons or rockets. This experiment provides an additional controllable factor that can yield more precise data; it also involves greater cost and hazard.

Acoustic Reconstruction of Lightning Channels

In the section on refraction we mentioned the utility of ray-tracing computer programs that could accurately calculate the curved path of an acoustic ray from its source to a receiver; the accuracy is limited to the precision with which we are able to define the atmosphere. An obvious application of thunder measurements is to invert this process; one measures thunder then traces it backward from the point of observation along the appropriate ray to its position at the time of the flash. Few (1970) showed that by performing this reverse-ray propagation for many sources in a thunder record it was possible to reconstruct in three dimensions the lightning channel producing the thunder signal. The sources in this case were defined by dividing the thunder record into short (~ 1/2 sec) intervals and associating the acoustics in a given time interval with a source on the channel.

Within each time interval the direction of propagation of the acoustic rays are found by cross correlating the signals recorded by an array of microphones. The position of the peak in the cross-correlation fraction gives the difference in time of arrival of the wave fronts at the microphones; from this and the geometry of the array, one calculates the direction of propagation. At least three noncollinear microphones are required. Close spacing of the microphones produces higher correlations and shorter intervals thus more sources; however, the pointing accuracy of a small array is less than that of a large array. Based on experiences with several array shapes and sizes, 50 m^2 has been adopted as the optimum by the Rice University Group (see Few, 1974a).

The reconstruction of lightning channels by ray tracing was described by Few (1970) and Nakano (1973). A discussion of the accuracy and problems of the technique is given in Few and Teer (1974) in which acousti-

cally reconstructed channels were found to agree closely with photographs of the channels below the clouds. The point is dramatically made in these comparisons that the visual part of the lightning channel is merely the "tip of the iceberg."

Nakano (1973) reconstructed, with only a few points per channel, 14 events from a single storm. Teer and Few (1974) reconstructed all events during an active period of a thunderstorm cell. MacGorman et al. (1981) similarly performed whole-storm analyses by acoustic channel reconstruction and compared statistics from several different storm systems. Reconstructed lightning channels by ray tracing have been used to support other electric observations of thunderstorms at the Langmuir Laboratory by Weber et al. (1982) and Winn et al. (1978).

A second technique for reconstructing lightning channels has been developed that is called thunder ranging. This technique was developed to provide a quick coarse view of channels (within minutes after lightning if necessary) as opposed to the ray-tracing technique, which is slow and time consuming. Thunder ranging requires thunder data from at least three noncollinear microphones separated distances on the order of kilometers. Experience with cross-correlation analysis of thunder signals has shown that the signals become spatially incoherent at separations greater than 100 m owing to differences in perspective and propagation path. However, the envelope of the thunder signals and the gross features such as claps remain coherent for distances on the order of kilometers. As discussed earlier these gross features are produced by the large-scale tortuous sections of the lightning channel. Thunder ranging works as follows: (1) The investigator identifies features in the signals (such as claps) that are common to three thunder signatures on an oscillograph. (2) The time lags between the flash and the arrival of each thunder feature at each measurement point are determined. (3) The ranges to the lightning channel segments producing each thunder feature are computed. (4) The three ranges from the three separated observation points for each thunder feature define three spheres, which should have a unique point in space that is common to all of them. (5) The set of points gives the locations of the channel segments producing the thunder features (see Few, 1974b; Uman et al., 1978).

The basic criticism of the thunder-ranging technique is that the selection of thunder features is the subjective judgment of the researcher; for many features the selection is unambiguous; other features, which are close together, may appear separated at one location and merged at another. The program developed by Bohannon (1978) included these uncertainties in the estimation of errors associated with such points. Most of the recent thunder research has used a combination of ranging and ray tracing.

The whole-storm studies in which an extended series of channels are reconstructed have proven to be the most valuable use of thunder data to date. They define the volume of the cloud actually producing lightning, the evolution of the lightning-producing volume with time, and the relationship of individual channels with other cloud observations such as radar reflectivity and environmental winds (Nakano, 1973; Few, 1974b; Teer and Few, 1974; Few *et al.*, 1977, 1978; MacGorman and Few, 1978; MacGorman *et al.*, 1981).

ELECTROSTATICALLY PRODUCED ACOUSTIC EMISSIONS

The concept of electrostatically produced acoustic waves from thunderclouds goes all the way back to the writings of Benjamin Franklin in the eighteenth century; Wilson (1920) provided a rough quantitative estimate of the magnitude of the electrostatically produced pressure wave. McGehee (1964) and Dessler (1973) developed quantitative models for this phenomenon—McGehee for spherical symmetry and Dessler for spherical, cylindrical, and disk symmetries. The theory developed by Dessler is of particular importance because it made specific predictions regarding the directivity and shape of the wave. The predictions were subsequently verified in part by Bohannon *et al.* (1977) and Balachandran (1979, 1983).

The charge in a thundercloud resides principally on the cloud drops and droplets. In a region of the cloud where the charge is concentrated producing an electric field E, the charged particles will experience an electric force, which is directed outward with respect to the charge center, in addition to the other forces expressed on them. These particles quickly (on the order of milli-

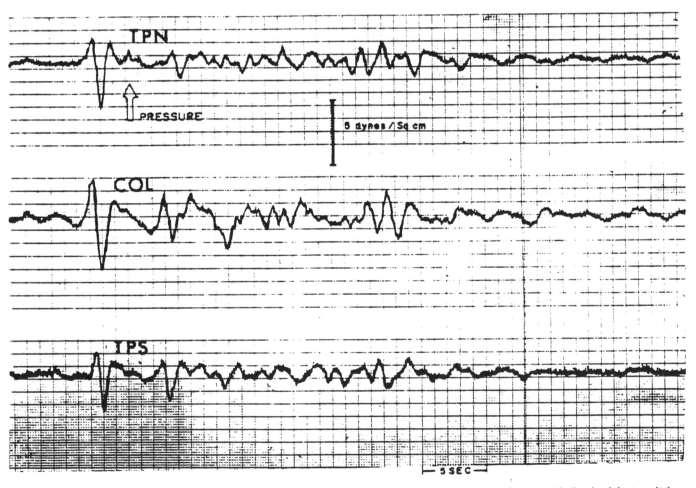

FIGURE 4.11 Low-frequency acoustic pulse thought to have been generated by an electrostatic pressure change inside the cloud during a lightning flash. The higher-frequency signals from thunder have been removed from this record. From Balachandran (1979) with permission of the American Geophysical Union.

seconds) come into dynamic equilibrium where the hydrodynamic drag force associated with their motion is balanced by the sum of all the externally expressed forces. When the electric field is quickly reduced by a lightning flash the cloud particles readjust to a new dynamic equilibrium. The change in the hydrodynamic drag force requires a change in the pressure distributions surrounding all the charged cloud particles; hence, the pressure in the volume containing the cloud particles is altered by the sudden reduction of the electric field. Since the electric force from a charge concentration is outward, the pressure inside the charged volume will be slightly lower than the surrounding air. When E is reduced by the lightning flash the charged volume produces a slight implosion; this radiates a negative wave. Few (1982) derived a general expression for the internal pressure gradient produced by the electrostatic force; when integrated the result is

$$P_E = P_0 - (n + 1) \frac{\epsilon_0(E_0^2 - E^2)}{2}. \quad (4.9)$$

In Eq. (4.9) the parameter n takes the value 0 for plane geometry, 1 for cylindrical geometry, and 2 for spherical geometry; P_0 and E_0 are the values at the edge of the charged volume.

The amplitude of this pressure signal is related to the electric field, the wavelength to the thickness of the charged region, and the directivity of the wave to the geometry to the source (Dessler, 1973). If the theory can be quantitatively verified, the signal can be used to determine remotely internal cloud electric parameters.

The experimental search for electrostatic pressure waves has been difficult. The wave is low frequency (~ 1 Hz), small amplitude (~ 1 Pa), and buried in large background pressure variations produced by wind, turbulence, and thunder. Prior to Dessler's prediction of beaming, one wondered why the signal was not more frequently seen in thunder measurements. Holmes *et al.* (1971) measured a low-frequency component in a few of their power spectra of thunder but found these components completely missing in others. Dessler showed that that signal would be beamed for cylindrical and disk geometry; the disk case would require that the detectors be placed directly underneath the charged volume for observation. This relationship has been observed by Bohannon *et al.* (1977) and by Balachandran (1979, 1983).

The electrostatic pressure wave predicted by the theory discussed above is a negative pulse. The measured acoustic signature thought to be the verification of the prediction actually exhibits a positive pulse followed by a negative pulse (see Figure 4.11). The negative pulse appears to fit the theory, but the theory is deficient in

that the positive component of the wave is not described. Recently, Few (1984) suggested that the diabatic heating of the air in the charged volume by positive streamers may be the source of the positive pulse.

Colgate and McKee (1969) described theoretically an electrostatic pressure pulse using this same mechanism but applied to a volume of charged air surrounding a stepped leader. This particular signature has not been experimentally verified because it has the regular thunder signal, which is 300 times more energetic, superimposed on it.

ACKNOWLEDGMENT

The author's research into the acoustic radiations from lightning has been supported under various grants and contracts from the Meteorology Program, Division of Atmospheric Sciences, National Science Foundation, and the Atmospheric Sciences Program, Office of Naval Research; their support is gratefully acknowledged.

REFERENCES

Arnold, R. T. (1982). Storm acoustics, in *Instruments and Techniques for Thunderstorm Observation and Analysis*, E. Kessler, ed., U.S. Department of Commerce, Washington, D.C., pp. 99-116.

Balachandran, N. K. (1979). Infrasonic signals from thunder, *J. Geophys. Res. 84*, 1735-1745.

Balachandran, N. K. (1983). Acoustic and electric signals from lightning, *J. Geophys. Res. 88*, 3879-3884.

Bass, H. E. (1980). The propagation of thunder through the atmosphere, *J. Acoust. Soc. Am. 67*, 1959-1966.

Bohannon, J. L. (1978). Infrasonic pulses from thunderstorms, M.S. thesis, Rice Univ., Houston, Tex.

Bohannon, J. L., A. A. Few, and A. J. Dessler (1977). Detection of infrasonic pulses from thunderclouds, *Geophys. Res. Lett. 4*, 49-52.

Brode, H. L. (1955). Numerical solutions of spherical blast waves, *J. Appl. Phys. 26*, 766.

Brode, H. L. (1956). The blast wave in air resulting from a high temperature, high pressure sphere of air, Rand Corp. Res. Memorandum RM-1825-AEC.

Colgate, S. A., and C. McKee (1969). Electrostatic sound in clouds and lightning, *J. Geophys. Res. 74*, 5379-5389.

Dessler, A. J. (1973). Infrasonic thunder, *J. Geophys. Res. 78*, 1889-1896.

Few, A. A. (1969). Power spectrum of thunder, *J. Geophys. Res. 74*, 6926-6934.

Few, A. A. (1970). Lightning channel reconstruction from thunder measurements, *J. Geophys. Res. 75*, 7517-7523.

Few, A. A. (1974a). Thunder signatures, *EOS 55*, 508-514.

Few, A. A. (1974b). Lightning sources in severe thunderstorms, in *Conference on Cloud Physics* (Preprint volume), American Meteorological Society, Boston, Mass., pp. 387-390.

Few, A. A. (1975). Thunder, *Sci. Am. 233*(1), 80-90.

Few, A. A. (1982). Acoustic radiations from lightning, in *Handbook of Atmospherics*, Vol. 2, H. Volland, ed., CRC Press, Inc., Boca Raton, Fla., pp. 257-289.

Few, A. A. (1984). Lightning-associated infrasonic acoustic sources, in

Preprints: VII International Conference of Atmospheric Electricity, American Meteorological Society, Boston, Mass., pp. 484-486.

Few, A. A., and T. L. Teer (1974). The accuracy of acoustic reconstructions of lightning channels, *J. Geophys. Res. 79*, 5007-5011.

Few, A. A., H. B. Garrett, M. A. Uman, and L. E. Salanave (1970). Comments on letter by W. W. Troutman, "Numerical calculation of the pressure pulse from a lightning stroke," *J. Geophys. Res. 75*, 4192-4195.

Few, A. A., T. L. Teer, and D. R. MacGorman (1977). Advances in a decade of thunder research, in *Electrical Processes in Atmospheres*, H. Dolezelak and R. Reiter, eds., Steinkopff, Darmstadt, pp. 628-632.

Few, A. A., D. R. MacGorman, and J. L. Bohannon (1978). Thundercloud charge distributions, inferences from the intracloud structure of lightning channels, in *Conference on Cloud Physics and Atmospheric Electricity*, American Meteorological Society, Boston, Mass., pp. 591-596.

Fleagle, R. G. (1949). The audibility of thunder, *J. Acoust. Soc. Am. 21*, 411.

Georges, T. M. (1982). Infrasound from thunderstorms, in *Instruments and Techniques for Thunderstorm Observation and Analysis*, E. Kessler, ed., U.S. Department of Commerce, Washington, D.C., pp. 117-133.

Hill, R. D. (1971). Channel heating in return-stroke lightning, *J. Geophys. Res. 76*, 637-645.

Hill, R. D. (1977). Thunder, in *Lightning*, R. H. Golde, ed., Academic Press, New York, pp. 385-408.

Holmes, C. R., M. Brook, P. Krehbiel, and R. A. McCrory (1971). On the power spectrum and mechanism of thunder, *J. Geophys. Res. 76*, 2106-2115.

Kinsler, L. E., and A. R. Frey (1962). *Fundamentals of Acoustics*, 2nd ed., Wiley, New York, 523 pp.

Krider, E. P., and C. Guo (1983). The peak electromagnetic power radiated by lightning return strokes, *J. Geophys. Res. 88*, 8471-8474.

Krider, E. P., G. A. Dawson, and M. A. Uman (1968). Peak power and energy dissipation in a single-stroke lightning flash, *J. Geophys. Res. 73*, 3335-3339.

Landau, L. D., and E. M. Lifshitz (1959). *Fluid Mechanics*, Pergamon Press, London, 536 pp.

MacGorman, D. R., and A. A. Few (1978). Correlations between radar reflectivity contours and lightning channels for a Colorado storm on 25 July 1972, *Conference on Cloud Physics and Atmospheric Electricity*, American Meteorological Society, Boston, Mass., pp. 597-600.

MacGorman, D. R., A. A. Few, and T. L. Teer (1981). Layered lightning activity, *J. Geophys. Res. 86*, 9900-9910.

McGehee, R. M. (1964). The influence of thunderstorm space charges on pressure, *J. Geophys. Res. 69*, 1033-1035.

Nakano, M. (1973). Lightning channel determined by thunder, *Proc. Res. Inst. Atmospherics* (Nagoya Univ.) *20*, 1-7.

Orville, R. E. (1968). A high-speed time-resolved spectroscopic study of the lightning return stroke, *J. Atmos. Sci. 25*, 827-856.

Otterman, J. (1959). Finite-amplitude propagation effect on shock-wave travel times from explosions at high altitudes, *J. Acoust. Soc. Am. 31*, 470-747.

Pierce, A. D. (1981). *Acoustics an Introduction to Its Physical Principles and Applications*, McGraw-Hill, New York, 642 pp.

Plooster, M. N. (1968). Shock waves from line sources, NCAR-TN-37, National Center for Atmospheric Research, Boulder, Colo., 83 pp.

Plooster, M. N. (1971a). Numerical simulation of spark discharges in air, *Phys. Fluid 14*, 2111-2123.

Plooster, M. N. (1971b). Numerical model of the return stroke of the lightning discharge, *Phys. Fluids 14*, 2124-2133.

Ribner, H. S., and D. Roy (1982). Acoustics of thunder: A quasilinear model for tortuous lightning, *J. Acoust. Soc. Am. 72*, 1911-1925.

Sakurai, A. (1954). On the propagation and structure of the blast wave, 2, *J. Phys. Soc. Japan 9*, 256-266.

Teer, T. L., and A. A. Few (1974). Horizontal lightning, *J. Geophys. Res. 79*, 3436-3441.

Uman, M. A. (1969). *Lightning*, McGraw-Hill, New York.

Uman, M. A., and R. E. Voshall (1968). Time interval between lightning strokes and the initiation of dart leaders, *J. Geophys. Res. 73*, 497-506.

Uman, M. A., A. H. Cookson, and J. B. Moreland (1970). Shock wave from a four-meter spark, *J. Appl. Phys. 41*, 3148-3155.

Uman, M. A., W. H. Beasley, J. A. Tiller, Y.-T. Lin, E. P. Krider, C. D. Weidman, P. R. Krehbiel, M. Brook, A. A. Few, Jr., J. L. Bohannon, C. L. Lennon, H. A. Poehler, W. Jafferis, J. R. Gluck, and J. R. Nicholson (1978). An unusual lightning flash at Kennedy Space Center, *Science 201*, 9-16.

Weber, M. E., H. J. Christian, A. A. Few, and M. F. Stewart (1982). A thundercloud electric field sounding: Charge distribution and lightning, *J. Geophys. Res. 87*, 7158-7169.

Weidman, C. D., and E. P. Krider (1978). The fine structure of lightning return stroke wave forms, *J. Geophys. Res. 83*, 6239-6247.

Wilson, C. T. R. (1920). Investigations on lightning discharges and on the electric field of thunderstorms, *Phil. Trans. R. Soc. London, Ser. A 221*, 73-115.

Winn, W. P., C. B. Moore, C. R. Holmes, and L. G. Byerly (1978). Thunderstorm on July 16, 1975, over Langmuir Laboratory: A case study, *J. Geophys. Res. 83*, 3079-3092.

Application of Advances in Lightning Research to Lightning Protection

5

MARTIN A. UMAN
University of Florida

INTRODUCTION

Significant advances in lightning protection have been made during the last decade. These advances have been a result of progress in two general areas of lightning research: (1) lightning phenomenology, including the technology for determining real-time strike locations, and (2) lightning physics, particularly the characteristics of return stroke currents and electromagnetic fields (see Krider, Chapter 2, this volume, for a description of the return-stroke phase of a lightning flash, as well as of the other salient events that make up the flash).

(1) By phenomenology, we mean those characteristics of thunderstorms that are associated with numbers of lightning events, as opposed to the physical properties of the individual events. A phenomenological parameter of particular interest is the average lightning flash density, that is, the number of lightnings per square kilometer per year (other units are possible) as a function of location. This parameter represents the starting point for almost all lightning protection designs (for example, the lightning overvoltage protection of utility power lines) because the number of lightning failures per year for which a system is designed is directly proportional to the number of ground flashes per unit area per year. Real-time identification of phenomenological parameters such as the total number of lightning events per

storm and the lightning flashing rate is now possible with newly developed detection equipment. This equipment also makes possible real-time decisions on utility system repair and repair preparation, early warning and detection of lightning-caused forest fires, and a variety of other warning functions in situations that allow protective action to be taken, such as launches at the NASA Kennedy Space Center.

(2) When an object (e.g., aircraft, building, power line, or person) is struck directly by lightning, or is exposed to the intense electromagnetic fields of a nearby flash, the potentially deleterious currents and voltages that appear in the object are determined by the physical characteristics of the lightning currents and fields and by the electric characteristics of the object that is struck. For example, it is thought that, to a first approximation, the voltages that are induced in electronics within an airborne metal aircraft that is struck by lightning are indirectly initiated by the fastest part of the current rate of rise. This fast change in current induces resonant oscillations on the metallic exterior of the aircraft (like a pestle striking a bell) that are then coupled inside the aircraft via holes or apertures, such as windows, in the conducting metal skin. Lightning protection is currently of considerable concern for the latest generation of military and commercial aircraft that operate with low-voltage computer circuits and have lightweight ep-

61

oxy surfaces (potential apertures) replacing the more-conventional conducting metal. For these types of aircraft and other similar advanced systems, the microelectronic components used are often more easily damaged by lightning-induced voltages and the shielding against the intrusion of those voltages is often less adequate than is the case for more conventional systems.

In the following three sections, we examine in more detail the recent and widespread use of lightning-detection techniques for protection; those properties of lightning that cause damage, the mechanisms of lightning damage, and new methods of protection; and some remaining questions that research can answer to facilitate additional improvements in lightning protection.

APPLICATIONS OF NEW LIGHTNING-DETECTION TECHNIQUES TO PROTECTION

In 1983 the Electric Power Research Institute (EPRI), the research arm of United States power utilities, funded a long-term study of lightning flash density in the United States for the purpose of making possible better lightning protection design for power lines. The EPRI research is being carried out using lightning-locating technology recently developed through basic research (Krider et al., 1980). For the initial part of the study a network of automatic lightning direction finding (DF) stations called the East Coast Network and operated by the State University of New York at Albany (SUNYA) (Orville et al., 1983) is being used. Future flash density studies can be expected to involve additional portions of the United States and perhaps Canada. As is clear from Figure 5.1 in which the East Coast Network is evident, over three quarters of the area of the United States and Canada is covered by DFs, a development that has occurred since 1976. In addition, lightning-locating systems of the DF type developed in the United States have been installed in Australia, Norway, Sweden, Mexico, South Africa, Japan, Hong Kong, and the People's Republic of China during the same time period.

The primary user of lightning-location data in the United States at present is the Bureau of Land Management (BLM), which is responsible for the majority of the DFs in the western United States and Alaska (Figure 5.1). The BLM and the Forest Services of most Canadian provinces utilize the time and location of lightning storms to determine when and where to look for forest fires. Early detection of these fires results in considerable savings in natural resources and in the cost of fighting the fires. BLM data are also disseminated in real time to all National Weather Service Offices in the western region via AFOS, to the National Severe Storms Forecast Center in Kansas City, to Vandenburg Air Force Base, and to Nellis Air Force Base. Data from the SUNYA East Coast Network are currently being displayed in real time at the FAA Washington Air Route Traffic Control Center (ARTCC) in Leesburg, Virginia, the National Weather Service Forecast Office in Albany, New York, and Langley Air Force Base in Hampton, Virginia.

In addition to applications-oriented research, operational forest fire management, and Weather Service storm warning, the newly developed lightning-locating equipment is used to warn of the approach of storms in a variety of practical applications where protective action can be taken. Examples range from power utility companies (e.g., Tampa Electric Company, China Power of Hong Kong) to missile launches (e.g., Kennedy Space Center, Vandenburg AFB) to sensitive military installation (e.g., Buckley Air National Guard Base, Colorado, Cudjoe Key AF Station, Florida). In addition, lightning maps from these lightning locating systems are becoming widely shown on TV weather shows, as they are often more meaningful to the typical viewer than the more conventional radar displays. An example of a 1-day lightning map from the Tampa Electric Company (Peckham et al., 1984) is shown in Figure 5.2.

AMELIORATION OF LIGHTNING DAMAGE

Mechanisms of Lightning Damage

The amount and type of lightning damage an object suffers is due to both the characteristics of the lightning discharge and the properties of the object. The physical characteristics of lightning of most interest are the currents and electromagnetic fields, particularly those from the return stroke since these are usually the largest; hence protection against the return stroke will usually protect against the currents and fields from other lightning processes.

Four properties of the return stroke current can be considered important in producing damage: (1) the peak current, (2) the maximum rate of change of current, (3) the integral of the current over time (i.e., the charge transferred), and (4) the integral of the current squared over time, the so-called action integral. Let us examine each of these properties and the type of damage that it can produce.

For objects that present a resistive impedance, such as a ground rod driven into the Earth, a long power line, or a tree, the peak voltage on the object will be proportional to the peak current. For example, a 50,000-A current injected into a 400-Ω power line produces a line voltage of 20,000,000 V. Such large voltages lead to

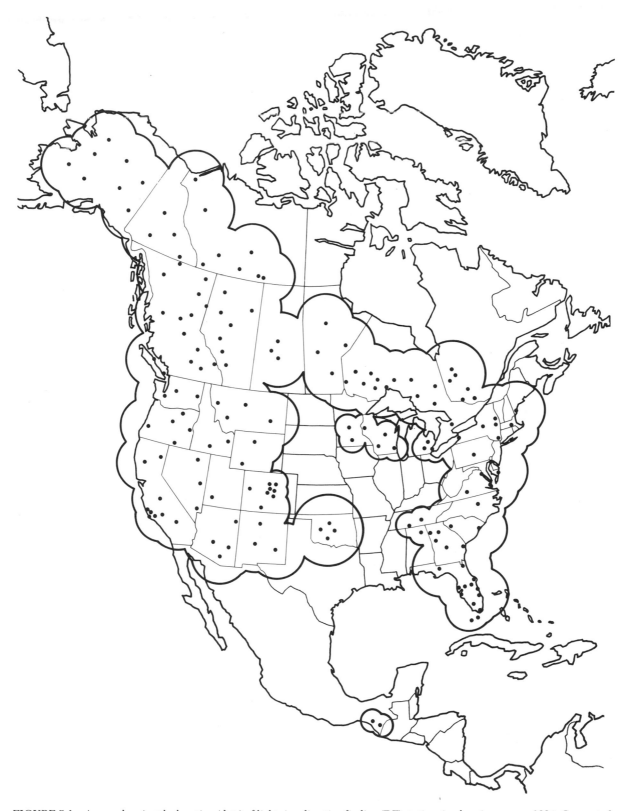

FIGURE 5.1   A map showing the location (dots) of lightning direction finding (DF) stations in place in summer 1984. Connected circles around each DF indicate area of lightning coverage. The area of coverage along the East Coast is the SUNYA East Coast Network.

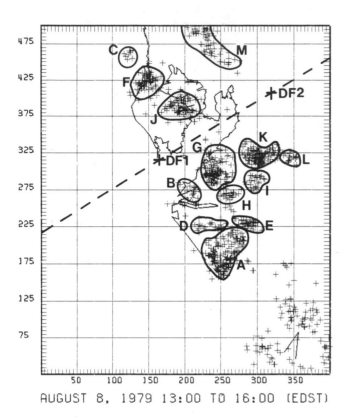

AUGUST 8, 1979 13:00 TO 16:00 (EDST)

FIGURE 5.2   A map of the cloud-to-ground lightning strikes in the Tampa Bay area for August 8, 1979. Individual storms have been circled. The two DF locations in the Tampa Electric Company's lightning location system are identified. Map scale is in thousands of feet.

electric discharges from the struck object to the ground through the air or through insulating materials. Such flashovers can, for example, short-circuit a power system or kill people that are standing close to the object that is struck. An example of discharges across the ground caused by the high voltage on a struck golf-course green marker is shown in Figure 5.3. The magnetic forces produced by the peak lightning currents are large and can crush metal tubes and pull wires from walls.

For objects that have an inductive impedance, such as wires in an electronic system, the peak voltage will be proportional to the maximum rate of change of the lightning current ($V = L\, di/dt$). For example, if 1 m of wire has an inductance L of 10^{-7} H and $di/dt = 10^{10}$ A/sec, 1000 V is generated across the wire. Voltages of this level often cause damage to solid-state electronic devices.

The heating or burn through of metal sheets such as airplane wings or metal roofs is, to first approximation, proportional to the lightning charge transferred (average current times time). Generally, large charge transfers are due to long-duration (tenths of a second to sec-

onds) lightning currents in the 100- to 1000-A range rather than to peak currents that have a relatively short duration. An example of a hole burned in an aircraft skin by lightning is shown in Figure 5.4, and some information on hole size versus charge transferred is given in Figure 5.5. A typical lightning transfers 20 to 30 C and extreme lightnings hundreds of coulombs, but, fortunately, the lightning does not often stay attached to one place on an aircraft in flight for the duration of that transfer.

The heating of many objects and the explosion of insulators is, to first approximation, due to the value of the action integral. In the case of wires, the action integral represents the heat that is generated by the resistive impedance of the wire. Some data on wire temperature rise for typical lightning action integrals is given in Figure 5.6. About 1 percent of negative strokes to ground have action integrals exceeding 10^6. About 5 percent of positive strokes is thought to exceed 10^7. In the case of a tree, this heat vaporizes the internal moisture of the wood, and the resultant steam pressure causes an explosive fracture. An example of typical tree damage from lightning is shown in Figure 5.7.

Two properties of the electromagnetic fields are sufficient to describe most of the important damage effects: the peak value of the field and the maximum rate of rise to this peak. For certain types of antennas or metal exposed to the lightning field, the peak voltage on the metal is proportional to the peak field. These antennas are commonly referred to as capacitively coupled. For other antennas, such as a loop of wire in an electronic circuit or an underground communication cable, the peak voltage is proportional to the maximum rate of change of the field.

New Results on Lightning Characteristics

Chapter 2 (this volume) by Krider describes the recent findings on lightning current and field characteristics. Much of this work has application to protection. A major step forward has been made in identifying the maximum rates of change of currents and fields, and it should be noted that these are now thought to be at least 10 times larger than was believed to be the case a decade ago. These recent results have important implications for the design of protection against damage that is caused by fast rates of change of currents and fields.

Chapter 3 (this volume) by Rust discusses positive lightning, recently identified and only partially characterized. Positive lightning apparently produces very large peak currents, charge transfers, and action integrals, much larger than the usual negative lightning. The Japanese report that their power systems are dis-

FIGURE 5.3   Lightning damage caused by a direct strike to a golf course green (photo courtesy of *Weatherwise*).

FIGURE 5.4   A lightning hole burned in the wing tip of a Boeing 707 (Uman, 1971).

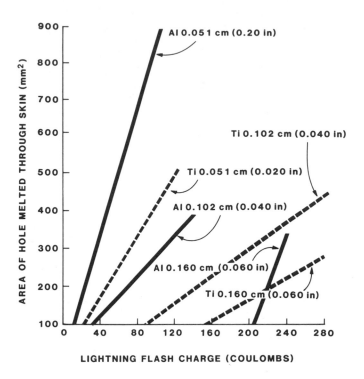

FIGURE 5.5   Area of holes melted through aluminum and titanium of various thicknesses by lightning charge (Fisher and Plumer, 1977).

rupted by a large fraction of the positive lightning strike, whereas only a small fraction of negative lightnings have this effect (Nakahori *et al.*, 1982).

Protection Techniques

There are two general types of lightning protection: (1) diversion and shielding and (2) limiting of currents and voltages. On a residential or commercial building, for example, the diversion of lightning currents to ground via a standard system of lightning rods, down leads, and grounds is sufficient to protect the building structure itself and to decrease by imperfect shielding potentially harmful effects to electronic equipment inside. An example of a diversion and shielding system is shown in Figure 5.8.

More complete protection of electronic equipment must include limiting of currents and voltages induced by the direct strike to the structure or by traveling waves into the structure on electric power, communication, or other wires connected to the outside world. The design of the current- and voltage-limiting system is obviously dependent on an understanding of the wave shapes of the deleterious signals that are to be controlled; and this in turn requires a knowledge of the lightning character-

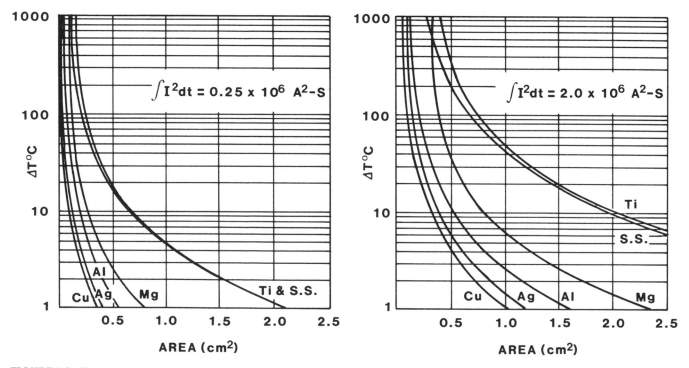

FIGURE 5.6   Temperature rise in various types of wires of various cross-sectional areas for two values of action integral (Fisher and Plumer, 1977).

FIGURE 5.7   Typical damage to a tree due to a direct lightning strike (Uman, 1971).

istics and how the properties of the system under consideration change these characteristics. Once such a determination is made, three general types of current- and voltage-limiting devices can be used for electronic or power systems: (1) voltage crowbar devices that reduce the voltage difference effectively to zero and short circuit the current to ground (the carbon block and gas tube arrestors used by the telephone company are good examples of crowbar devices); (2) voltage clamps such as recently developed solid-state metal oxide varistors (MOVs) or Zener diodes, which do not allow the voltage to exceed a given value; and (3) electric filters that reflect or absorb the higher and generally more damaging frequencies in the lightning transient. Frequently, all three of these forms of protection are used together in a coordinated way. Examples of some of these protective devices are shown in Figure 5.9.

In recent years, a systematic approach has been developed that allows an optimal lightning protection system to be designed for most structures. This new technique is called "topological shielding," and it uses both diversion and shielding and the limiting of currents and voltages discussed above. The technique consists of nesting shields and "grounding" each shield to the one enclosing it. All incoming wires are connected to the outside of each successive shield by a transient protective device, and therefore, at each successively inner shield, the voltage and power levels to be protected against are reduced. In Figure 5.10, we illustrate the principles of topological shielding. In Figure 5.10a, the equivalent

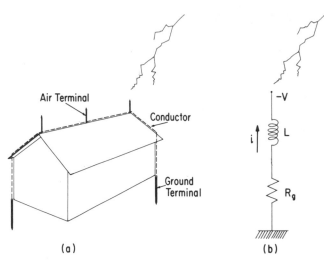

FIGURE 5.8   A standard lightning protection system for a small structure.

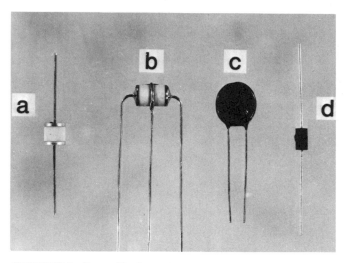

FIGURE 5.9   Examples of typical lightning protective devices: a and b are sealed spark gaps (crowbar devices); c is a solid-state metal oxide varistor; and d is a solid-state Zener diode. The diameter of c is about 1 inch.

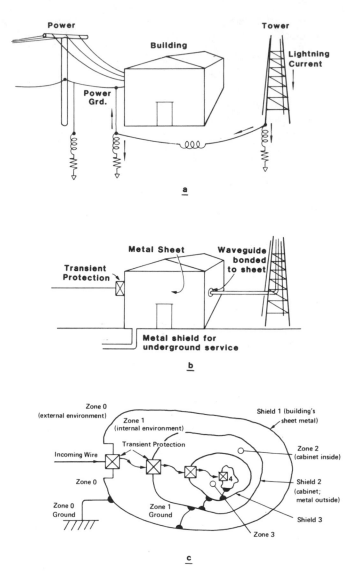

FIGURE 5.10   A diagram illustrating the principles of topological shielding. a, Equivalent electric circuit for a grounded building served by power lines and a communications tower. b, External view of building after topological shielding. c, Schematic of the topological shielding.

circuit is shown for the grounding of a building associated with a communications tower. Figure 5.10b shows an external view of the building after topological shielding, and Figure 5.10c shows a schematic of the topological shielding technique.

FUTURE RESEARCH NEEDED FOR IMPROVEMENTS IN PROTECTION

The detailed physics of how lightning strikes a structure, power line, or aircraft is still not well understood. The approaching lightning leader is not influenced by the object to be struck until it is perhaps a few tens to hundreds of meters away. At that time, an upward-moving spark leaves the object to be struck eventually and similar sparks may also leave nearby objects. The upward-moving spark connects to the downward-moving leader attaching the leader to ground. (See Krider, Chapter 2, this volume, for a discussion of the attachment process.) When this process is better understood through basic research, we should be able to determine with higher probability what will and what will not be struck and to provide better lightning protection accordingly. For example, the positioning of overhead ground wires above transmission lines should be able to be optimized.

More information is needed about the character of lightning currents, particularly those in processes other than return strokes. Is there, for example, an upper limit on the maximum rate of change of current? We need more data on positive lightning to be able to characterize all aspects of it in a statistical way. Only then can it be taken account of properly in protection design.

Much work needs to be done on the interaction of lightning currents and fields with objects like aircraft. For example, how are aircraft resonances affected by channel attachment? Computer models are now being developed with which to study these problems even in the presence of nonlinear discharge properties.

CONCLUSIONS

Basic research over the last decade has made possible impressive improvements in lightning protection. Those related to lightning detection and to the specification of current and electromagnetic-field wave shapes have been discussed in this chapter. As with all research, new discoveries raise new questions. With the present interest in lightning among scientists, due partly to recent successes and partly to the important unsolved problems, we can expect continued progress in lightning protection during the next decade.

REFERENCES

Fisher, F. A., and J. A. Plumer (1977). Lightning protection of aircraft, *NASA Reference Publication 1008.*

Krider, E. P., R. C. Noggle, A. E. Pifer, and D. L. Vance (1980). Lightning direction-finding systems for forest fire detection, *Bull. Am. Meteorol. Soc. 61,* 980-986.

Nakahori, K., T. Ogawa, and H. Mitani (1982). Characteristics of winter lightning currents in Hokuriku District, *IEEE Trans. PAS-101,* 4407-4412.

Orville, R. E., R. W. Henderson, and L. F. Bosart (1983). An east coast lightning detection network, *Bull. Am. Meteorol. Soc. 64,* 1029-1037.

Peckham, D. W., M. A. Uman, C. E. Wilcox, Jr. (1984). Lightning phenomenology in the Tampa Bay area, *J. Geophys. Res. 89,* 11789-11805.

Uman, M. A. (1971). *Understanding Lightning,* BEK, Pittsburgh, Pa.

The Role of Lightning in the Chemistry of the Atmosphere

6

WILLIAM L. CHAMEIDES
Georgia Institute of Technology

ABSTRACT

The high temperatures in and around the discharge tube of a lightning stroke cause the dissociation of the major atmospheric constituents N_2, O_2, CO_2, and H_2 and the formation of trace species such as NO, CO, N_2, OH, N, O, and H. As this cylinder of hot air cools, the levels of these trace species drop. However, if the cooling is sufficiently rapid the concentrations of these trace species can be "frozen-in" at levels significantly above their ambient, thermochemical equilibrium abundances, thereby leading to a net source of these gases to the background atmosphere. It is estimated that about 3 tg of N yr^{-1} as NO are produced in the present-day atmosphere by lightning through this process. Other gases produced by lightning are CO and N_2O in the Earth's atmosphere; HCN in the Earth's prebiological atmosphere; CO, NO, and O_2 in the cytherian atmosphere; and CO, N_2, and a variety of hydrocarbons in the jovian atmosphere. The major uncertainty in quantifying the role of lightning in the chemistry of the terrestrial atmosphere, as well as that of other planetary atmospheres, arises from the lack of accurate statistics on the energy and frequency of lightning. The role of coronal discharges in the chemistry of clouds also needs to be investigated.

INTRODUCTION

In addition to the spectacular visual and aural effects that accompany a lightning flash, intense chemical reactions occur, which, on a relatively short time scale, can radically alter the chemical composition of the air in and around the discharge tube and, on longer time scales, can ultimately affect the composition of the atmosphere as a whole. The short-term chemical changes associated with lightning have been well documented by spectroscopic studies of lightning strokes (cf., Salanave, 1961; Uman, 1969). For instance, the strong

emissions from neutral atomic nitrogen (N I), singly ionized atomic nitrogen (N II), neutral atomic oxygen (O I), and singly ionized atomic oxygen (O II) typically observed from the hot core of discharges, are indicative of the widespread dissociation of atmospheric N_2 and O_2 and the subsequent ionization of their atomic daughters. Other prominent spectroscopic features are the emission lines from CN and H, species arising from the dissociation of CO_2 and H_2O.

For the most part the large changes in the chemical composition of the air in and around the discharge tube can be related to the rapid variations in temperature in

70

this region. The lightning bolt and associated shock wave produce a cylinder of very hot air within which chemical reactions between the atmospheric gases proceed rapidly to bring the mixture into thermochemical equilibrium. Immediately after the energy deposition, the temperature in the discharge tube approaches 30,000 K and the gas is a completely ionized plasma. As the gas cools by hydrodynamic expansion and turbulent mixing, the equilibrium composition of the gas changes from a plasma to a mixture of neutral atoms such as N and O and then to a mixture of molecular species and ultimately as the temperatures return to ambient to a mixture of N_2, O_2, H_2O, and CO_2 much like the background composition of the atmosphere. This variation in the equilibrium composition of air as a function of temperature is illustrated in Figure 6.1; note that as the temperatures fall below 5000 K the equilibrium shifts from N, O, H, and CO to NO, OH, and CO and then to N_2, O_2, H_2O, and CO_2.

If the gas around the lightning discharge was always to remain in thermochemical equilibrium, the net effect of lightning on the atmospheric composition would be negligible; once the temperature of the gas returned to its ambient level, its composition would be essentially the same as that of the background atmosphere's, and thus there would be no net production or destruction of atmospheric chemical species by lightning discharges. On the other hand, it is well known that laboratory sparks can have significant effects on the composition of air; most notable is the fixation of atmospheric nitrogen (N_2) by sparks to produce nitric oxide (NO). Given the basic equivalence between laboratory sparks and lightning discharges it would seem reasonable to expect that lightning also affects the composition of air. In fact, the knowledge that NO is produced by laboratory sparks led von Liebig to propose in 1827 that the NO_3^- typically observed in rainwater arises from the fixation of atmospheric N_2 by lightning discharges. This nineteenth century hypothesis of von Liebig's has only recently been qualitatively confirmed by direct observations of enhanced levels of NO and NO_2 in and around active thunderclouds (Noxon, 1976, 1978; Davis and Chameides, 1984) and in the vicinity of a cloud-to-ground lightning flash (Drapcho et al., 1983).

The identification of the mechanisms responsible for the net production of trace species such as NO by lightning and the quantification of their source rates on a global scale define the current frontier in the field of the chemistry of atmospheric lightning and will, therefore, be the major subject of this review. The discussion begins by focusing on the production by lightning of atmospheric NO, a species of special interest because of its central role in the photochemistry of the atmosphere

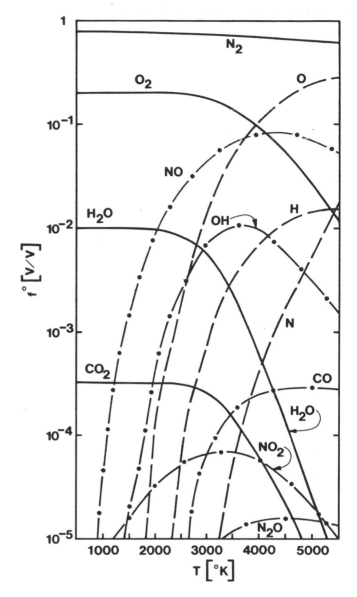

FIGURE 6.1 Equilibrium volume mixing ratios of selected atmospheric species as a function of temperatures in heated tropospheric air.

(cf., Crutzen, 1983). Following the discussion of NO production by lightning, a more general presentation will be given of the production of other trace species in both the present and the prebiological, terrestrial atmosphere, as well as in other planetary atmospheres. A brief discussion is then presented on the possible effect of electrical discharges on the chemistry of cloudwater and the generation of acids in precipitation. Finally, a brief outline of the needs for future work in this area is presented.

NO PRODUCTION BY LIGHTNING

Similar to the Z'elovich mechanism for the fixation of nitrogen in explosions (Z'elovich and Raizer, 1966), the production of NO in lightning discharges is believed to be driven by high-temperature chemical reactions within a rapidly cooling parcel of air; the rapid cooling causes NO levels above its thermochemical abundance to be "frozen" into the gas. A simple physical analogy to this chemical production mechanism is that of dropping a bead through a column of rapidly cooling water in a gravitational field. Because the bead wants to minimize its potential energy with respect to the gravitational field, the bead will tend to fall to the bottom of the water column. If, however, the water were to cool so rapidly that it froze before the bead reached the bottom of the column, the bead would be frozen in the column at a position of higher potential energy and would be prevented from reaching its energetically preferred position at the bottom of the column.

In the case of NO production in lightning, the high temperatures in and surrounding the discharge channel result in a series of chemical reactions that both produce and destroy NO. NO production is initiated by the thermal dissociation of O_2.

$$O_2 \longleftrightarrow O + O \qquad \text{(Reaction 6.1)}$$

followed by the production of NO via the reaction chain

$$O + N_2 \longrightarrow NO + N \qquad \text{(Reaction 6.2)}$$

and

$$N + O_2 \longrightarrow NO + O. \qquad \text{(Reaction 6.3)}$$

In competition with these NO-producing reactions are

$$NO + N \longrightarrow N_2 + O \qquad \text{(Reaction 6.4)}$$

and

$$NO + O \longrightarrow N + O_2, \qquad \text{(Reaction 6.5)}$$

which convert NO back to N_2 and O_2 as well as

$$NO \longleftrightarrow N + O, \qquad \text{(Reaction 6.6)}$$

the thermal dissociation of NO itself, and

$$NO + NO \longrightarrow N_2O + O, \qquad \text{(Reaction 6.7)}$$

the formation of N_2O from NO.

The equilibrium NO concentration, f_{NO}^0, is the NO level at which NO-producing and NO-destroying reactions are in balance. As illustrated in Figure 6.2, f_{NO}^0 is a strong function of temperature. As the temperature rises above 1000 K the dissociation of N_2 and O_2 causes an increase in the NO equilibrium level. At about 4000 K,

f_{NO}^0 peaks at a value approaching 10 percent. For higher temperature, N and O atoms become increasingly more stable relative to NO (See Figure 6.2) and f_{NO}^0 decreases.

Thus if NO were always to maintain thermochemical equilibrium, its concentration would reach a maximum when the temperature in and around the discharge tube was ~ 4000 K and would then decrease to a negligibly small value as the heated air cooled to ambient temperatures. However, similar to the equilibrium NO concentration, the time, τ_{NO}, required to establish thermochemical equilibrium for NO also varies with temperature. As illustrated in Figure 6.2, this time becomes increasingly longer as temperature decreases because the reactions acting to establish equilibrium become slower. (In this figure, τ_{NO} was calculated by summing the loss frequencies for NO due to Reactions 6.4-6.7.) Whereas only a few microseconds are required for NO to equilibrate at 4000 K, equilibrium requires milliseconds at 2500 K, a second at 2000 K, and approximately 10^3 years at 1000 K. Hence, as the air cools, a temperature is eventually reached at which the rates of reaction

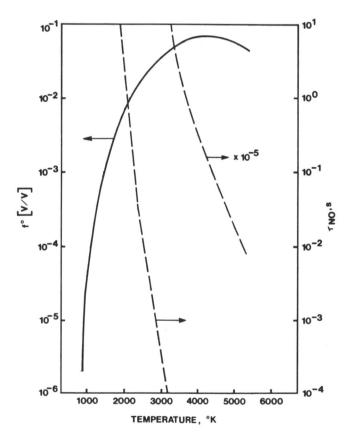

FIGURE 6.2   The NO equilibrium volume mixing ratio f^0, represented by the solid curve, and the NO chemical lifetime τ_{NO}, represented by the dashed curve, as a function of temperatures in heated tropospheric air. (After Borucki and Chameides, 1984.)

become too slow to keep NO in equilibrium. Instead of falling to the thermochemical equilibrium concentration of the ambient temperature, a higher NO level becomes frozen into the gas. This higher concentration, which corresponds to the NO equilibrium level at the temperature at which the NO concentration departs from equilibrium, is called the "freeze-out" temperature.

The freeze-out temperature of NO, T_F, is approximately determined by the relationship

$$\tau_T(T_F) = \tau_{NO}(T_F), \qquad (6.1)$$

where τ_T is the characteristic cooling time of the heated air. When $T > T_F$, then $\tau_{NO} < \tau_T$ and the chemical reactions are sufficiently rapid to keep NO in the thermochemical equilibrium. However, for $T < T_F$, $\tau_{NO} > \tau_T$ and chemical reactions are too slow to adjust to the rapidly decreasing T; NO, therefore, freezes out with a mixing ratio $f_{NO}^0(T_F)$. Although a lower abundance of NO is favored thermodynamically at low T, the kinetics are too slow for readjustment. P, the net yield of NO produced by this process, is then approximated by

$$P(NO) = f_{NO}^0(T_F)M(T_F)E_0^{-1} \text{ molecules J}^{-1}, \quad (6.2)$$

where M is the number of molecules per meter heated to, or above, T_F in the region where NO is being produced for a discharge energy of E_0 (in units of J/m). Thus it is necessary to determine values for T_F and M that, when combined with the results of Figure 6.2, will allow an estimate of $P(NO)$ from Eq. (6.2).

Once $P(NO)$ is obtained, the global rate of NO production by lightning, $\phi(NO)$, can be estimated from

$$\phi(NO) = P(NO) \cdot R$$

$$\cdot \frac{(14 \text{ g/mole}) (10^{-12} \text{ tg/g}) (3.16 \times 10^7 \text{ sec/yr})}{(6.02 \times 10^{23} \text{ molecules/mole})} \quad (6.3)$$

in units of teragrams (tg) (i.e., 10^{-12} g or 10^6 metric tons) of N per year, where R is the number of joules dissipated globally by lightning per second. Because of the current interest in developing global budgets for the flow of fixed nitrogen and nitrogen oxides through the atmosphere, reasonably accurate estimates for $\phi(NO)$ are desirable. A brief discussion of how the parameters needed to solve $\phi(NO)$ are calculated is presented below.

Estimate of P(NO)

Following the approach of Borucki and Chameides (1984), we infer values for T_F and M that are needed to

calculate $P(NO)$ in Eq. (6.2) from the laboratory study of linear discharge channels by Picone et al. (1981). This study indicated that lightning-like discharge channels cooled with a τ_T of about 2.5×10^{-3} sec. For this choice of τ_T, T_F and $f_{NO}^0(T_f)$ can be estimated from Figure 6.2 to be about 2660 K and 0.029, respectively. Furthermore, using the result of Picone et al. (1981) that in spark discharges 1 J of energy is required to heat each 1 cm^3 of air to a temperature of 3000 K and assuming that the gas cools from 3000 K to the freeze-out temperature of 2660 K by mixing with the ambient atmosphere, it can be inferred that

$$M(2660 \text{ K})/E_0 = 3.2 \times 10^{19} \text{ molecules/J}. \quad (6.4)$$

Substituting the above values for $f_{NO}^0(T_F)$ and $M(T_F)/E_0$ into Eq. (6.2),

$$P(NO) = (0.29)(3.2 \times 10^{18})$$

$$= 9.2 \times 10^{16} \text{ molecules/J}. \quad (6.5)$$

A comparison of the above-estimated NO yield with those of previous investigators is presented in Table 6.1 and indicates a rather good agreement with a wide variety of theoretical calculations, laboratory spark experiments, and atmospheric measurements. The largest discrepancy appears to be with the NO yield attributed to Drapcho et al. (1983). The yield of Drapcho et al. was based on their observation of a sudden increase in NO and NO$_2$ levels in the vicinity of a cloud-to-ground discharge; given the many assumptions necessary to infer a yield from this observation the disparity between the yield of Drapcho et al. and the others in Table 6.1 is not very surprising.

The Global Dissipation Rate, R

The rate at which energy is dissipated by lightning globally can be expressed as a function of two other parameters, i.e.,

$$R = E_F \cdot F, \qquad (6.6)$$

where E_F is the average number of joules dissipated per lightning flash and F is the number of lightning flashes occurring globally per second. Borucki and Chameides (1984) recently examined the existing data base on lightning flashes to estimate these parameters. Combining optical and electrical measurements of the energy of a single stroke, observations of the number of strokes per flash, as well as measurements of the distribution of energy among the first and subsequent strokes, E_F was estimated to be about 4×10^8 J/flash with a factor of 2.5 uncertainty. From satelliteborne optical detection sys-

TABLE 6.1   Estimates of NO Yield from Lightning Discharge

	$P(NO)$ (molecules/J)	Investigator
A. This work	$(9 \pm 2) \times 10^{16}$	Based on the calculations of Borucki and Chameides (1984)
B. Theoretical calculation	$3 \times 10^{16\,a}$	Tuck (1976)
	$(3\text{-}7.5) \times 10^{16}$	Chameides et al. (1977)
	$(4\text{-}6) \times 10^{16\,a}$	Griffing (1977)
	$80 \times 10^{16\,b}$	Hill et al. (1980)
	16×10^{16}	Hill et al., as corrected by Borucki and Chameides (1984)
	$(8\text{-}17) \times 10^{16}$	Chameides (1979)
C. Laboratory spark experiment	$(6 \pm 1) 10^{16}$	Chameides et al. (1977)
	$(5 \pm 2) \times 10^{16}$	Levine et al. (1981)
	$(2 \pm 0.5) \times 10^{16}$	Peyrous and Lapeyre (1982)
D. Atmospheric measurement	$(20\text{-}30) \times 10^{16\,a}$	Noxon (1976)
	$(25\text{-}2500) \times 10^{16\,a}$	Drapcho et al. (1983)

[a]The NO yields obtained by these investigators were expressed as molecules/flash. These yields were converted to units of molecules/Joule by assuming $E_F = 4 \times 10^8$ J/flash.

[b]Derived from Hill et al. (1980) by dividing their NO yield (6×10^{25} molecules/flash) by their energy per flash [$(1.5 \times 10^4$ J/m) $(5 \times 10^3$ m/flash)].

tems, a value of 100 flashes/sec was assigned to F by Borucki and Chameides (1984) with an uncertainty factor of 25 percent. Substituting these parameters into Eq. (6.6), R was thus estimated to be about 4×10^{10} W with a possible range of $(1.3$ to $12) \times 10^{10}$ W.

The Global NO Production Rate, $\phi(NO)$

The above estimates for R and $P(NO)$ can be combined in Eq. (6.3) to yield a global NO production rate of ~ 2.5 tg of N/yr. However, it should be noted that this number is highly uncertain; Borucki and Chameides (1984) in a similar analysis arrived at a possible range in $\phi(NO)$ from 0.8 to 8 tg of N/yr. By far the largest source of uncertainty arises from the uncertainty in E_F, the energy dissipated by a lightning flash. A comparison between the estimate for the global fixation rate calculated here and those of previous investigators is presented in Table 6.2. For the most part our result is consistent with, although somewhat smaller than, the other estimates.

Biological processes fix atmospheric N_2 at a rate of about 200 tg of N/yr and anthropogenic fixation (primarily due to the synthesis of fertilizers) occurs at a rate of about 60 tg of N/yr (Burns and Hardy, 1975). Thus it would appear that, at present, lightning is responsible for at most a few percent of the Earth's total nitrogen fixation. On the other hand, lightning appears to represent one of, if not the, major natural source of NO_x to the atmosphere. The other natural sources of atmospheric NO_x include stratospheric oxidation of N_2O at a rate of 0.6 tg of N/yr (Levy et al., 1980); oxidation of NH_3, which is not well known but could be important (Mc-

TABLE 6.2   Estimates of the Amount of Nitrogen Fixed by Lightning

Investigator	Nitrogen Fixed per Year (tg)[a]
Tuck (1976)	4.2
Chameides et al. (1977)	30 to 40
Chameides (1979)	35 to 90
Dawson (1980)	3
Hill et al. (1980)	4.4
Levine et al. (1981)	1.8
Kowalczyck and Bauer (1982)	5.7
Ehhalt and Drummond (1982)	5
Peyrous and Lapeyre (1982)	9
Logan (1983)	8
Drapcho et al. (1983)	30
Present result [based on calculations of Borucki and Chameides (1984)]	Best Estimate: 2.6 Range: 0.8 to 8

[a]1 tg = 10^{12} g = 10^6 metric tons.

Connell, 1973); and NO emissions from soils as a result of microbial activity at a rate of $(1$ to $10)$ tg of N/yr (Galbally and Roy, 1978; Lipschultz et al., 1981). As noted earlier, qualitative confirmation of the importance of lightning as a natural source of atmospheric NO_x has been obtained from a variety of NO and NO_2 measurements (Noxon, 1976, 1978; Drapcho et al., 1983; Davis and Chameides, 1984), which reveal anomalously high concentrations of NO and NO_2 in air within and above clouds in remote regions of the globe. In regions strongly affected by anthropogenic activities, however, this natural NO source is swamped by NO production from the burning of fossil fuel and biomass at a combined rate of about 30 tg of N/yr (Crutzen, 1983). It is for this reason that NO_x levels in urban areas are some 10^4 to 10^5 times higher than in remote regions of the marine atmosphere (McFarland et al., 1979) and in conjunction with the anthropogenic release of non-methane hydrocarbons is the cause of photochemical smog and related air pollution problems. Nevertheless, it is interesting to note that even the extremely low levels of NO_x that are characteristic of the remote troposphere are believed to have a significant effect on the chemistry of the atmosphere, catalyzing the photochemical production of tropospheric O_3 and enhancing OH levels (Logan et al., 1981; Davis and Chameides, 1984). To the extent that lightning is responsible for the NO_x levels in the remote troposphere, it would appear that lightning plays an important role in the photochemistry of the atmosphere.

OTHER TRACE GASES PRODUCED BY LIGHTNING

While NO has received the bulk of the attention with regard to production by lightning because of its importance in the photochemistry of the atmosphere, research has revealed that a myriad of other trace gases in addition to NO can be generated by lightning in a variety of interesting environments. These gases and their yields in electrical discharges as determined by both theoretical calculations and laboratory experiments are listed in Table 6.3. The comparison between experimentally and theoretically derived yields is quite good over a wide range of gases and atmospheric compositions.

The production of HCN in a reducing, prebiological terrestrial atmosphere is of particular interest because it has been proposed that lightning-produced HCN was an

TABLE 6.3   Calculated and Experimentally Derived Yields of Trace Gases in Various Atmospheres[a]

Species	Calculated Yield (molecules/J)	Experimental Yield (molecules/J)	Reference
A. Present-Day Terrestrial Atmosphere			
NO	9×10^{16}	$(2\text{-}6) \times 10^{16}$	1, 2, 3, 4
CO	$(0.1\text{-}5) \times 10^{14}$	1×10^{14}	5, 6
N_2O	$(3\text{-}13) \times 10^{12}$	4×10^{12}	6
B. Reducing Prebiological Terrestrial Atmosphere (95% N_2, 5% CH_4)			
HCN	$(6\text{-}17) \times 10^{16}$	$\sim 10^{17}$	7, 8, 9, 10
C. Cytherian Atmosphere (95% CO_2, 5% N_2)			
CO	$(1\text{-}1.4) \times 10^{17}$	3×10^{16}	6, 11, 12
NO	$(5\text{-}6) \times 10^{15}$	4×10^{15}	11, 12, 13
$(O_2 + O)$	$(6\text{-}9) \times 10^{16}$	—	11
D. Jovian Atmosphere (99.95% H_2, 0.05% CH_4)			
CO	5×10^{15}	—	14, 15
N_2	5×10^{14}	—	14, 15
HCN	9×10^{13}	—	14, 15
C_2H_2	3×10^{13}	—	14, 15
C_2H_4	2×10^{12}	—	14, 15
HCHO	8×10^{11}	—	14, 15
CO_2	3×10^{11}	—	14, 15
C_2H_6	4×10^{11}	—	14, 15
E. Titan Atmosphere (97% N_2, 3% CH_4)			
HCN	1.2×10^{17}	—	16
C_2N_2	2.5×10^{14}	—	16
C_2H_2	7.5×10^{15}	—	16
C_2H_4	5×10^{11}	—	16

[a]References:
1. Borucki and Chameides (1984)
2. Chameides et al. (1977)
3. Levine et al. (1981)
4. Peyrous and Lapeyre (1982)
5. Chameides (1979)
6. Levine et al. (1979)
7. Chameides and Walker (1981)
8. Sanchez et al. (1967)
9. Bar Nun and Shaviv (1975)
10. Bar Nun et al. (1980)
11. Chameides et al. (1979)
12. Bar Nun (1980)
13. Levine et al. (1982)
14. Lewis (1980)
15. Bar Nun (1975)
16. Borucki et al. (1984)

organic precursor that ultimately led to the chemical evolution of life on Earth (Miller and Urey, 1959). Both laboratory and theoretical calculations indicate that in a highly reducing atmosphere, rich in hydrocarbons, lightning could have produced HCN in copious quantities, possibly large enough to allow HCN levels in ponds and ocean water to build to levels large enough to trigger the formation of peptide chains and similar precursors to amino acids. On the other hand, the calculations of Chameides and Walker (1981) indicate that the HCN yield rapidly decreases as the atmosphere becomes less reducing. For an atmosphere where C is primarily in the form of CO, the HCN yield decreases by about 3 orders of magnitude from that of a hydrocarbon atmosphere, and it decreases by an additional 3 orders of magnitude for a CO_2 atmosphere. Thus in order to better understand the role of lightning in the evolution of life, studies are needed to better determine the relative amounts of CH_4, CO, and CO_2 in the primitive atmosphere.

THE POSSIBLE ROLE OF ATMOSPHERIC DISCHARGES IN CLOUD CHEMISTRY

In recent years the growing concern over the possible deleterious effects of acidic precipitation on lakes, forest ecosystems, and crops has led to an increased interest in the chemistry of clouds, a region where acids can be efficiently generated and incorporated into rainwater. One aspect of cloud chemistry that has yet to be adequately studied is the role of atmospheric electrical phenomena in acid generation in electrified clouds. One possible effect of electrical discharges on cloud chemistry is briefly described below.

Suppose, under the appropriate conditions, continuous, low-level positive point coronal discharges from droplets occurred in a cloud. These discharges would cause 1 electron to be deposited on the droplet and 1 positive ion (most often O_2^+) to be produced in the gas phase for each ion pair produced (Loeb, 1965). The electrons deposited on the droplet would be incorporated into the droplet and become hydrated electrons [i.e., $(e^-)_{aq}$]. In the presence of dissolved O_2, these hydrated electrons would rapidly form O_2^- via

$$(e^-)_{aq} + (O_2)_{aq} \longrightarrow O_2^-. \qquad \text{(Reaction 6.8)}$$

The O_2^- species is related to the aqueous-phase HO_2 radical by the acid-base equilibrium reaction

$$HO_2 \longleftrightarrow O_2^- + H^+. \qquad \text{(Reaction 6.9)}$$

The O_2^+ ions produced in the gas phase would lead to the eventual formation of an OH radical and a hydrated oxonium ion, H_3O^+ (Good et al., 1970). Heterogeneous scavenging of the $H_3O^+ \cdot nH_2O$ ion and its incorporation into the droplets to form H^+ would maintain the nominal charge neutrality of the droplets. A sizable fraction of the gas-phase OH radicals produced by the discharge would also be scavenged, either as OH or HO_2 in the gas phase, and incorporated into the droplet representing an additional radical source to the aqueous phase. Calculations similar to those of Chameides (1984) indicate that about 1.5 aqueous-phase HO_2 free radicals would be produced for each ion pair generated.

The HO_2 radicals thus produced in the aqueous phase would rapidly react to form dissolved H_2O_2 via reactions such as

$$HO_2 + O_2^- \xrightarrow{\text{H}^+} H_2O_2 + O_2. \qquad \text{(Reaction 6.10)}$$

Since aqueous-phase H_2O_2 is believed to be, in many cases, the most important oxidant of dissolved SO_2 in cloud droplets leading to the production of sulfuric acid (Martin, 1983), it is conceivable that this electrical process could play a significant role in the generation of acids in clouds.

CONCLUSION

The agreement between theoretical calculations and experimental determinations of chemical yields from discharges for a wide range of gaseous species and a wide range of atmospheres suggests that the basic chemical mechanism by which trace species are produced by lightning is fairly well understood. However, in order to infer global production rates from these chemical yields, accurate values for the rate at which energy is dissipated by lightning is needed. Because these dissipation rates are not well known (uncertainty factors of 10 for the Earth and much larger for other planets are estimated), the role of lightning in the global budgets of species such as NO remains uncertain. To reduce this uncertainty, studies are needed to characterize more accurately the energy and frequency of lightning strokes on a global scale.

Another area where research is needed concerns coronal discharges and their role as local sources of trace species. Mechanisms exist, for instance, by which positive-point corona from cloud droplets could lead to enhanced generation of sulfuric acid in cloudwater. To determine if this and similar processes occur at a significant rate, the magnitude of low-level coronal currents in clouds needs to be more accurately established.

REFERENCES

Bar Nun, A. (1975). Thunderstorms on Jupiter, Icarus 24, 86-94.
Bar Nun, A. (1980). Production of nitrogen and carbon species by thunderstorms on Venus, Icarus 42, 338-342.
Bar Nun, A., and A. Shaviv (1975). Dynamics of the chemical evolution of Earth's primitive atmosphere, Icarus 24, 197-211.

Borucki, W. L., and W. L. Chameides (1984). Lightning: Estimates of the rates of energy dissipation and nitrogen fixation, *Rev. Geophys. 22*, 364.

Borucki, W. L., C. P. McKay, and R. C. Whitten (1984). Possible production by lighting of aerosols and trace gases in Titan's atmosphere, *Icarus 60*, 260-274.

Burns, R. C., and R. W. Hardy (1975). *Nitrogen Fixation in Bacteria and Higher Plants*, Springer-Verlag, Berlin.

Chameides, W. L. (1979). The implications of CO production in electrical discharges, *Geophys. Res. Lett. 6*, 287-290.

Chameides, W. L. (1984). The photochemistry of a remote marine stratiform cloud, *J. Geophys. Res. 89*, 47-39-4755.

Chameides, W. L., and J. C. G. Walker (1981). Rates of fixation by lightning of carbon and nitrogen in possible primitive atmospheres, *Origins of Life 11*.

Chameides, W. L., D. H. Stedman, R. R. Dickerson, D. W. Rusch, and R. J. Cicerone (1977). NO_x production in lightning, *J. Atmos. Sci. 34*, 143-149.

Chameides, W. L., J. C. G. Walker, and A. F. Nagy (1979). Possible chemical impact of planetary lightning in the atmospheres of Venus and Mars, *Nature 280*, 820-822.

Crutzen, P. J. (1983). Atmospheric interactions—Homogeneous gas reactions of C, N, and S containing compounds, in *The Major Biogeochemical Cycles and Their Interactions*, B. Bolin and R. B. Cook, eds., SCOPE, Paris.

Davis, D. D., and W. L. Chameides (1984). The atmospheric chemistry of electrified clouds, presented at VII International Conference on Atmospheric Electricity, Albany, N.Y.

Dawson, G. A. (1980). Nitrogen fixation by lightning, *J. Atmos. Sci. 37*, 174-178.

Drapcho, D. L., D. Sisterson, and R. Kumar (1983). Nitrogen fixation by lightning activity in a thunderstorm, *Atmos. Environ. 17*, 729-734.

Galbally, I. E., and C. R. Roy (1978). Loss of fixed nitrogen from soils by nitric oxide exhalation, *Nature 275*, 734-735.

Good, A., A. Durden, and P. Kebarle (1970). Mechanism and rate constants of ion-molecule reactions leading to formation of $H^- \cdot (H_2O)_n$ in moist oxygen and air, *J. Chem. Phys. 52*, 222-229.

Griffing, G. W. (1977). Ozone and oxides of nitrogen during thunderstorms, *J. Geophys. Res. 82*, 943-950.

Hill R. D., R. G. Rinker, and H. Dle Wilson (1980). Atmospheric nitrogen fixation by lightning, *J. Atmos. Sci. 37*, 179-192.

Levine, J. S., R. E. Hughes, W. L. Chameides, and W. E. Howell (1979). N_2O and CO production by electric discharge: Atmospheric implications, *Geophys. Res. Lett. 6*.

Levine, J. S., R. S. Rogowski, G. L. Gregory, W. E. Howell, and J. Fishman (1981). Simultaneous measurements of NO_x, NO, and O_3 production in a laboratory discharge: Atmospheric implications, *Geophys. Res. Lett. 8*, 357-360.

Levine, J. S., G. L. Gregory, G. A. Hervey, W. E. Howell, W. J. Borucki, and R. E. Orville (1982). Production of nitric oxide by lightning on Venus, *Geophys. Res. Lett. 9*, 893-896.

Levy, H., II, J. D. Mahlman, and W. J. Moxim (1980). Stratospheric NO_y: A major source of reactive nitrogen in the unpolluted troposphere, *Geophys. Res. Lett. 7*, 441-444.

Lewis, J. S. (1980). Lightning synthesis of organic compounds on Jupiter, *Icarus 42*, 85-95.

Lipschultz, F., O. C. Zafirious, S. C. Wofsy, M. B. McElroy, F. W. Valois, and S. W. Watson (1981). Production of NO and N_2O by soil nitrifying bacteria: A source of atmospheric nitrogen oxides, *Nature 294*, 641-643.

Loeb, L. B. (1965). *Electrical Coronas: Their Basic Physical Mechanisms*, Univ. of California Press, Berkeley, 694 pp.

Logan, J. A. (1983). Nitrogen oxides in the troposphere: Global and regional budgets, *J. Geophys. Res. 88*, 10785-10807.

Logan, J. A., M. J. Prather, S. C. Wofsy, and M. B. McElroy (1981). Tropospheric chemistry, *J. Geophys. Res. 86*, 7210-7254.

Martin, L. R. (1983). Kinetic studies of sulfite oxidation in aqueous solution, in *Acid Precipitation*, J. G. Calvert, ed., Ann Arbor Science, Ann Arbor, Mich.

McConnell, J. C. (1973). Atmospheric ammonia, *J. Geophys. Res. 78*, 7812-7821.

McFarland, M. C., D. Kley, J. W. Drummond, H. L. Schmeltekopf, and R. H. Winkler (1979). Nitric oxide measurements in the equatorial pacific region, *Geophys. Res. Lett. 6*, 605-608.

Miller, S. L., and H. C. Urey (1959). Organic compounds synthesis on the primitive Earth, *Science 130*, 245-251.

Noxon, J. A. (1976). Atmospheric nitrogen fixation of lightning, *Geophys. Res. Lett. 3*, 463-465.

Noxon, J. A. (1978). Tropospheric NO_2, *J. Geophys. Res. 83*, 3051-3057.

Peyrous, W., and R. M. Lapeyre (1982). Gaseous products created by electrical discharges in the atmosphere and condensation nuclei resulting from gaseous phase reactions, *Atmos. Environ. 16*, 959-968.

Picone, J. M., J. P. Boris, J. R. Greig, M. Rayleigh, and R. F. Fernster (1981). Convective cooling of lightning channels, *J. Atmos. Sci. 38*, 2056-2062.

Salanave, L. E. (1961). The optical spectrum of lightning, *Science 134*, 1395-1399.

Sanchez, R. A., J. P. Ferris, and L. E. Orgel (1967). Studies in prebiotic synthesis. II. Synthesis of purine precursors and amino acids from aqueous hydrogen cyanide, *J. Mol. Biol. 30*, 223-253.

Tuck, A. F. (1976). Production of nitrogen oxides by lightning discharges, *Q. J. R. Meteorol. Soc. 102*, 749-755.

Uman, M. A. (1969). *Lightning*, McGraw-Hill, New York, 264 pp.

Z'elovich, Y. B., and Y. P. Raizer (1966). *Physics of Shock Waves and High-Temperature Hydrodynamic Phenomena*, Academic Press, New York.

II
CLOUD AND THUNDERSTORM ELECTRICITY

Thunderstorm Origins, Morphology, and Dynamics

7

EDWIN KESSLER
NOAA National Severe Storms Laboratory

INTRODUCTION

Thunderstorms involve rapid vertical rearrangement of deep air layers. Large processes promote and shape vertical and horizontal air motions, and processes within storms control development of rain, hail, and strong local winds. Espy (1841) first presented the concept that thunderstorm circulations are driven largely by the heat latent in water vapor, released during condensation. During 1946-1947, a U.S. federally funded project investigated thunderstorms in Ohio and in Florida. Findings of that project (Byers and Braham, 1949) were the first accurate description of thunderstorm details, and project methods were a principal part of the foundation of subsequent studies. A modern comprehensive and abundantly illustrated textbook was prepared by Ludlam (1980). Thunderstorm processes are presented in much detail, with many references, in Volume 2 of a recent three-volume treatise on thunderstorms (Kessler, 1985), and the lightning process is detailed elsewhere in this volume.

VERTICAL CONVECTION AND THE GREENHOUSE EFFECT

Vigorous vertical air currents and thunderstorms are a consequence of excessive warmth and moisture at low altitudes. In a global sense this stratification of properties stems largely from the radiative properties of atmospheric constituents. The atmosphere is largely transparent to solar radiation, which is concentrated in the yellow region of the spectrum. Most of the solar energy not reflected by clouds passes through the atmosphere unabsorbed and is converted to heat at the ground or in surface waters. The absorbing surface also radiates, but because its temperature is much lower than the Sun's, terrestrial radiation is concentrated in infrared wavelengths. Atmospheric carbon dioxide, water vapor, and some other trace constituents are significant absorbers of much of the outgoing infrared emission, so this outgoing radiation is substantially blocked and further heats the air at low altitudes.

The average temperature at low altitudes provides long-term equality between the rate at which heat is carried away and the rate at which it is received. Removal processes include some radiative losses through spectral windows between absorbing bands of carbon dioxide and water vapor and transport of heat by air parcels in motion. This latter process is known as convection. Heat gained at the surface is carried by convection to altitudes where the radiative process is more effective because there is less absorbing medium above. Figure 7.1 shows that, globally, thunderstorms are most frequent in low latitudes where a larger surplus of heat

81

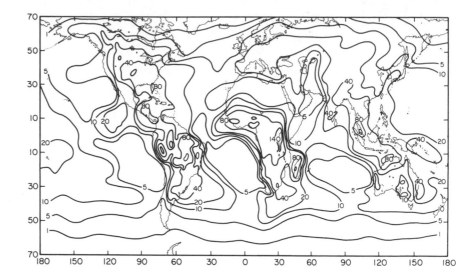

FIGURE 7.1 Annual number of thunderstorm days. Adapted from World Meteorological Organization (1953).

(due to radiative imbalance) is transported aloft by convection.

The process of vertical convection within an air mass contrasts with a global process driven by horizontal temperature differences on a very large scale. Thus, equatorial regions are warmer than polar regions, and there are various circulation regimes that function to reduce this temperature contrast. These circulations, primarily horizontal, associate with some vertical air motions as well, and they tend sometimes to enhance and sometimes to diminish the conditions that favor the intense vertical circulations of thunderstorms.

THE SIZE OF THUNDERSTORMS

About 85 percent of the atmospheric mass is contained in the lowest 13 km—this is the layer that generally participates in the vertical circulation of a typical thunderstorm. The typical horizontal dimension of a thunderstorm is similar to its vertical dimension. On the other hand, the storms of winter, essentially embodying horizontal air motions, respond to variations of temperature in the horizontal plane extending over hundreds or even thousands of kilometers; these storms are correspondingly larger horizontally than vertically.

The vertical circulation of air in a thunderstorm is a result of density differences. When the air is relatively warm, it is relatively expanded and weighs less than a parcel of the same size in its surroundings. The warm parcel is subject to pressure forces; these vary with height according to the weight of air in the large environment, and they are not fully balanced by the warm parcel weight. The imbalance is represented by a net upward force, Archimedean buoyancy, which causes the warm parcel to rise. As it rises, the air ahead of it

must move out of the way. The greater the horizontal extent of a rising parcel, the farther the displaced air must move and the more resistance there is to the parcel's motion. Very large buoyant parcels cannot rise rapidly because of the large volume of air that must be simultaneously displaced horizontally. Thus, vertical motions are faster in smaller rising parcels. On the other hand, the smallest buoyant parcels tend to be more eroded by diffusive processes operative on their boundaries. In the final analysis, more definite conclusions as to scale rest largely on empirically determined characteristics of diffusion and consideration of equations of three-dimensional motion.

Figure 7.2 shows a zone of severe thunderstorms extending hundreds of miles across the central United States. The large horizontal extent of the zone is related to the large scale of a disturbance engendered primarily by horizontal temperature contrasts. Along the zone, numerous boundaries between individual thunderstorms are evident. They mark the dimensions of local storms driven primarily by vertical temperature contrasts, locally enhanced by processes intrinsic to the larger-scale disturbance.

Conditioning the Air for Overturning

A proclivity for overturning is enhanced as the lower atmosphere is warmed. This occurs daily after sunrise, and afternoons are often marked by puffy clouds indicative of rising thermal currents. Seasonally, temperature rises as day length increases, and as the Sun rises higher in the sky, so spring and early summer usually present more frequent and more vigorous showers and thunderstorms than does fall. In other words, the thundery weather that tends to be frequent around the vernal

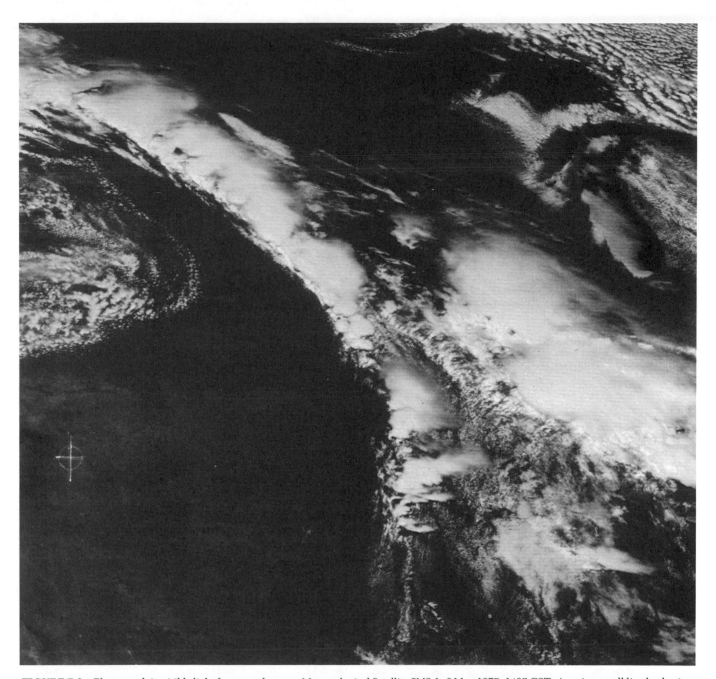

FIGURE 7.2   Photograph in visible light from synchronous Meteorological Satellite SMS-1, 6 May 1975, 1405 CST. A major squall line harboring tornadoes lies across the central United States. Lakes Superior, Huron, and Michigan are in upper right. Photo courtesy of NOAA, National Environmental Satellite Service.

equinox in mid-latitudes is stimulated by the relative coolness aloft that remains from winter.

A rising air parcel encounters decreasing air pressure and expands. This expansion represents work done on the environment and is accompanied by a decline in temperature. A rising parcel continues to rise as long as it remains warmer than its environment—this means that the ambient temperature must decline with height as fast as or faster than the temperature of the rising parcel declines as its altitude increases. A shower or thunderstorm occurs only when there is a sufficient decline of ambient temperature with height in a deep layer.

There would be no rain without moisture in the air,

but moisture represents energy as well—a large amount is latent in water vapor and enters the atmosphere as surface waters evaporate. Most of the air that harbors thunderstorms is conditioned during days or weeks over warm seas. During the westward journey of an air mass in the North Atlantic trade-wind zone, for example, the depth of the moist layer increases daily and is often about 2000 m thick when the air turns northward and enters the United States. The low-altitude moist layer may be surmounted by a very dry layer. This stratification is produced even as heat and moisture are supplied from below, while weak subsidence, associated with the same subtropical high-pressure area that engenders the trade winds, maintains dry air at middle levels, protected from incursions of moisture from below by a temperature inversion. Dry air aloft may also be produced by other processes. For example, precipitation that accompanies ascent of air on the windward side of a mountain range leaves the air drier on the leeward side. A warm temperature and large quantity of water vapor in an air mass at low altitudes and only a small amount of water vapor aloft, is a significant combination (known as a convectively unstable condition) whose potency is most realized when a large-scale disturbance causes generalized horizontal convergence and rising air motion. In such a case, the moist air at low altitudes cools less rapidly than the air above because the low-level air gains the heat latent in the water vapor as that vapor condenses. A result is that the temperature decline with height increases, i.e., the lapse rate becomes much steeper. Then rising parcels within the air mass and overturning within the air mass can become a violent process.

CONDENSATION AND PRECIPITATION IN RELATION TO AIR MOTION

As an ascending air parcel cools below the temperature at which its vapor is saturated, condensation occurs on nuclei, usually motes of sea salt, nitrates, or sulfates, numbering typically about 1000/cm³ over land. There is an initial selection or activation process whose details depend on the nature and number of the nuclei and on the ascent rate, reflected in extent of vapor supersaturation. Following activation, only the selected nuclei continue to grow and the balance evaporate. Owing to differences in initial size and composition, growth of the different nuclei proceeds at different rates. If ice coexists with supercooled liquid water, i.e., liquid at temperature below the normal freezing point, there is growth of the ice particles by deposition of vapor diffused from the liquid particles, which tend to evaporate. Brownian motion, turbulence, and differential fall speeds among the particles all contribute to coalescence and further broadening of the size distribution. After some tens of minutes, unless reduced in size by evaporation in dry air mixed into the cloud from its environment, some of the particles have appreciable fall speeds. Whether ice or liquid, they now begin to sweep out the smaller particles in their paths and fall ever more rapidly to the ground.

Convective precipitation is typically intermittent. Its fluctuating character at a place is related not only to the passage of discrete shower cells overhead but also to a dual effect of the condensation process. The condensation process releases latent heat, which tends to reduce air density and stimulate the updraft, but the condensation process also burdens the rising air with the weight of the thousandfold-denser condensate particles, which may be water or ice. Unless the updraft speed is markedly stronger than the fall speed of precipitation particles, precipitation accumulates in the updraft until the positive effects of thermal buoyancy are quite overwhelmed. Thus an initial updraft usually becomes a downdraft after about 30 minutes, as indicated in Figure 7.3. The downdraft can be quite strong, especially when the weight of precipitation is augmented by a cooling that accompanies evaporation of cloud and precipitation into dry air mixed into the cloud at middle levels. Such downdrafts necessarily become divergent horizontally as they near the Earth's surface, where they represent a significant threat to aviation.

In some severe storms, the weight of accumulated condensation products contributes to a splitting process that leads to separate storms, moving to the left and right of the direction of the vertically averaged wind. In the northern hemisphere, the rightward-moving storms are more likely to harbor tornadoes; damaging hail is often found in left movers.

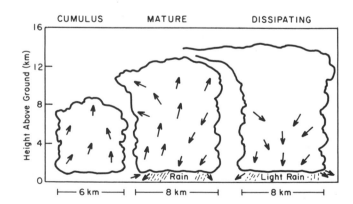

FIGURE 7.3   Three stages in the life cycle of an ordinary thunderstorm cell, as observed by the Thunderstorm Project. After Byers and Braham (1949).

STRONG CONVECTION

Development of a severe convective system depends on a stratification of temperature, moisture, and horizontal wind that produces extreme buoyancy forces tending to accelerate the motion of both ascending and descending air parcels. A steepening of the temperature-lapse rate is not alone sufficient, because there are many situations in which the energy of overturning can be dissipated in minor events about as fast as it becomes available through solar heating or other processes. Regimes of frequent events of weak to moderate intensity characterize air masses that are moist and weakly unstable through great depths.

Usually a combination of the following conditions is associated with extreme convection in middle latitudes:

(a) The airmass is convectively unstable. That is, there is a considerable lapse rate of temperature, and moisture is abundant only at low altitudes.

(b) A disturbance in the larger-scale flow (e.g., a short-wave trough in the westerly current aloft, often having a marked low-pressure system at the Earth's surface) provides generalized lifting, which causes the convective instability to become realized.

(c) The moist lower layer is separated from the dry zone by a stable layer or even a temperature inversion, which inhibits early overturning within the airmass and premature loss of potential energy.

(d) There usually is differential temperature advection (warm air advection at low altitudes and, occasionally, cold air advection aloft).

(e) A marked increase of the wind with height has a dual enhancing effect. First, warm air from convective towers is carried rapidly downstream by the strong winds aloft, preventing its local accumulation aloft and permitting longer duration of local convection; second, variation of the wind with altitude can cause the updraft column to slope with height and contribute to an organization of the flow that enables the intrinsic coolness of the air at middle levels to be manifested in overall storm energy. In brief, the precipitation formed and carried to great heights in strong updrafts can descend into intrinsically cold middle-level air, hastening that air's descent. In this case, where precipitation does not descend in the updraft in which it was formed, the storm may acquire a quasi-steady character.

The conditions described above are often associated with weather-map features like those shown in Figure 7.4.

Figure 7.5 shows a schematic vertical cross section through a quasi-steady severe storm that might form under the conditions depicted in Figure 7.4. The more de-

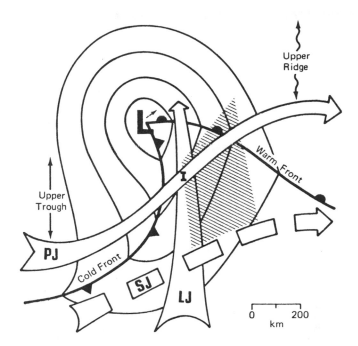

FIGURE 7.4   Idealized sketch of a middle-latitude weather situation especially favorable for development of severe thunderstorms. Thin lines denote sea-level isobars around a low-pressure center with cold and warm fronts. Broad arrows represent low-level jetstream (LJ), polar jet (PJ), and subtropical jet (SJ). The LJ advects moisture-rich air from subtropical regions to provide the basic fuel for convection. Severe storms (hatched area) are most likely to start near I and gradually shift toward the east while building southward. Severe thunderstorms also occur with many variations on this basic theme. From Barnes and Newton (1985).

tailed plan view of such a storm (Figure 7.6) illustrates two downdrafts. The forward-flank downdraft in the rain area is largely an effect of the weight of condensation products; the rear-flank downdraft is thought to owe its existence in part to a barrier effect on the ambient winds produced by ascent of air from low altitudes in a shearing environment. Another cause of the rear-flank downdraft is evaporative cooling of air, dry and intrinsically cold at heights of 3 to 4 km, by precipitation descending into it from greater altitudes. Finally, Brandes (1984) proposed a mechanism by which this rear-flank downdraft would be stimulated by formation of a tornado or other vortex at low altitudes.

Figures 7.5 and 7.6 show asymmetries that are critical features of persistent severe-traveling-storm complexes; Figure 7.3, in contrast, presents more symmetrical features, especially in its first and third frames. South of the severe-storm center in Figure 7.6, where a mesocyclone and possibly a tornado are located, a line of convective clouds (the flanking line) marks the intersection of air descended from middle levels with warm air rushing to-

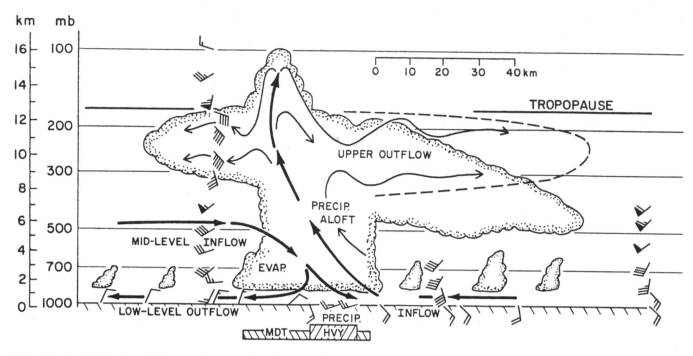

FIGURE 7.5　Profile of squall-line thunderstorm (based on series of radar observations, fivefold vertical exaggeration) as it passed Oklahoma City on May 21, 1961. Winds (full barb, 5 m/sec; pennant, 25 m/sec) are plotted relative to squall-line orientation; a shaft pointing upward is parallel to the squall line (azimuth 225°). In sounding behind storm, balloon passed through anvil outflow, and winds shown are not representative of the environment. Arrows indicate main branches of airflow relative to squall line that was moving toward right at 11 m/sec. On the right, dashes outline the supposed air plume; the radar-detected cloud plume at lower elevations consists of small precipitation particles that have partly fallen out of the air plume while drifting downwind from the storm core. Simplified from Barnes and Newton (1985).

ward the storm along the ground. At the Earth's surface, the intersection is called the gust front. It can itself be hazardous to aircraft because of remarkable turbulence and wind shears, and its horizontal winds are sometimes strong enough to do considerable damage to trees and buildings.

HAIL

At temperatures below the melting point of ice, solid and liquid phases can coexist, but the liquid phase is metastable and starts to freeze with release of latent heat in the presence of a suitable nucleus or on contact with an ice surface.

Growth of hail to large sizes occurs when strong and enduring updrafts, with temperature below the melting point of ice, bear liquid cloud particles in addition to some icy motes. The liquid particles start to freeze when contacted by the ice particles, and the latter thereby grow; they descend to the ground when their fall speed exceeds the rising speed of the enveloping air current.

The release of latent heat that attends freezing causes a rise of temperature toward 0°C; a growing hailstone

thus tends to be somewhat warmer than its environment. (This condition of relative warmth is associated with an important electrification process treated in Chapters 9 and 10, this volume.) In the presence of much liquid water and little supercooling, the hailstone may enter a stage of wet growth, i.e., one wherein all the impacting liquid is not frozen immediately but becomes absorbed into a previously developed porous structure or is somewhat shed. In this growth regime, the frozen material exhibits a clear structure of large crystals, readily distinguishable from each other under polarized light. At sufficiently low temperatures and water contents, liquid cloud and small raindrops impinging on the growing hailstones freeze so quickly that air bubbles remain entrapped within and between the globules. The variable growth process is revealed by the concentric translucent or opaque layers and nearly transparent zones that appear in hailstone sections (Figure 7.7).

The magnitude and distribution of vertical and horizontal air currents and the associated content of supercooled liquid water determine the growth and trajectory of hailstones. If the updraft speed has a maximum

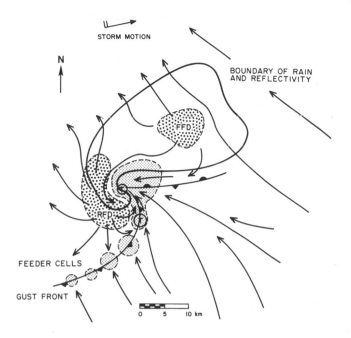

FIGURE 7.6 Schematic plan view of a tornadic thunderstorm at the surface. The heavy solid line encompasses radar echo. The wavelike gust front structure is depicted by a solid line and frontal symbols. Surface positions of the updraft are finely stippled; forward-flank downdraft (FFD) and rear-flank downdraft (RFD) are coarsely stippled; arrows represent associated streamlines (relative to the ground). Likely tornado locations are shown by encircled Ts. From Davies-Jones (1985).

value of 30 m/sec, and the cloud content is 4 g/m^3, a hailstone can grow to about 7.5 cm in diameter before arriving at the ground, 25 minutes after being selected for such growth by having fortuitously attained a larger size than its neighbors.

The rapid ascent of small particles in sufficiently strong updrafts gives insufficient time for the growth of any of them to large size. The development of hail in such strong updrafts may depend on the insertion into the updraft of hail embryos formed nearby and cycled into the updraft column by virtue of their descent from higher levels into horizontally convergent regions below.

Major tornadoes are usually accompanied by large hailstones. However, we find that tornadoes are usually absent from the storm class that includes the most damaging hailstorms. Doppler-radar observations of the airflows show that the hailstorm updrafts are not so strong as updrafts in tornadic storms, but the updrafts in major hailstorms cover a substantially larger area.

TORNADOES

Major tornadoes are most often associated with thunderstorms of a type illustrated in cross section in Figure 7.6, and they are identified by a rapidly rotating funnel-shaped cloud that marks the condensation boundary of in-spiraling air at low altitudes undergoing adiabatic expansion and cooling. Tornadoes are most severe and least uncommon in the United States, but they occur occasionally in India, Australia and New Zealand, South Africa, Argentina, Japan, and several countries of western Europe. In Mississippi, the state in the United States most subject to tornado deaths and damage, statistics on annual damage and storm frequency suggest that about 1/1000 to 1/300 of the area is affected by tornadoes each year, with tornado winds of 50 m/sec or more and significant damage to structures. The maximum rate of pressure change may be between 50 and 100 mbars/sec, and the maximum wind of a major tornado is about 100 m/sec. The frequency and intensity of these storms and the areas visited vary from year to year in association with irregular departures from seasonal norms of other quantities such as temperature and moisture.

Radar data show that tornadoes start at middle levels (about 5 km), to the rear side of pre-existing cyclonic circulations about 3 to 10 km in diameter, and develop downward and upward on a time scale of about half an hour. At the ground, tornadoes appear on or near a boundary between rising warm moist air, within which the release of latent heat of condensation is the storm's principal source of energy, and air descending from middle levels where it is intrinsically cold, and cooled sensibly by the evaporation of precipitation into it.

The most critical dynamical aspect of tornadoes involves the concentration of rotation within them. Various investigations during the past 10 years have established that two processes have direct importance in this concentration. The air's angular momentum is a consequence of the Earth's rotation and of various weather systems. The conservation of angular momentum accompanying horizontal convergence and ascent of air has long been appreciated. Such conservation is manifested, for example, by the increased rotation rate as a skater's arms are brought in from an outstretched position. The second process, more recently detailed, involves the rotation or twisting of horizontal vorticity toward the vertical plane. Horizontal vorticity is represented by the vertical variation of the horizontal wind (vertical wind shear), already cited as significant for severe-storm development through its role in facilitating removal of condensation products from the updraft. The twisting process is commonly effective on an

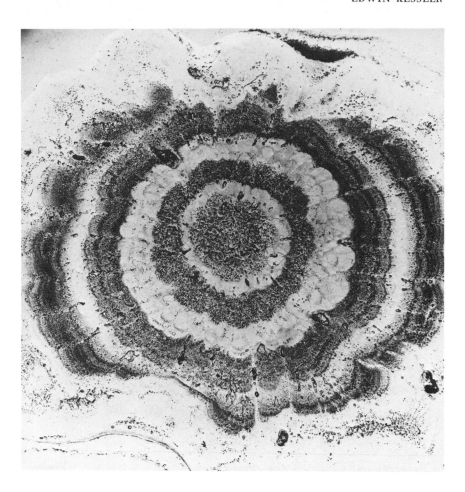

FIGURE 7.7 Cross section through hailstone. Alternating clear and opaque layers
mark growth under different conditions of
temperature and rate of accretion of liquid
water. Photo courtesy of Charles Knight and
Nancy Knight, National Center for Atmospheric Research.

updraft boundary when the winds at low altitude relative to the storm motion veer markedly with height.

The properties related to these processes—vorticity,
circulation, and angular momentum—are interrelated
through considerations of area and distance. Thus the
circulation in a region is the average velocity along the
closed curve that defines the region times the length of
the curve. This is the same as the average component of
vorticity normal to the same region times the area of the
region. The angular momentum of a particle or air parcel is measured about a reference point as the distance
from the point times the velocity normal to the line connecting the particle and reference point.

CONCLUDING REMARKS

The thunderstorm entity is a result of thermodynami-
cal, microphysical, and electric processes. All processes
interact and must be observed contemporaneously in order to be well understood. The summary presented in
the foregoing pages represents remarkable progress in
understanding, a result of public investment in a focused use of new tools during the past two or three decades. Meteorological radar, artificial satellites, and
marvelous new technologies for data processing, computing, and communicating have all been critically important aids and have facilitated impressive new methods for early identification of storms, dissemination of
information about their location, movement and intensity, significant reduction in death rates from tornadoes,
and marked decline in the rate of weather-related aircraft accidents. Now there are a host of new methods for
study of the lightning process. As past is prologue, we
confidently expect these new aids to contribute much
toward clarification of the function of electric processes
in severe-storm evolution and toward diminishing the
still significant lightning hazard.

ACKNOWLEDGMENT

The author thanks Evelyn Horwitz, who typed this
paper through several drafts, and Lindsay Murdock,
who improved the paper editorially.

REFERENCES

Barnes, S. L., and C. W. Newton (1985). Thunderstorms in the synoptic setting, in *Thunderstorm Morphology and Dynamics*, E. Kessler, ed., 2nd edition, Univ. of Oklahoma Press, Norman, pp. 75-112.

Brandes, E. A. (1984). Relationship between radar derived thermodynamic variables and tornadogenesis, *Mon. Weather Rev. 112*, 1033-1052.

Byers, H. R., and R. R. Braham (1949). *The Thunderstorm*, Report of The Thunderstorm Project, U.S. Weather Bureau, U.S. Government Printing Office, Washington, D.C., 287 pp.

Davies-Jones, R. P. (1985). Tornado dynamics, in *Thunderstorm Morphology and Dynamics*, E. Kessler, ed., 2nd edition, Univ. of Oklahoma Press, Norman, pp. 197-236.

Espy, J. P. (1841). *The Philosophy of Storms*, Charles C. Little and James Brown, Boston, Mass., 552 pp.

Kessler, E., ed. (1985). *Thunderstorm Morphology and Dynamics*, 2nd edition, Univ. of Oklahoma Press, Norman, 415 pp.

Ludlam, F. H. (1980). *Clouds and Storms*, Pennsylvania State University Press, University Park, Pa., 405 pp.

World Meteorological Organization (1953). *World Distribution of Thunderstorm Days*, WMO no. 21, part 2, Geneva, Switzerland, 71 pp. and maps.

The Electrical Structure of Thunderstorms

8

PAUL R. KREHBIEL
New Mexico Institute of Mining and Technology

Thunderstorms and the lightning that they produce are inherently interesting phenomena that have intrigued scientists and mankind in general for many years. A number of theories have been proposed to explain how thunderstorms become electrified, and many field and laboratory experiments have been conducted to determine the electrical nature of storms and to test the electrification theories. Through this effort we are beginning to understand how electric charge is distributed in storms, but the mechanisms that cause their electrification continue to elude scientists and remain the subject of considerable inquiry and debate.

The basic difficulty in determining how thunderclouds become electrified lies in the fact that they are large, complex, and short-lived phenomena that need to be examined both as a whole and in detail to understand how they function. The electrical processes are intimately related to the cloud dynamics or motions and to the microphysics of the cloud, namely, to the populations and interactions of the precipitation, cloud droplets, ice crystals, and other particles that make up the cloud. These important aspects of storms are themselves incompletely known or understood, yet a detailed comprehension of them is necessary to understand the electrification processes.

Attempts to simulate possible electrification processes in the laboratory or by theoretical modeling have been helpful in evaluating some theories but have not demonstrated the efficacy of any particular mechanism. This is because thunderstorm conditions are inherently difficult to simulate and are insufficiently understood for us to be confident that we are simulating them properly.

At present, further progress in understanding the electrification of thunderstorms, and indeed in understanding their dynamics and precipitation processes as well, requires simultaneous observations of their dynamical, microphysical, and electrical properties. This need has been increasingly recognized in recent years and has given rise to a number of cooperative studies of storms. The cooperative studies employ the latest techniques for internally and remotely probing storms and rely on the combined expertise of university and national laboratory researchers to conduct and analyze the observations. The studies typically use instrumented aircraft and balloons to penetrate the storms, multiple radar systems to measure precipitation strengths and velocities, and ground-based instrument networks for measuring meteorological and electrical quantities.

A few research programs have focused on the electrification question, including the ongoing studies at the Langmuir Laboratory for Atmospheric Research in the mountains of central New Mexico and the Thunderstorm Research International Program (TRIP) in Florida and New Mexico. These and other studies have steadily improved our ability to observe the electrical characteristics of thunderclouds both from the ground and inside the storm. Most electrification studies have lacked particle observations inside the storm, a shortcoming that has been partially addressed in recent years.

The study of thunderstorms and their electrification is important not only because of their instrinsic scientific interest but for other reasons as well. A significant fraction of the Earth's rainfall in temperate climates comes from electrified clouds, and it is possible or likely that the precipitation processes in these storms are influenced by their electrification. Also, electrified and lightning-producing storms may play an important role in the chemical reactions responsible for the production of acid rain. Finally, vigorous downbursts, which recently have been identified as dangerous to aircraft in the vicinity of airports, may be accompanied by an electrical signature that could aid in their detection and the warning of hazardous conditions.

In this chapter we discuss what is currently known about the electrification of thunderstorms and what is not known and indicate some directions for research efforts over the next decade or so.

ELECTRICAL STRUCTURE AND DEVELOPMENT

The distribution and motion of electric charge in and around a thunderstorm is complex and changes continuously as the storm evolves. Nevertheless, we have a rudimentary picture of how charge is distributed in an already-electrified storm. This is depicted in Figure 8.1 on a photograph of a small thunderstorm over Langmuir Laboratory.

The interior of the storm contains a dipolar charge distribution consisting of positive charge in the upper part of the cloud and negative charge below the positive. These are the dominant accumulations of charge in the storm and are called the "upper positive" and "main

FIGURE 8.1   An isolated thundercloud over Langmuir Laboratory in central New Mexico and a rudimentary picture of how electric charge appears to be distributed inside and around the thundercloud, as inferred from in-cloud and remote observations.

negative" charges, respectively. The upper positive charge attracts negative ions to the top of the cloud from the electrically conducting clear air around the storm. The ions, which are produced by cosmic radiation, attach to small cloud particles at the edge of the cloud, forming a negative screening layer that partially cancels or screens the interior positive charge from an outside observer. The main negative charge causes point discharge or corona from trees, vegetation, and other pointed or exposed objects on the ground below the storm, which leaves positive charge in the air above the Earth's surface.

Positive charge is also found beneath and inside the base of the cloud below the main negative charge; this is called the "lower positive charge." Two sources for the lower positive charge are the corona from the ground, which may be carried upward into the cloud by the updraft, and positive ions generated by cosmic rays below and around the cloud base, which are attracted to the cloud by the main negative charge. Additional lower positive charge is carried by descending precipitation and occurs in localized regions known as "lower positive charge centers" (LPCCs). Several hypothetical LPCCs are depicted in Figure 8.1; they may be caused by a subsidiary charging process in the cloud.

The above description presents a simplified picture of how charge is distributed in a thunderstorm; the actual charge distribution is more complicated than this and needs to be better understood before we can answer the question of how thunderstorms become electrified. For example, it is necessary to know what types of particles carry the charges and how these particles move. Information on these and other questions is being obtained both from in-cloud and remote observations, as we discuss in this review.

Charge accumulates in the upper positive and main negative regions as a result of the charging mechanisms until the electric stresses are such that a lightning discharge occurs. Two primary forms of discharges are cloud-to-ground and intracloud lightning. Cloud-to-ground (CG) is the most familiar and spectacular form of lightning; it usually occurs between the main negative charge and ground and lowers negative charge to ground along one or more distinct and highly luminous channels. Some CG flashes lower positive charge to ground (see Chapter 3, this volume); these are called positive CG flashes and have been difficult to distinguish from normal CG discharges until recent years. They are of interest both because they are different from normal-polarity CG flashes and because they are often more damaging to objects that they strike.

Intracloud (IC) lightning is usually confined to the cloud interior and diffusely illuminates the cloud, being visible primarily at night. Intracloud lightning often occurs as a primarily vertical discharge between the main negative and upper positive charge regions of the storm. Horizontal IC lightning is also common, particularly in large storm systems where the lightning may propagate over distances of 100 km or more. These extensive discharges may have CG components or may be initiated by a CG discharge.

For studying the processes of electrification and electrical breakdown, the most interesting parts of a lightning discharge are inside the cloud where they are obscured from direct optical observation. Clouds and precipitation are transparent at microwave and longer wavelengths and to other kinds of signals, however, and several techniques that sense these signals are beginning to provide us with important information on what lightning looks like inside a storm. The techniques locate the lightning channels and charges and are discussed later in this chapter.

The charges of the storm itself can be sensed by measuring the electric field that they produce. The electric field indicates the strength and direction with which the storm charges attract or repel other charges and can be measured at the ground or in the air outside or inside the cloud. In clear-sky (fair-weather) conditions the atmospheric electric field at the Earth's surface has a negative value of about 100 to 200 V/m. This is caused by the fact that the ionosphere is charged positively to a potential of about 300,000 V with respect to the Earth's surface. (In turn the ionospheric potential is believed to result from the global thunderstorm activity.) Beneath a thunderstorm the electric field at the ground is often substantially larger, up to 10,000 V/m or more, and tends to be reversed in sign from fair-weather conditions.

Figure 8.2 shows a recording of the electric field versus time measured on the ground beneath a thunderstorm over central New Mexico. The storm went through its complete life cycle over the recording instrument, and the electric field record illustrates different stages in its electrical activity. As the storm became electrified, the buildup of negative charge in its base caused the electric field at the ground to reverse sign from the fair-weather (negative) value and to increase rapidly in magnitude. This is called the initial electrification of the storm. In-cloud measurements have indicated that the initial electrification can occur in a relatively short time, on the order of 5 to 10 minutes or perhaps less. The initial electrification is usually considered to end with the occurrence of the first lightning discharge, which marks the beginning of the active or lightning-producing stage.

The active stage can last from a few minutes to an hour or more depending on the size and convective vigor

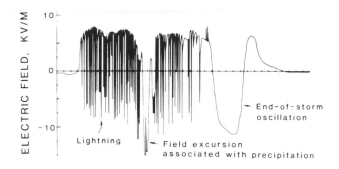

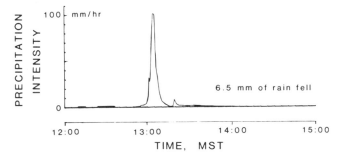

FIGURE 8.2 The atmospheric electric field and precipitation intensity on the ground beneath an isolated, stationary thunderstorm in central New Mexico (adapted from Moore and Vonnegut, 1977). See text for description.

of the storm. During this time lightning discharges suddenly decrease the electric field while the charging process steadily increases the field. As observed at the ground, the electric field jumps from positive to negative values and then grows back to positive values. The sign reversal indicates the presence of positive corona charge above the measuring instrument, which dominates the field for a short time after the discharge. As the storm charges build back up to the next lightning flash, point discharge limits the electric field at the ground to some maximum value indicated by the upper envelope of the record (about 8000 V/m in the case of Figure 8.2). Electric-field meters flown from balloons several hundred meters above the ground do not show field reversals during lightning and usually do not show limiting field values (Standler and Winn, 1979; Holden et al., 1983). This confirms that both effects are associated with corona from the ground and indicates that most of the corona charge resides in a relatively shallow "blanket" close to the Earth's surface.

The pronounced excursion of the electric field to neg-

ative values in the middle of the active stage was associated with the arrival of a downdraft and a transient burst of precipitation at the observing location. This is a common feature of thunderstorm observations and is called a field excursion associated with precipitation. In this example the charge carried by the precipitation as it arrived at the ground was measured to be weak and negative, i.e., of the wrong sign (and insufficient in magnitude) to have caused the field excursion. This is often the case and is called the mirror-image effect (Chalmers, 1967). The precipitation is believed to capture point discharge ions produced during the field excursion and to return them to earth. On the other hand, balloonborne measurements of lower positive charge centers carried by precipitation have been correlated with the subsequent occurrence of a field excursion at the ground, of the right sign to be explained by the precipitation charge (Marshall and Winn, 1982; Holden et al., 1983). It is uncertain whether field excursions are usually caused by descending, charged precipitation (whose sign may be reversed close to the earth by the capture of point discharge ions) or whether the downdrafts that accompany the precipitation carry or reveal other charge that causes the excursion. In any event, it has been suggested that the field excursions could help to detect downbursts in storms that are responsible for aircraft accidents on takeoff and landing (C. B. Moore, New Mexico Institute of Mining and Technology, private communication; Lhermitte and Williams, 1985a).

During the final or dissipating stage of the storm the lightning activity died out and the electric field exhibited a large, sustained swing to negative values and back, called an end of storm oscillation (EOSO). EOSOs are observed directly beneath a dissipating storm and are associated with the storm's physical collapse. The field at the ground is dominated for relatively long periods of time by positive charge overhead, and this is found to be a favored time for the occurrence of positive CG lightning. (None occurred in the storm of Figure 8.2, however.) It is not understood what the charge distribution is during an EOSO or how it changes to produce the field reversals; what little information we have is discussed at the end of this chapter.

The electrification of a storm is cellular in nature, i.e., it is associated with the development of individual convective cells within the overall storm. All but the simplest of storms are multicellular, with the lifetime of an individual cell being about 10 or 15 minutes. Some severe storms of the Great Plains develop into large, highly organized systems called supercell storms. These and other large storms appear also to have a dipolar charge structure, but little is known about the details of their electrification.

THE MAIN NEGATIVE AND UPPER POSITIVE CHARGES: SOME OBSERVATIONAL EVIDENCE

The dipolar structure of the storm interior was first inferred in England during the 1920s by Wilson (1920, 1929). Wilson observed that the electric-field change produced by intracloud lightning reversed sign with increasing distance from storms, as if the lightning were discharging an upper positive and lower negative charge. (Earlier work by Wilson on the properties of atmospheric ions led him to develop the cloud chamber for studying high-energy particles, for which he was awarded a Nobel Prize in 1927.) Subsequent field studies in England and New Mexico between about 1935 and 1955 confirmed this basic picture of the storm charges and indicated that the main negative charge resided at altitudes where the ambient temperature is less than 0°C (Simpson and Scrase, 1937; Workman *et al.*, 1942; Reynolds and Neill, 1955). The observations by Simpson and Scrase in England also revealed the presence of lower positive charge below the main negative-charge region. Field studies during the past 15 yr have further confirmed and refined these early results, as discussed below. In particular, the studies have indicated that the main negative charge is found in a relatively narrow range of altitudes at temperatures that vary between about 0 and −25°C.

Figure 8.3 presents a modern-day equivalent of Wil-

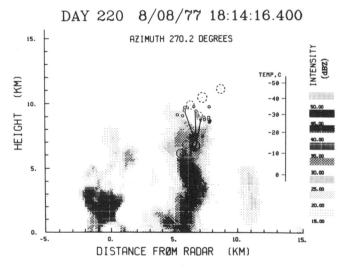

DAY 220   8/08/77   18:14:16.400

FIGURE 8.3   Vertical cross section of the radar echo from a small Florida thunderstorm at the time of the first lightning flash in the storm and the centers of charge transferred by the first lightning flash. The lightning was an intracloud discharge that effectively transported negative charge from within the precipitation echo centered at about 7-km altitude (−15°C) to above the detectable precipitation (from Krehbiel *et al.*, 1984b).

son's result. It shows the centers or sources of charge for the first, intracloud lightning discharge in a small Florida storm, superimposed on a vertical cross section of the radar echo from the storm's precipitation. The charge centers were determined from simultaneous measurements of the electric-field change produced by the lightning at eight locations on the ground beneath and around the storm. The flash effectively removed negative charge from within the precipitation echo between about 6- and 7.5-km altitude and transported the charge upward in the cloud, to above the detectable precipitation. Data from a higher-power, Doppler radar observing the same storm showed that a weaker radar echo extended up to and above the upper charge centers, indicating that the lightning remained within the cloud. The air temperature outside the cloud at the level of the negative-charge centers was between −10 and −15°C.

Comparison of the lightning and radar data in three dimensions has shown that the lightning occurred in the part of the storm having the greatest vertical extent of precipitation. Additional comparison with the Doppler observations of precipitation velocities has indicated that the negative charge sources of the lightning were located adjacent to and in the updraft of the storm. The initial charge centers coincided with a localized region of stronger precipitation that was falling toward earth on one side of the updraft, and the subsequent charge centers were displaced through the updraft toward its opposite side.

Figure 8.4 shows a vertical sounding of the electric field in a small New Mexico storm, obtained with a balloonborne instrument that measured the corona current from a 1-m-long vertical wire. The corona current reversed sign as the instrument ascended through the negative-charge region between 6- and 7-km altitude [above mean sea level (MSL)] and reversed sign again as it ascended through the upper positive charge, above 9-km MSL. The temperature at the altitude of the negative-charge region was between about −5 and −10°C. No lightning was produced by the storm. Soundings through lightning-producing storms also indicate a dipolar charge structure but are complicated by the large-amplitude field changes of the lightning discharges. The soundings can be made with more sophisticated instruments that sense the electric field directly and in three dimensions (e.g., Winn *et al.*, 1981). This can be done from balloons or on aircraft, and the observations show that the fields and charges have a more detailed structure than suggested by Figure 8.4.

The electric-field measurements indicate that the volume density of electric charge is on the order of 1-10 coulombs/km^3 inside storms. This results in total charge amounts of a few coulombs to a few hundred coulombs

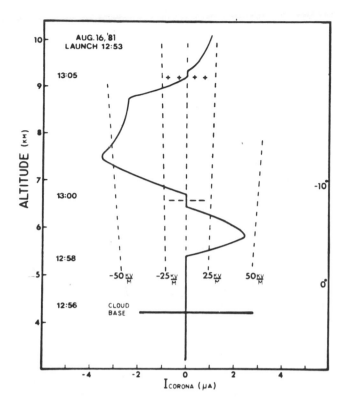

FIGURE 8.4 Vertical sounding of the corona current from a 1-m-long wire carried by balloon through a storm over Langmuir Laboratory in New Mexico (Byrne *et al.*, 1983). The wire was vertically oriented, and the corona current record is indicative of the vertical component of the electric field in the storm.

or more depending on the size and age of the storm. Greater charge densities may exist in localized regions of a storm (e.g., Winn *et al.*, 1974).

The above results indicate the dipolar nature of thunderstorms and illustrate two techniques for studying their charge structure. Of particular importance from these and similar types of observations are the findings (1) that net negative charge is distributed horizontally within a storm [rather than in vertical columns as had been inferred by Malan and Schonland (1951)] and (2) that the negative charge is found at similar temperature values in different sizes and types of storms. These results have been inferred from a combination of lightning and in-cloud measurements and are illustrated in Figure 8.5. The negative-charge sources of both cloud-to-ground and intracloud lightning are found to be displaced horizontally within a storm and are found to be at similar temperature levels in Florida storms and New Mexico storms. The similarity of the lightning-charge heights and temperatures is particularly significant in view of the fact that Florida storms have substantially greater depth of cloud and precipitation below the 0°C

level, and often above it as well. New Mexico storms are drier and generally smaller than Florida storms, having cloud bases just below the 0°C level. The results suggest that the charging processes are the same in the two types of storm and operate at temperatures less than 0 or −10°C. They also suggest that the part of a storm warmer than 0°C is not directly involved in the electrification.

The lightning-charge observations are supported by electric-field soundings through storms, which indicate that the main negative-charge region is relatively shallow, on the order of a kilometer deep, and is laterally extensive (e.g., Winn *et al.*, 1981; Byrne *et al.*, 1983). The altitude of the negative-charge region from sounding observations tends to be lower than that inferred from lightning-charge observations, and there is some indication that the negative-charge region may be systematically higher in Florida storms than in New Mexico storms (Williams, 1985). The latter difference could result from the greater water content or size of Florida storms. But any such differences in electrical structure need to be substantiated by more direct observational comparisons.

The main negative charge appears not only to be distributed horizontally in a storm but to remain at approximately constant altitude or temperature as the storm grows. This is indicated by the results of Figure 8.6, which shows the heights of the charge centers for the first 15 lightning discharges in the small Florida storm of Figure 8.3. As the storm grew vertically, the positive (upper) charge centers of the intracloud lightning flashes increased from 10- to 14-km altitude (−30 to −60°C), but the negative-charge centers remained at about 7-km altitude (−15°C). Sequences of radar pictures like the one shown in Figure 8.3 confirm the upward growth of the storm and show that it occurred at the same rate as for the positive-charge centers of the lightning (8 m/sec). This agrees with the idea that the upper positive charge resides on small particles that are carried by the updraft into the upper part of the storm.

The apparent altitude stability of the negative charge is remarkable in view of the fact that convective storms are characterized by substantial upward and downward motions of both air and precipitation. The storm charges are carried on cloud and precipitation particles and must follow the motions of the particles until their charge somehow changes. As time-resolved observations become available such charge motions will undoubtedly be found; indeed there is some evidence for them in the variability of electric-field data from storm to storm. Possible explanations for the otherwise horizontal and stable nature of the main negative charge are that the charging process operates only at certain tem-

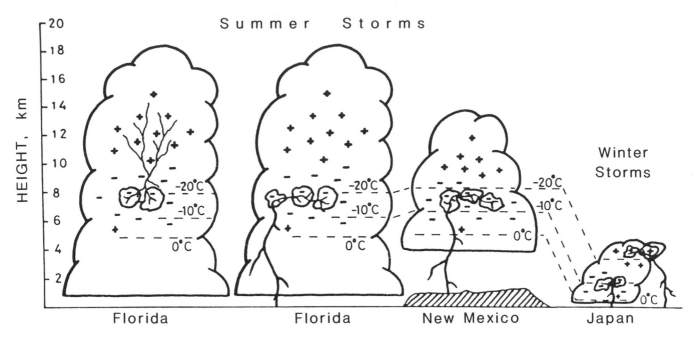

FIGURE 8.5   Illustration of how the negative-charge centers of cloud-to-ground lightning are at similar temperature levels in New Mexico and Florida storms, even though the latter have much greater extent of cloud and precipitation below the 0°C level and often above this level as well (adapted from the original by M. Brook, expressing the results of Jacobson and Krider, 1976; Krehbiel *et al.*, 1979; Krehbiel, 1981; Brook *et al.*, 1982). The negative charge centers of intracloud lightning are also at similar altitudes and temperatures even though intracloud discharges extend upward in the cloud rather than downward. Preliminary studies of lightning in Japanese winter storms suggest that the negative charge is at lower altitude but similar temperature values as in the summer storms.

peratures (or pressures) or that the dynamics of the storm causes negatively charged particles to accumulate at the observed altitude. In any event the fact that net negative charge tends to be observed over a limited vertical extent indicates that other processes operate to change or mask the particle charges as they emerge from the negative-charge region.

Preliminary studies of lightning in Japanese winter storms have indicated that these shallow but vigorously convective and strongly sheared storms also have a dipolar charge structure in which negative charge is at a lower altitude (but at similar temperature values) as in summer thunderstorms (Brook *et al.*, 1982). These results are also illustrated in Figure 8.5; if confirmed by additional observations they suggest that temperature or the storm dynamics, rather than absolute altitude or pressure, are the important factors in the electrification.

FIGURE 8.6   The altitude of the lightning charge centers for the first 15 discharges in the small Florida storm of Figure 8.3. The upper positive-charge centers of the intracloud flashes increased in altitude as the storm grew, while the negative-charge centers remained at constant altitude. Two cloud-to-ground discharges occurred toward the end of the sequence.

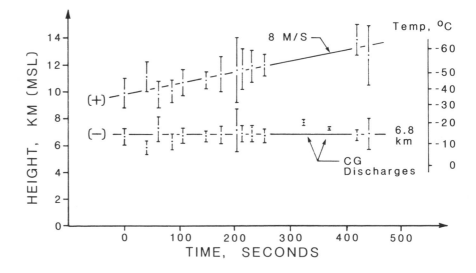

INITIAL ELECTRIFICATION

Observations of the onset of electrification in storms are consistent with the above picture of thunderstorm charges and provide additional insight into the electrification problem. In particular, it is found that a storm does not become strongly electrified until its radar echo extends above a certain altitude threshold and is growing vertically. The threshold altitude depends somewhat on the sensitivity of the radar but is about 8 km above MSL in the summer months, corresponding to an air temperature of about $-20°C$. Kasemir and Cobb (see Cobb, 1975; Illingworth and Latham, 1977) reported a similar threshold effect from aircraft measurements near the tops of Florida clouds; electric-field values of 1 kV/m were not detected until the radar echo top grew above about $-5°C$.

A recent set of observations that illustrate the onset of electrification in a storm is presented in Figure 8.7 [J. E. Dye (National Center for Atmospheric Research) and W. P. Winn and C. B. Moore (New Mexico Institute of Mining and Technology), private communications]. The figure shows the height of the precipitation echo versus time in a small storm near Langmuir Laboratory and an electric-field record from a ground station 5 km distant from the storm. Electrification was not detected at the ground until about 12:40 MST, shortly after the radar echo began a sustained period of growth above 8-km altitude. More sensitive measurements from an instrumented aircraft penetrating the cloud at 7-km altitude ($-15°C$) showed weak electrification (on the order of 100 V/m electric-field perturbations) during a pass between 12:34 and 12:36 and strong electrification on re-entering the cloud at 12:45. Other measurements from a sailplane at 4-km altitude inside the cloud indicated weak electrification starting at about 12:31, probably associated with the earlier convective surge at 12:25. The sailplane spiraled upward in the storm updraft and measured 1 kV/m maximum fields by 11:40 at 6-km altitude. The field appeared to originate from negative charge in nearby precipitation of 40-dBZ reflectivity (Dye *et al.*, 1985). The first lightning discharge occurred at 12:44 when the echo top had reached 10-km altitude. By this time, moderately strong (40-dBZ) echoes had developed up to 8-km altitude and were beginning to subside. Equally strong precipitation developed during the earlier convective surge, but the earlier cell had less convective energy and did not become strongly electrified.

The above example illustrates graphically the importance of convective growth in the electrification of a storm. This fact has been recognized for a number of years and is generally accepted (e.g., Workman and Reynolds, 1949; Reynolds and Brook, 1956; Moore *et al.*, 1958). The convective growth is often retarded by stable air or by strong winds at mid-altitudes in the atmosphere and is usually preceded by a succession of convective surges or turrets before one or more of these succeed in penetrating the stable layer. The example also indicates that moderately strong precipitation had developed in the storm before to its electrification. That precipitation must be present and must develop above a certain altitude or temperature threshold is a consistent feature of field observations that is being documented for an increasing number of storms in New Mexico, Florida, and Montana (Reynolds and Brook, 1956; Holmes *et al.*, 1977; Lhermitte and Krehbiel, 1979; Krehbiel *et al.*, 1984a; Dye *et al.*, 1985, 1986).

DISCUSSION

The above results and others like them indicate that the electrification process operates at temperatures of less than 0 or $-10°C$. In addition, they indicate that convection and precipitation somehow combine to cause the electrification.

One of the biggest questions and sources of debate among thunderstorm researchers has been whether the kinds of precipitation and cloud particles that grow in convective storms cause their electrification or whether the convective motions themselves directly electrify the storm without involving or requiring precipitation. Historically, observations have led many scientists to assume or favor the precipitation explanation, and the recent radar and electrical observations described above continue to fuel this idea. The temperature values at which electrification is observed have caused many researchers to focus on frozen precipitation as a primary agent in the electrification process. Other observations, discussed below, have raised questions about precipitation theories and cause some researchers to look toward a convective explanation.

Chapters 9 and 10 (this volume) discuss the various theories and mechanisms that have been proposed to explain how thunderstorms become electrified. Precipitation theories hypothesize that the relatively large precipitation particles acquire negative charge, in most cases by colliding with or shedding smaller cloud particles. The cloud particles acquire a corresponding positive charge and are carried by the updraft into the upper part of the storm, whereas the precipitation may rise or fall with respect to the ground depending on the relative magnitudes of its fall speed and the updraft. Negative and positive charges are segregated onto large and small particles, respectively, and are separated by the action of gravity to electrify the storm. In convection theories,

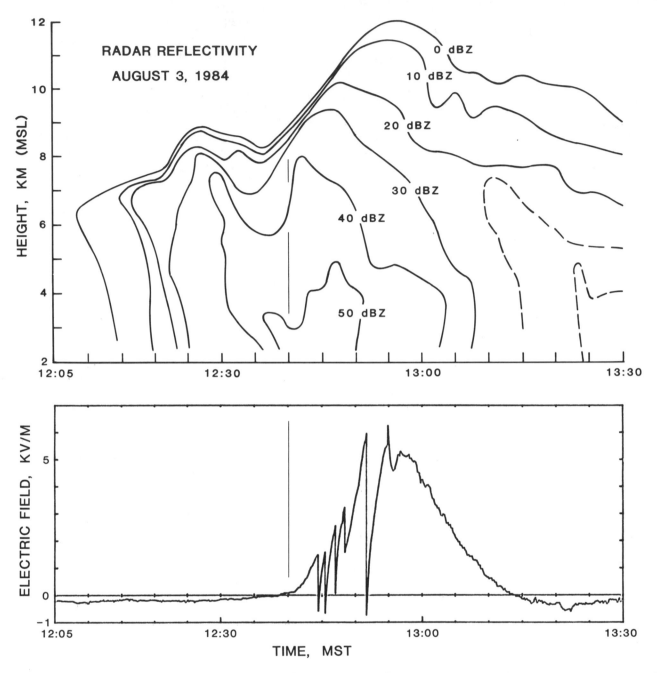

FIGURE 8.7   The radar reflectivity of precipitation versus height and time in a small storm near Langmuir Laboratory on August 3, 1984, and a record of the electric field at the ground 5 km from the storm. The electrification was associated with convective growth above 8-km altitude (about −20°C) and with the development of moderately strong precipitation up to this altitude. An initial convective surge between 12:20 and 12:25 produced only weak electrification, as measured by instrumented aircraft inside the storm. [Unpublished data from J. E. Dye (National Center for Atmospheric Research) and C. B. Moore and W. P. Winn (New Mexico Institute of Mining and Technology).]

positive and negative charges are spatially segregated and the energy of electrification is derived directly from the convective motions of the storm, which transport charges of opposite sign away from each other. The charges are expected to reside primarily on small cloud particles, with the net charge on precipitation being ei-

ther small or of the same sign as that on the cloud particles.

In-cloud observations at the level of the main negative charge show that the cloud contains a mixture of particle sizes and types. All or most of the precipitation particles are frozen and are in the form of graupel or hail. The precipitation particles coexist with a large number of small, unfrozen cloud droplets that are carried above the 0°C level by the updraft. The droplets remain in a supercooled liquid state until they contact an ice surface, whereupon they freeze and stick to the surface in a process called riming. (Alternatively, the supercooled droplets freeze spontaneously at sufficiently low temperature.) Riming is the dominant growth process of graupel and dry hail and entraps significant amounts of air, giving the particles a milky appearance. The riming process is also responsible for the dangerous ice loads that aircraft develop in flying through convective clouds above the 0°C level.

Storms that produce snow, such as winter storms and the dissipating parts of summer storms, can be strongly electrified but tend to produce only occasional lightning or none at all. This suggests that snow, whose crystals grow directly from water vapor in the air, does not in itself cause the electrification of active thunderclouds. The primary differences between winter or dissipating storms and active thunderclouds is that the latter are more strongly convective, develop greater vertical extents, and produce graupel or hail rather than snow.

A number of laboratory studies since the 1950s have shown that rebounding collisions between hail pellets and small ice particles cause charge of the appropriate sign to be transferred between the particles (e.g., Reynolds *et al.*, 1957; Takahashi, 1978; Gaskell and Illingworth, 1980; Jayaratne *et al.*, 1983). This charging process operates in the correct temperature range and is considered by some researchers to be the most promising of the precipitation mechanisms at present (e.g., Latham, 1981; Illingworth, 1985). But the laboratory-observed charging is able to account for the observed electrification only when the precipitation rates are high, on the order of 30 mm/h, and when the ice crystals are relatively abundant, 10-50 per liter or more (e.g., Illingworth, 1985; Williams, 1985). A precipitation rate of 30 mm/h corresponds to a radar echo of about 40 dBZ if the precipitation is frozen. Radar echoes of this strength have been observed during the initial electrification of Florida storms (Lhermitte and Krehbiel, 1979; Krehbiel *et al.*, 1984a) and recently in New Mexico storms (Figure 8.7; Dye *et al.*, 1986). Earlier observations of New Mexican storms have indicated that they can become electrified when their radar echoes are weaker—33 to 35 dBZ or perhaps less (Moore, 1963; Holmes *et al.*, 1977). These echo strengths correspond to frozen precipitation rates of about 10 mm/h or less. While precipitation rates can be estimated remotely using radar, the populations of small ice crystals can be determined only from in-cloud measurements and vary greatly with the particular conditions and with altitude.

Concentrations of 10-50 per liter are large but have been observed. As noted by Dye *et al.* (1986), however, few measurements have been made in the conditions and locations of interest.

The above discussion points to a central issue of thunderstorm studies, namely, whether sufficient precipitation is present and involved in enough charging interactions to account for the initial electrification. There has been much discussion of this issue in the scientific literature (e.g., Moore, 1976a, 1976b, 1977; Mason, 1976; Illingworth and Latham, 1977; Illingworth, 1985; Williams, 1985). An increasing number of field studies are indicating that the initial electrification occurs during the growth of precipitation in an updraft, where the conditions would be conducive to an ice-based precipitation charging mechanism. (These are cited at the end of the preceding section.) Recent results from these studies indicate that the electric fields inside the cloud appear to originate from regions of stronger radar reflectivity at the negative-charge level and indicate negative charge in those regions (Dye *et al.*, 1986). But observations in already-electrified storms show that the electrification is more widespread than the strong precipitation echoes (Krehbiel, 1981; Winn *et al.*, 1981; Weber *et al.*, 1982). In addition, estimates of the energy available from falling precipitation indicate that the energy may only be comparable with the electrical energy of some storms, particularly at altitudes where the electrification occurs. In this case a precipitation mechanism would have to be highly efficient if it were to cause the electrification (Williams and Lhermitte, 1983).

Similar issues and questions exist with regard to convection theories of electrification. The convective energy of a storm is easily sufficient to account for the storm's electrical energy, but it has not been shown that the convective motions transport charge in a manner and in amounts required to explain the electrification.

There are some reports of lightning in clouds whose tops have not reached the 0°C level and that therefore cannot contain frozen precipitation (see Moore, 1976a, for a summary). These are called warm clouds, and the occurrence of lightning within them is a phenomenon that needs to be better documented and studied. Warm-cloud lightning appears to be uncommon, however, even though warm clouds in tropical climates can be strongly convective and can produce heavy rainfall. This, coupled with the observation that thunderstorms in temperate climates become electrified only when they grow above the 0°C level, leads many researchers to consider that warm-cloud electrification is an anomaly that is explained by a different mechanism than that which electrifies colder clouds.

If a precipitation mechanism operates to electrify

storms, negatively charged precipitation and positively charged cloud particles would overlap for some distance above the main negative charge, producing a semineutral but segregated reservoir of charge between the main negative- and upper positive-charge regions. This is the situation depicted in Figure 8.1. The existence of such a reservoir was first postulated by Wilson (1920, 1929) to explain the large apparent separation of the positive and negative charges. Net negative charge would be observed only at and below the lower boundary of the neutral region, and this would partly explain why the main negative charge appears to be distributed horizontally in a storm. The presence of such a reservoir has not been demonstrated by direct observations. But a charge reservoir almost certainly exists in some form no matter what the charging process, owing to the large distances and volumes through which charge must be transported. Such reservoirs would provide inertia to the charging process and would help to explain why lightning often occurs at nearly regular time intervals in a storm. Observations of the nature of the reservoirs would greatly aid our understanding of the charging processes.

PARTICLE-CHARGE OBSERVATIONS

To sort out how the electrification occurs it is essential to know the charge carried by the different types and sizes of particles in the cloud. Precipitation theories predict that the main negative charge of the storm resides on precipitation particles, and it has been of interest to test this prediction by direct measurement. Such measurements have become possible in recent years using instruments that sense the charge on individual precipitation particles. The instruments have been used in several programs since 1978, sometimes in conjunction with particle size measurements, and show that precipitation carries a mixture of positive and negative charges (Gaskell et al., 1978; Christian et al., 1980; Marshall and Winn, 1982; Gardiner et al., 1984). The magnitudes of the individual charges are relatively large, and their signs are sometimes predominantly negative; but the fraction of charged particles is small, and the inferred volume charge densities may or may not be adequate to account for the observed electrification. Few measurements have been obtained in the interesting parts of a storm, i.e., at temperatures of less than $-10°C$ and in updrafts. The particle-charge measurements are made from aircraft or below balloons and are difficult to obtain. First there are the logistical problems of being in the right place at the right time; then there are experimental problems of measuring weak charges in an icing and strongly electrified environment.

As further observations are obtained we can expect better answers to the question concerning precipitation charge. [Marshall and Marsh (1985) recently reported measurements of precipitation charges within the main negative-charge region of a storm in which all the precipitation particles whose charge was great enough to be detected by their instrument were negatively charged, in amounts that appeared to be sufficient to account for the field gradient in the negative-charge region.] Still unknown, however, will be the amounts and sign of charge carried by the large number of smaller particles that coexist with the precipitation but that are below the detection limit of present instruments. Cloud particles have a much greater charge-carrying capacity per unit volume of cloudy air than precipitation particles, and it is important to know how much charge they carry. No good technique exists for doing this in the uncontrolled and hostile environment of an active thunderstorm.

The in-cloud observations show that millimeter-size precipitation particles sometimes carry sufficient charge so that the electrical force on them would be comparable with the gravitational force in the strong-field regions of a storm. These particles would be expected to exhibit measurable velocity changes after nearby lightning. But attempts to detect such velocity changes using Doppler radars have been unsuccessful in most instances (Zrnic et al., 1982; Williams and Lhermitte, 1983). These results indicate that only a fraction of the precipitation particles are highly charged, in agreement with the in-cloud observations. If the energy considerations mentioned earlier were to require that the precipitation be efficiently charged, these results would indicate that convective motions are important in charging a storm (Williams, 1985). The fact that velocity changes are observed occasionally indicates that precipitation is strongly and efficiently charged at some locations and times.

Measurements of the charge on precipitation arriving at the Earth's surface show that it often has the same polarity as the point discharge being given off from the ground. This is the mirror-image effect mentioned earlier and indicates that the precipitation charges have been modified by the capture of point discharge ions as the precipitation falls to earth. Below cloud base or in the bases of clouds, precipitation is often observed to be positively charged and occurs in localized regions referred to as lower-positive-charge centers (Simpson and Scrase, 1937; Rust and Moore, 1974; Winn et al., 1981; Marshall and Winn, 1982; Holden et al., 1983). One explanation for these observations has been that the precipitation captures positively charged cloud droplets while falling through cloud base. However, positively charged precipitation is found well inside the cloud, up to and above the 0°C temperature level (Moore, 1976b; Marshall and Winn, 1982). These observations are not simply explained and pose an obvious problem for precipitation theories of electrification, since it is necessary to explain how the sign of the precipitation charge could be reversed within a relatively short distance of (or even inside) the main negative-charge region.

Several ideas have been proposed to explain the positive charge on precipitation in the interior of a storm (Not all of these are aimed at salvaging the precipitation hypothesis.) One is based on laboratory observations that the charge transferred during collisions between hail pellets and ice crystals changes sign at temperatures warmer than about $-10°C$ (e.g., Takahashi, 1978; Gaskell and Illingworth, 1980; Jayaratne et al., 1983). The reversed, positive charging is somewhat stronger than the negative charging observed at lower temperatures. This explanation may be correct, but it becomes more difficult or complicated for similar theories to explain the main negative charge as well. Specifically, it becomes necessary to explain why the main negative charge is laterally extensive and a major reservoir of charge in the storm, while the lower positive charge is more pointlike and a lesser reservoir of charge.

Another explanation for the lower-positive-charge centers is that they are caused by lightning (Marshall and Winn, 1982; Holden et al., 1983). Lightning flashes do not neutralize the storm charge on a fine scale, as the term discharge may imply. Rather, they deposit positive charge along a finely branched structure in or near the main negative-charge region of the storm (Vonnegut and Moore, 1960). If some of the lightning tendrils terminated on precipitation particles, these particles would tend to be left with a positive charge after the lightning regardless of any initial charge that they might have carried. This explanation would account for the localized nature of the lower-positive-charge centers and could operate in storms. But recent observations have shown that positively charged precipitation is present in storms before the first lightning (T. C. Marshall, University of Mississippi, private communication). Similar findings can be inferred from the observations of Simpson and Scrase (1937) and probably from other studies of non-lightning-producing storms as well. In this event another mechanism must be operating to charge the precipitation positively, and a lightning mechanism is not necessarily required.

A third possibility is that an inductive process might operate to charge the precipitation positively (e.g., Chiu, 1978). An inductive mechanism differs from the noninductive hail/ice-crystal mechanism described earlier in that the charge transferred during rebounding collisions depends on the strength and direction of the ambient electric field. In-cloud observations and laboratory studies have caused the inductive theory to lose favor (Gaskell et al., 1978; Christian et al., 1980;

Gaskell, 1981), but it would be premature to rule out inductive charging altogether given our general inability to explain the various electrification processes.

In evaluating observations of precipitation charge it is important to differentiate between measurements made before and those made after the onset of lightning. Lightning drastically complicates the detailed distribution of charge in a storm, to the point that the particle charges may not reflect the processes that caused the storm to become electrified. This occurs for several reasons. First, the lightning does not neutralize the storm charge on a fine scale, as discussed above, but deposits its charge along a myriad of channels and streamer paths and on particular particles. Second, the lightning subjects the cloud to large electric stresses by virtue of the fact that its channels extend across millions of volts of electric potential. The stresses undoubtedly cause transient corona discharges from particles in the vicinity of the lightning channels, which effectively erase the original charges and leave the particles with an unrelated residual charge (Dawson, 1969; Griffiths and Latham, 1972; Griffiths, 1976). Laboratory studies have shown that the residual charges are large in magnitude and variable in sign. Both the nonuniform charge deposition and corona effects would cause particles near the lightning channels to become strongly charged and would tend to leave other particles unaltered, giving rise to a mixture of particle charges, as is observed.

Lightning also generates vast numbers of ion pairs, some of which will be separated in the strong electric field to increase temporarily the electrical conductivity of the cloud. This could discharge particles that are charged, or randomly charge other particles, further complicating matters.

In addition to complicating the charge picture, lightning-induced or other corona could sustain or enhance the electrification of a storm if the corona from precipitation had a systematic sign (Dawson, 1969; Krehbiel, 1984). Laboratory studies have shown that corona from liquid surfaces is preferentially positive above 4-5 km altitude in the atmosphere and becomes more intense at lower pressures (higher altitudes) (Dawson, 1969). Such corona would result in positive ions and negatively charged precipitation. The positive ions would attach to nearby cloud droplets, producing a segregated distribution of negative precipitation and positive cloud particles similar to that of a collisional charging process. Systematic positive corona could occur, for example, from the liquid surfaces of wet or riming hail or from the bottom surfaces of (liquid or frozen) precipitation above the negative-charge region.

It has been proposed that the strong and uneven charging that occurs in the vicinity of lightning channels could enhance the rate of precipitation formation after a lightning discharge (Vonnegut and Moore, 1960) and that this may be responsible for the sudden bursts or gushes of rain that are commonly observed from thunderstorms. Some observations have been reported in support of this idea (e.g., Moore *et al.*, 1964; Syzmanski *et al.*, 1980), but the possible effects that lightning may have on the cloud microphysics and precipitation formation remain unclear and are in need of continued study. At the very least, the fact that the storm produces strong electric fields makes it easier for precipitation to form, in that the electric forces increase the probability that particles will coalesce after coming in contact (Rayleigh, 1879; Goyer *et al.*, 1960; Moore *et al.*, 1958).

CHARGE MOTIONS AND CURRENTS

The charging mechanism constitutes a source of current that continually increases the electrical energy of the storm, while lightning intermittently reduces the storm's electrical energy. The charging current is on the order of 1 A and ranges from inferred values of about 0.1 A during the initial stages of small storms (Krehbiel, 1981) to values greater than 1 A for large storm complexes. Currents of comparable average magnitude flow between the cloud and ground and between the upper atmosphere and cloud top. The current at the cloud top results from the flow of negative screening charge to the upper cloud boundary. The current between the cloud and ground is due to the action of CG lightning and point discharge from the Earth's surface and, like each of the other currents, predominantly lowers negative charge (or raises positive charge). The various currents are illustrated in Figure 8.8. Their net effect is to transfer negative charge from the upper atmosphere to the

Earth's surface. This transfer, from approximately 1000 storms that are estimated to be in progress at any time over the entire planet, is thought to be the reason that the ionosphere is maintained at a potential of several hundred thousand volts with respect to the Earth's surface.

The vast majority of the electrical charges inside a storm reside on cloud or precipitation particles. Free charge or ions, such as are generated by cosmic radiation, corona, or lightning, collide with and attach to cloud particles within about a second after being produced. This immobilizes the charge and makes the cloud a good electrical insulator. The currents that charge the storm should not be weakened significantly by leakage currents between the charge regions, except those involving the motion of charged cloud particles, which are expected to have small effect. The storm charges and electric fields are therefore expected to build up in a manner that reflects the strength of the charging process, until lightning occurs, after which the buildup repeats.

In-cloud electric-field measurements with slowly moving balloonborne instruments have shown that the electric field increases in an approximately linear manner with time between lightning discharges (Winn and Byerley, 1975). This indicates that the current source that charges the storm is relatively constant between discharges and is independent of the magnitude of the field. Theories that employ positive feedback to electrify a storm, such as an inductive theory, predict that the charging current would increase exponentially with time and thus would not appear to be operating. This may not be a reliable prediction or test of the mechanism type, however, if the charging process establishes and

FIGURE 8.8 Diagrammatic illustration of the electric currents for precipitation and convection scenarios of electrification. In both cases the thunderstorm transfers negative charge from the upper atmosphere to the Earth's surface. The arrows point in the direction of positive current flow; the short half-arrows indicate when this current is caused by negative charge flowing in the opposite direction. Intracloud lightning opposes the charging process in the precipitation scenario but provides a parallel charging path to the upper atmosphere in the convection scenario.

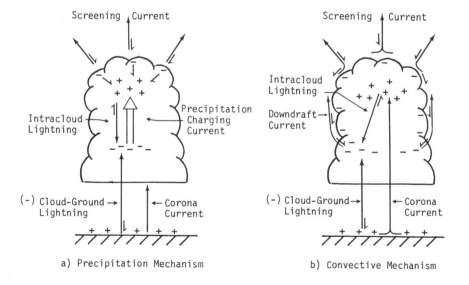

a) Precipitation Mechanism   b) Convective Mechanism

draws on a reservoir of charge as discussed earlier. In this case the charging current and field buildup would depend on the rate at which charge emerges from the reservoir, which will tend to be field independent.

A major unresolved question in understanding the electrical behavior of storms concerns the role and fate of the upper screening current in the electrical budget of the storm (Vonnegut, 1982). In the absence of convective motions and turbulence the flow of negative screening charge would completely shield the interior positive charge in a few tens of seconds. Airborne electric-field observations outside the tops of growing clouds show that a screening charge does form, but not to completion. Negative charge continues to be attracted to the cloud surface at rates that are comparable with the charging current of the storm (Gish and Wait, 1950). The question is what happens to this charge. The convection hypothesis of electrification postulates that the screening charge is carried downward by convective overturning to the level of the main negative charge and that this is the primary source of the main negative charge (Vonnegut, 1953). This charge transport would be generative, i.e., negative charge would be carried downward away from the upper positive charge, increasing the electrical energy of the storm. An alternative possibility is that turbulent mixing folds the screening charge into the upper positive charge of the storm, which would be a dissipative process.

Regardless of its eventual fate, the substantial flow of negative screening charge to the upper part of the cloud appears not to be matched by the flow of positive charge to cloud base, causing the storm as a whole to build up a net negative charge with time. The buildup is alleviated intermittently by negative cloud-to-ground lightning, and this is undoubtedly the reason why most CG lightning has a negative polarity. The buildup also increases the dominant effect of negative charge on the electric field at the ground, which is alleviated by positive corona from the ground. The convection hypothesis postulates that the positive corona charge is carried into the upper part of the cloud by the updraft, which feeds the cloud with low-level moisture, and that this is the primary source of the upper positive charge. Other possibilities are that much of the corona charge is carried into the main negative-charge region and is dissipated there or that most of it remains near the ground.

The strength of the point discharge current beneath storms has been estimated both from ground-based electric-field observations and from measurements of the corona current given off by vegetation beneath storms. Over a typical area of 10 km^2 the total corona current is estimated to be about 0.1 A (Livingston and Krider, 1978; Standler and Winn, 1979). This is comparable with the charging current at the beginning of a storm but is less than the average current of cloud-to-ground lightning in the active stage of a storm.

Recent experiments designed to test electrification ideas have attempted to influence or alter the electrification of a storm by releasing charge into the bases of growing clouds prior to their electrification (Vonnegut et al., 1984; Moore et al., 1985). In these experiments, several kilometers of cable and fine wire are strung over mountainous terrain and maintained at a high positive or negative potential. Natural clouds grow over a fair-weather supply of positive space charge near the Earth's surface, which tends to be ingested into the cloud along with surface moisture. By maintaining the wires at a high negative potential the researchers hope to give off sufficient negative corona charge to override the natural supply of positive charge and to prime the cloud with negative charge. If a convective mechanism operates to initiate the electrification, or if the electrification is influenced by the direction of the initial electric field inside the storm (as in the case of an inductive precipitation mechanism), such priming should invert the polarity of the electrification, i.e., produce a storm having an upper negative- and main positive-charge structure. The results of the experiments are that storms developing above negative-charge releases are anomalous in that the field at the ground is often dominated by positive charge overhead, which lightning acts to remove. There is incomplete and conflicting information on the question of whether the polarity of the main storm charges was inverted. One alternative possibility is that the experiment modifies primarily the subcloud and cloud-base charges. The success of the experiments in at least partially altering the electrical structure of storms makes them intriguing subjects for continued field programs.

Because the interior storm charges reside on cloud or precipitation particles, their motion is the same as the particle motions and can be investigated using Doppler radars when the particles are large enough to be detected by radar. A single Doppler radar measures the component of the particle velocity along the direction of the radar beam; a network of three or more Doppler radars is needed to determine the particle velocities in three dimensions. Three-dimensional measurements of particle velocites have become possible only recently (e.g., Lhermitte and Williams, 1985b) but are a key element in furthering our understanding of thunderstorms. A continuing problem in their determination is the rapidly changing nature of convective storms, which requires that the storm be scanned as rapidly as possible. Multiple Doppler radars have been used to study the electrification of storms in Florida and New Mexico; one case study has shown how the onset of lightning in a cell was correlated with the development of an updraft and precipitation within the cell (Lhermitte and Krehbiel, 1979).

Particle velocity measurements provide only a part of the information needed to estimate the electrical currents of the storm; also needed is some knowledge of the charge distribution or amounts of charge carried by the particles. Determination of the charge distribution is in itself a central problem of electrification studies, for which there are unfortunately no radarlike instruments. The charge information must be determined from in-cloud measurements, which are necessarily limited in scope, or inferred from other information, such as that obtained from other storms or from lightning.

Attempts to determine the charge structure of storms from remote measurements of the total electric field have given a qualitative picture of the storm charges but have not been successful at estimating their amounts or locations. There are several reasons for this, having to do with the facts that (a) the conductivity of the atmosphere increases exponentially with altitude and causes the upper charges of the storm to be masked or screened, (b) the overall charge distribution is complex and not uniquely defined by electric-field measurements, and (c) total electric-field measurements are strongly affected by local charges. These problems are alleviated somewhat by measuring the time rate of change of the electric field, which is related to the time rate of change of the charges, or to the storm currents. Such measurements have formed the basis of a new approach for estimating the storm currents, in which the displacement current associated with a time-varying electric field is added to other measurements of the local corona, conduction, and rain currents (Krider and Musser, 1982). The sum of these currents has been termed the Maxwell current after the British physicist who first described the significance of the displacement current. An example of displacement current measurements is shown in Figure 8.9. The displacement current density values can be integrated over the area affected by the storm to estimate the charging current of the storm; this gives results that are in reasonable agreement with the charging current values inferred from lightning data. The pattern of Maxwell current values, either at the ground or aloft, can in principle be used to locate and quantify the different currents of the storm, in much the same way that the lightning charges can be located. But this possibility has yet to be realized, in part because of the problems enumerated above for interpreting total electric-field measurements.

A totally different approach for determining the storm currents would involve measuring the pattern of magnetic fields that they produce. This approach has not been feasible owing to the difficulty of measuring the weak fields and to the presence of the geomagnetic field, but such an approach may become practical in the future.

LIGHTNING AND THE STORM ELECTRIFICATION

The study of lightning is an important part of thunderstorm investigations. Lightning is of interest not only as a phenomenon in itself but as an indicator and significant modifier of the storm's electrification. Lightning generates, deposits, and redistributes substantial amounts of free charge within a storm, and this greatly complicates the storm's electrification. In the process, lightning may also enhance the electrification or the formation of precipitation within the storm. But little is known even about what lightning looks like inside a

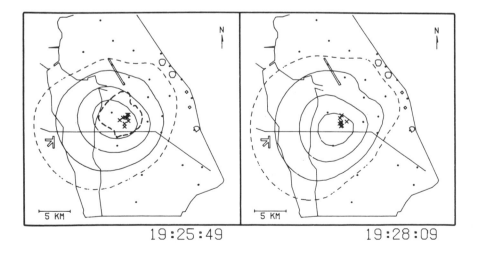

FIGURE 8.9 Contours of constant displacement current density at the ground beneath a thunderstorm on July 11, 1978 at Kennedy Space Center, Florida. Observations from two 5-minute time intervals are shown; contours are at 0.5 nA/m^2 intervals. The heavy dashed contour shows the detectable radar echo at 7.5-km altitude; the x's mark the negative-charge centers of lightning discharges. The areal integral of the displacement current was about 0.4 A in each instance. (Krider and Blakeslee, 1985.)

19:25:49   19:28:09

storm, much less about its detailed behavior or about the possible effects that it may have.

The study of lightning as a phenomenon in itself is the subject of the first five chapters in this volume. One question of interest here concerns how lightning is initiated. Measurements of the electric field inside storms give maximum (large-scale) values typically between 1×10^5 and 2×10^5 V/m (e.g., Winn et al., 1974, 1981). Winn et al. (1974) reported one measurement of 4×10^5 V/m. These values are 3 to 10 times smaller than the field strength required to break down clear air at the same altitude. Hydrometeors concentrate the field onto their surface by a factor of 3 or more, and this leads scientists to think that the breakdown is initiated at particle surfaces—by corona that somehow develops into a full-scale discharge (e.g., Loeb, 1953; Richards and Dawson, 1971; Crabb and Latham, 1974; Griffiths and Phelps, 1976).

The manner in which lightning is initiated is an unanswered and intriguing question, but however this happens it is most likely to occur in a strong-field region of the storm. In-cloud measurements like that shown in Figure 8.4 indicate that the electric field is strongest on the periphery of the main negative-charge region. Lightning radiation studies indicate that discharges indeed tend to be initiated at these altitudes in a storm (Proctor, 1981, 1983). There is also some evidence that IC flashes begin at slightly higher altitude than CG flashes (Taylor, 1983). This suggests that discharges that are initiated above the negative charge region tend to become IC flashes, while those that are initiated below the negative charge tend to become CG flashes.

Although highly variable, intracloud lightning generally outnumbers cloud-to-ground lightning in a storm, often by a factor of 5 or 10 to 1 or so, and it is of interest to ask why this happens. The charging process establishes the main negative and upper positive charges as the primary charges of the storm, and this may cause the electric field to be stronger above the main negative charge than below it. Also, the decrease in atmospheric pressure with altitude favors the occurrence of IC flashes, in that the critical field required for discharges to form and to propagate is smaller at higher altitudes. Finally, it may be that there are a greater number of initiation events above the main negative charge than below it.

The occurrence of CG flashes is thought to be aided by the presence of the lower positive charge, which increases the electrical energy below the negative-charge region, and by the tendency (mentioned earlier) for a storm to acquire a net negative charge with time.

The past 15 years have seen major advances in techniques for remotely sensing lightning inside a storm. In particular, radio-frequency radiation from the lightning may be located using one of several direction-finding or time-of-arrival techniques (Proctor, 1971, 1983; Taylor, 1978; Warwick et al., 1979; Hayenga and Warwick, 1981; Taylor et al., 1984; Richard et al., 1986). The charge centers of the lightning can be located from simultaneous measurements of the lighting electric-field change at a number of ground locations (Figure 8.3; Jacobson and Krider, 1976; Krehbiel et al., 1979). The hot lightning channels are readily detected by radar at 10-cm wavelength or longer (e.g., Holmes et al., 1980; Mazur et al., 1985), and the main channels can be reconstructed from recordings of the thunder that they produce (e.g., Teer and Few, 1974; Winn et al., 1978; Christian et al., 1980; MacGorman et al., 1981; Chapter 3, this volume). Finally, changes in the electrical forces on charged cloud particles during lightning cause low-frequency changes in the atmospheric pressure, called infrasound, which can be detected and used to estimate the charge heights (e.g., Wilson, 1920; Bohannon et al., 1977; Balachandran, 1983; Few, 1985).

Figure 8.10 shows two examples of lightning data that complement the electrical observations discussed earlier. Figure 8.10(a) shows the height of the radiation sources from lightning as a function of time in a Florida storm. Although not resolved in the figure, the radiation occurred in distinct bursts from individual discharge events. Only a few radiation sources were located during each discharge, but the results give a useful picture of the overall lightning activity in the storm. Events with sources located below 7-8-km altitude were usually CG discharges; the large number of remaining events were IC discharges. Of particular interest in the figure are the sequences of increased lightning activity whose sources moved upward with time. These were associated with the electrification of new convective cells in the storm and provide another indication that the electrification is associated with vertical growth. The fact that the sequences start above about 8-km altitude reflects the existence of an altitude threshold for the electrification. The discharge rate during the most intense sequence reached 37 per minute.

Similar observations have been reported by Lhermitte and Krehbiel (1979), who found a discharge rate of 60 per minute in a relatively small cell of a storm. Such high discharge rates are not unusual for large storms, but their occurrence in small, individual cells of normal-sized storms is a new finding. The high-rate discharges have been shown to transfer relatively small amounts of charge (Krehbiel et al., 1984b), indicating that the high-rate sequences result from a large number of initiating events rather than from superelectrification

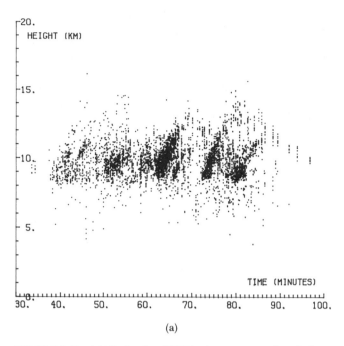

(a)

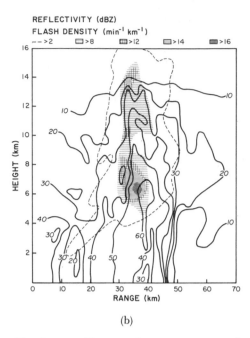

(b)

FIGURE 8.10   (a) The height of VHF radiation sources from lightning versus time in a Florida storm. The upward-moving sequences of enhanced lightning activity were associated with the electrification of new, growing cells, while the increase in the number of sources above 8 km reflects the altitude dependence of the electrification (Krehbiel *et al.*, 1984b). (b) A vertical cross section of the radar reflections from precipitation (solid contours) and from lightning (hatched areas, dashed contours) during a 5-minute time interval in a squall line near Wallops Island, Virginia (Mazur *et al.*, 1984). The greatest number of lightning echoes were observed at altitudes that correspond with the charge centers of intracloud and cloud-to-ground discharges (Figures 8.3 and 8.6) and may indicate the locations of the main negative and upper positive charges in the storm.

of the cell. High-rate sequences have been observed only in subsequent cells of already-electrified Florida storms, but it appears that they are a common feature of such storms. This suggests that initiation events are somehow enhanced in subsequent cells. It would not be surprising if high-rate sequences of small discharges are found in storms at other locations as well. For instance, Taylor *et al.* (1983) reported the occurrence of minor discharges in large Oklahoma storms.

Figure 8.10(b) shows observations of radar echoes from lightning during a 5-minute time interval in a squall line over the East Coast of the United States. The echo locations are superimposed on measurements of the precipitation reflectivity in the storm. The lightning echoes were detected by a UHF radar operating at 70-cm wavelength; the precipitation reflectivities were determined using a separate radar at 10-cm wavelength. The lightning echoes were located most often in strong precipitation on the leading edge of a well-developed cell at 30- to 40-km range from the radars. The largest number of echoes were observed between 5- and 8-km altitude and vertically above this from 10- up to 14-km altitude. These altitudes correspond to the heights of the positive- and negative-charge centers of lightning in other storms (e.g., Figures 8.3 and 8.6), suggesting that

the echoes are strongest in the vicinity of the lightning charge centers. This is where the discharges are expected to be most highly branched.

The increasing ability of researchers to sense lightning inside thunderclouds has raised questions about the extent to which lightning indicates or reflects the electrification of a storm (Vonnegut, 1983b). Once initiated, the lightning channels and charges themselves influence the continued propagation of a discharge, enabling the discharge to develop in a manner that can be unrelated to the storm charges and fields. While it is necessary to be cautious in making inferences about the electrification from lightning observations, some evidence exists that suggests that lightning can be a reasonable indicator of the storm charges. For example, it has been found that the negative charge sources of the CG lightning in the storm of Figure 8.6 coincided with those of the IC lightning that immediately preceded and followed the CG discharges, even though the two types of discharge developed in opposite vertical directions. This result, which is illustrated in Figure 8.5, suggests that the negative-charge sources for the lightning coincided with main negative charge in the storm.

The question of how the lightning and storm charges are related has also been investigated by studying the behavior of electrical discharges in plexiglass (Williams et al., 1985). These laboratory discharges appear to simulate the large-scale behavior of lightning in clouds. High-energy electrons are deposited in a controlled manner within the plexiglass and the resulting space charge is discharged by mechanical disruption at some point on the surface. The dendritic or finely branched structure of the discharges follows the pattern of space charge within the plexiglass, suggesting that real lightning may do the same in storms.

Assuming that lightning tells us something about the electrification, one question of interest has been how the lightning channels and charges are related to precipitation in the storm, as revealed by radar. If a precipitation mechanism were operating to electrify the storm, one would expect the lightning and precipitation to be correlated in some manner. Not surprisingly in phenomena as complex as thunderstorms and lightning (and as complicated to study), a wide variety of observations have been found. These range from observations that lightning and precipitation are correlated (e.g., Larsen and Stansbury, 1974; Krehbiel et al., 1979; Taylor et al., 1983; Figures 8.3 and 8.10b), to observations that lightning avoids regions of strong precipitation (MacGorman, 1978; Williams, 1985), to other observations of precipitation echoes that develop after nearby lightning (Moore et al., 1964; Szymanski et al., 1980). There is some uncertainty and debate as to what the various observations mean. In the author's opinion, one of the most striking results has been the degree to which the lightning charges correlate with radar echoes from precipitation.

LIGHTNING EVOLUTION

We finish this review with a brief and simple description of how lightning appears to evolve with time in a thunderstorm. This is depicted in Figure 8.11 and provides a framework for understanding some of the wide variety of lightning observations. In addition it gives further insight into the electrical nature of storms. The description is based on a number of different studies and observations of lightning (e.g., Ligda, 1956; Teer and Few, 1974; Krehbiel et al., 1979, 1984a; Krehbiel, 1981; Proctor, 1981, 1983; Rust et al., 1981; Fuquay, 1982; Taylor, 1983).

In response to the dipolar structure of the storm, the initial lightning discharges are usually intracloud flashes that transport charge vertically between the main negative- and upper positive-charge regions (Figures 8.11a and 8.3). The first cloud-to-ground discharge usually follows an initial sequence of intracloud flashes (Figures 8.11b and 8.6), but simple CG flashes some-

times begin the lightning activity. The latter situation occurs presumably because conditions somehow favor the initiation of CG discharges. CG flashes consist of a number of discrete strokes down the channel to ground; the early CG flashes are simple in that they produce one or only a few strokes.

The initial lightning activity is associated with the cell having the greatest vertical development in the storm. Other cells do not generate lightning until they develop vertically above 7-8-km altitude MSL (in summertime), even though the subsequent cells may have stronger precipitation echoes within them than within the initial, lightning-producing cell.

As additional cells become electrified, the IC flashes remain basically vertical but become broader in horizontal extent and can exhibit a pattern of cross-discharging between cells (Figure 8.11c). The CG flashes produce a larger number of discrete strokes whose negative-charge sources progress horizontally through the precipitating part of the storm (Figure 8.11d). For some still-unknown reason, the CG flashes can initiate a continuing current or arc-type discharge down the channel to ground from within the horizontally extensive negative charge. The continuing currents can last for a few tenths of a second and produce a persistent luminosity that is sometimes detectable visually.

In large storm complexes having a number of cells, the intracloud and cloud-to-ground discharges can have large horizontal extents, corresponding to the horizontal dimensions of the storm system. Because the horizontal dimensions can be much greater than a storm's vertical dimension, the discharges become primarily horizontal in nature.

As the storm grows, its top reaches the base of the stratosphere or is sheared off by high-level winds to form an anvil cloud (Figure 8.1). The anvil cloud is composed of small ice crystals that carry part of the upper positive charge and is penetrated by intracloud discharges from the active region of the storm (Figure 8.11e). The anvil clouds commonly extend tens or hundreds of kilometers downwind from the parent storm. Cloud-to-ground discharges have also been observed to emanate from anvil clouds, well away from the active region of the storm.

As older cells dissipate, predominantly horizontal intracloud lightning occurs between negative charge in still-active cells and apparent positive charge at about the same level in the dissipating part of the storm (Figure 8.11f). In propagating storms the dissipating part trails the active part and can have substantial horizontal extent. The horizontal discharges within them are correspondingly extensive and are observed to propagate over distances of 50 to 100 km (Ligda, 1956; Proctor, 1983). The radar echo from within the dissipating part

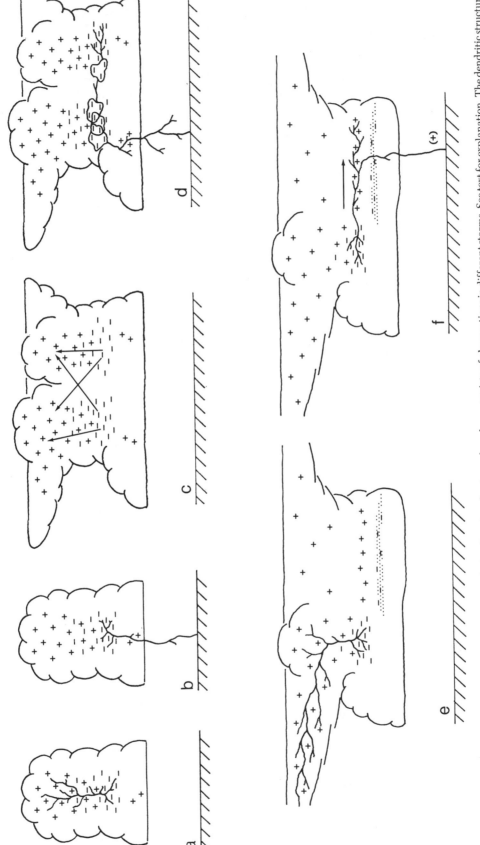

FIGURE 8.11   The apparent evolution of lightning with time in a thunderstorm, based on a variety of observations in different storms. See text for explanation. The dendritic structure of the lightning has been guessed in all cases except for the multicellular intracloud discharge of part c. The dotted region in the dissipating part of the storm in parts e and f represents the radar brightband from melting snowflakes.

of a storm is characterized by a horizontal *brightband* at and just below the 0°C level, caused by melting snow. The horizontal discharges appear to propagate just above the level of the brightband and effectively remove positive charge at this level (Krehbiel, 1981). The lightning is observed to repeat at intervals of a few minutes or more and occasionally produces positive strokes to ground. The repetitive nature of the discharges suggests that a widespread, low-rate charging process is operating to regenerate the positive charge and that the charging process is associated with the production or fall of snow.

Independent evidence for the existence of a positive-charge layer is found in electric-field soundings through dissipating storms. Only a few such soundings have been made, even though dissipating storms have a stratiform, slowly changing structure that is relatively easy to probe. One sounding obtained by researchers in France is shown in Figure 8.12 (Chauzy *et al.*, 1980). The balloonborne instrument passed through negative charge within the radar brightband and positive charge 1 km above the brightband. Although the charge distribution was still dipolar, positive charge was found at the level

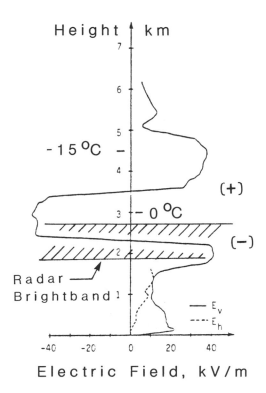

FIGURE 8.12 Sounding of the vertical electric field in the dissipating part of a large frontal storm in France, indicating the presence of a positive-charge layer just above the 0°C level and a negative-charge layer within the radar brightband from the storm (adapted from Chauzy *et al.*, 1980).

where negative charge is found in the active part of a storm. Similar results were obtained by Simpson and Scrase (1937) in the dissipating parts of English storms. The sounding and lightning observations agree, but further observations are needed to check their validity.

The above observations do not explain how the charge structure of the storm changes to produce the end-of-storm electric-field oscillation discussed in connection with Figure 8.2. Moore *et al.* (1958) interpreted their observations of the oscillations as being due to the subsidence that occurs as the storm dissipates, which reveals the upper positive charge and transports it downward toward the ground [see also Moore and Vonnegut (1977)]. The observations of Williams (1981) support the idea that the field reversals are associated with downward motion of charge during subsidence, but the nature and source of the charges still remain to be determined.

SUMMARY AND CONCLUSIONS

The ability of scientists to observe and study thunderstorms has increased greatly over the past decade or two, and this has brought their study to a particularly exciting stage. A number of ideas have been proposed over the years to explain how thunderstorms become electrified, and it is now becoming possible to test the various ideas by direct measurement. Thunderstorms provide a difficult environment for measurements, but scientists are increasingly able to probe them with instruments that reliably measure electric-field profiles and particle charges and sizes, as well as air temperature, cloud water content, and other parameters. At the same time, remote-sensing techniques are providing increasingly detailed pictures of the storm as a whole. For example, networks of Doppler radars are able to measure the three-dimensional particle motions at different locations in the storm, and the lightning channels and charges are able to be located in space and time.

The major ingredients for a thunderstorm continue to be vigorous convection and the formation of precipitation at altitudes where the air temperature is colder than 0°C. Strong electrification does not occur until the cloud and precipitation develop above a threshold altitude that is 7-8 km above MSL in the summer months, corresponding to an air temperature of −15 to −20°C. The main negative charge resides at and below this altitude at temperatures that are remarkably similar within a given storm and in different kinds of storms.

A central issue of thunderstorm studies is whether the electrification is caused by the gravitational fall of charged precipitation or whether it results primarily from the convective transport of charges by the air motions of the storm. Precipitation theories predict that the main negative charge is carried by precipitation particles, and this is being tested by in-cloud measurements of the charges carried by precipitation. Laboratory studies continue to point to rebounding collisions between hail and small ice crystals as a mechanism that can charge precipitation negatively and possibly explain the electrification. This mechanism is expected to operate in storms, but it has not been shown that enough precipitation is present and involved in enough charging interactions to account for the electrification. Compounding this difficulty are observations that electrification is more widespread than strong precipitation in a storm and that precipitation below the main negative-charge region is often observed to be positively charged. In convection theories the charges reside primarily on small cloud particles, which can carry much more charge per unit volume of cloudy air than precipitation. But little is known about the amounts and motions of the cloud particle charges and whether they would combine to produce charge accumulations consistent with observations.

Once a storm becomes strongly electrified and starts to produce lightning, it is likely that additional charging processes occur that complicate or possibly enhance the electrification. In particular, the lightning itself could have such a role. This could explain some of the complexity of the electrical observations and would make it more difficult to sort out the various charging processes. In order to study the initial electrification processes, it is necessary that observations be made before the onset of strong electrification and of lightning. This need has been recognized but greatly increases the logistical difficulties of studying storms.

By the same reasoning, it is possible that the primary electrification mechanism changes once a storm becomes strongly electrified. For example, precipitation could initiate the electrification, and then the larger convective energies of the storm could continue the electrification. Or the electrification could be sustained and enhanced if the corona from precipitation had a systematic sign. Already there is some evidence that a different, lower-rate mechanism operates to electrify dissipating storms or parts of storms.

Our understanding of the electrification processes remains limited by the need for better observations of the electrical and physical characteristics of actual thunderstorms. This need has guided thunderstorm research for a number of years and involves several parallel and interacting efforts: field programs, data analysis, and the development of observational techniques. Field programs provide the basic experimental data and allow scientists to test new instruments and observational

techniques. Data analysis extracts the scientific information from the field programs and provides feedback for future studies.

Instrumental development and testing can be done in limited field programs, but significant advances in understanding the electrification processes require focused, cooperative field programs that bring together the best available observational techniques. A substantial amount of data is already in hand from recent field programs of this type whose continued analysis will provide further insights into the electrification problem. But too many questions remain unaddressed in the measurements of those studies for them to hold the answers to the problems. And, as is usual in science, the results of one set of observations and experiments often raise new questions and avenues of investigation. A prime example of this is the recent experiments that have attempted to invert the electrical polarity of a storm.

Much of our information about thunderstorm electrification has come from the study of relatively small, isolated storms such as those that form over the mountains of the southwestern United States or above the sea-breeze convergence in southeastern coastal areas. These storms provide relatively stationary and predictable targets for study and remain attractive subjects for field programs. Although relatively small, the storms are not simple, and we have much to learn from them. As their study has demonstrated, however, it is important that different types and sizes of storms be studied and compared. In particular, it is important that electrical studies be made of severe storms, propagating squall lines, tropical storms (both ice-free and ice-containing), and winter storms.

This review has concentrated primarily on the scientific observations and issues related to the problem of thunderstorm electrification. Other recent reviews on the same subject have been made by Moore and Vonnegut (1977), Illingworth and Krehbiel (1981), Latham (1981), Vonnegut (1982), Lhermitte and Williams (1983), Illingworth (1985), and Williams (1985). Another whole review could be devoted to a description of the techniques that are used to study thunderstorms and their electrification. Many of the techniques are new and are still under development and have been used in cooperative studies for only one or a few thunderstorm seasons. Other techniques have yet to be used in electrification studies—for example, the differential reflectivity polarization radar technique (Bringi et al., 1984). (This technique measures the difference in precipitation reflectivity for vertical and horizontal polarizations and is able to distinguish between ice and liquid water in clouds.) Much could be learned by bringing existing techniques together and applying them to the same storms. In some instances a brute-force approach would also be helpful, for example, in obtaining successive or simultaneous balloon soundings of the electric-field profile in a storm.

Still other observational needs provide challenging problems for development—for example, measurement of the net charge density in storms or of the charge carried by small cloud particles.

Even after comprehensive observations have been obtained on storms, it is entirely possible that the specific mechanism or mechanisms that caused their electrification will continue to elude precise definition. In the case of a precipitation mechanism, it would be extremely difficult to catch particular charging events "in the act." This would need to be done in the controlled environment of the laboratory, simulating cloud conditions as closely as possible. Specific areas of interest in present and future laboratory studies are (1) contact electrification processes at ice surfaces and (2) corona discharges from precipitation. Computational models, both simple and complex, will continue to be useful in interpreting field and laboratory observations and in predicting the ability of particular mechanisms to electrify a storm. Such modeling will rely heavily on parameterizations of observational data, however, since the electrification results from a cascade of physical processes each of which are inherently difficult to simulate in their own right.

In conclusion we note that, although the problems of thunderstorm electrification are difficult and complex, their solution is becoming possible and is a prized goal of scientists.

ACKNOWLEDGMENTS

The comments and reviews by William Winn, Earle Williams, Don MacGorman, Charles Moore, and Marx Brook significantly improved this review. The author is also indebted to numerous other colleagues for their insights and discussions.

REFERENCES

Balachandran, N. K. (1983). Acoustic and electric signals from lightning, *J. Geophys. Res.* 88, 3879-3884.

Bohannon, J. L., A. A. Few, and A. J. Dessler (1977). Detection of infrasonic pulses from thunder, *Geophys. Res. Lett.* 4, 49-52.

Bringi, V. N., T. A. Seliga, and K. Aydin (1984). Hail detection with a differential reflectivity radar, *Science* 225, 1145-1147.

Brook, M., M. Nakano, P. Krehbiel, and T. Takeuti (1982). The electrical structure of the Hokuriku winter thunderstorms, *J. Geophys. Res.* 87, 1207-1215.

Byrne, G. J., A. A. Few, and M. E. Weber (1983). Altitude, thickness and charge concentration of charged regions of four thunderstorms during TRIP 1981 based upon in situ balloon electric field measurements, *Geophys. Res. Lett.* 10, 39-42.

Chalmers, J. A. (1967). *Atmospheric Electricity*, 2nd ed., Pergamon, Oxford.

Chauzy, S., P. Raizonville, D. Hauser, and F. Roux (1980). Electrical and dynamical description of a frontal storm deduced from LANDES 79 experiment, *Proceedings 8th International Conference on Cloud Physics*, American Meteorological Society, Boston, Mass., pp. 477-480.

Chiu, C. S. (1978). Numerical study of cloud electrification in an axisymmetric, time-dependent cloud model, *J. Geophys. Res.* 83, 5025-5049.

Christian, H., C. R. Holmes, J. W. Bullock, W. Gaskell, A. J. Illingworth, and J. Latham (1980). Airborne and ground-based studies of thunderstorms in the vicinity of Langmuir Laboratory, *Q. J. R. Meteorol. Soc.* 106, 159-174.

Cobb, W. E. (1975). Electric fields in Florida cumuli, *EOS 56*, 990.

Crabb, J. A., and J. Latham (1974). Corona from colliding drops as a possible mechanism for the triggering of lightning, *Q. J. R. Meteorol. Soc.* 100, 191-202.

Dawson, G. A. (1969). Pressure dependence of water-drop corona onset and its atmospheric importance, *J. Geophys. Res.* 74, 6859-6868.

Dye, J. E., J. P. Winn, C. B. Moore, and J. J. Jones (1985). The relationship between precipitation and electrical development in New Mexico thunderstorms, *EOS 66*, 840.

Dye, J. E., J. J. Jones, W. P. Winn, T. A. Cerni, B. Gardiner, D. Lamb, R. L. Pitter, J. Hallett, and C. P. R. Saunders (1986). Early electrification and precipitation development in a small, isolated Montana cumulonimbus, *J. Geophys. Res.* 91, 1231-1247.

Few, A. A. (1985). The production of lightning-associated infrasonic sources in thunderclouds, *J. Geophys. Res.* 90, 6175-6180.

Fuquay, D. M. (1982). Positive cloud-to-ground lightning in summer thunderstorms, *J. Geophys. Res.* 87, 7131-7140.

Gardiner, B., D. Lamb, R. Pitter, and J. Hallett (1984). Measurements of initial electric field and ice particle charges in Montana summer thunderstorms, *J. Geophys. Res.* 90, 6079-6086.

Gaskell, W. (1981). A laboratory study of the inductive theory of thunderstorm electrification, *Q. J. R. Meteorol. Soc.* 107, 955-966.

Gaskell, W., and A. J. Illingworth (1980). Charge transfer accompanying individual collisions between ice particles and its role in thunderstorm electrification, *Q. J. R. Meteorol. Soc.* 106, 841-854.

Gaskell, W., A. J. Illingworth, J. Latham, and C. B. Moore (1978). Airborne studies of electric fields and the charge and size of precipitation elements in thunderstorms, *Q. J. R. Meteorol. Soc.* 104, 447-460.

Gish, O. H., and G. R. Wait (1950). Thunderstorms and the Earth's general electrification, *J. Geophys. Res.* 55, 473-484.

Goyer, G. G. *et al.* (1960). Effects of electric fields on water-drop coalesence, *J. Meteorol.* 17, 442-445.

Griffiths, R. F. (1976). Corona charging of frozen precipitation, *J. Atmos. Sci.* 33, 1602-1606.

Griffiths, R. F., and J. Latham (1972). The emission of corona from falling drops, *J. Meteorol. Soc. Jpn.* 5, 416-422.

Griffiths, R. F., and C. T. Phelps (1976). A model for lightning initiation arising from positive streamer development, *J. Geophys. Res.* 81, 3671-3676.

Hayenga, C. O., and J. W. Warwick (1981). Two-dimensional interferometric positions of VHF lightning sources, *J. Geophys. Res.* 86, 7451-7462.

Holden, D. N., G. R. Holmes, C. B. Moore, W. P. Winn, J. W. Cobb, J. E. Griswold, and D. M. Lytle (1983). Local charge concentrations in thunderclouds, in *Proceedings in Atmospheric Electricity*, L. H. Ruhnke and J. Latham, eds., Deepak Publ., Hampton, Va., pp. 179-183.

Holmes, C. R., C. B. Moore, R. Rogers, and E. Szymanski (1977). Radar study of precipitation development in thunderclouds, in *Electrical Processes in Atmospheres*, H. Dolezalek and R. Reiter, eds., Steinkopff, Darmstadt, pp. 623-627.

Holmes, C. R., E. W. Szymanski, S. J. Szymanski, and C. B. Moore (1980). Radar and acoustic study of lightning, *J. Geophys. Res. 85*, 7517-7532.

Illingworth, A. J. (1985). Charge separation in thunderstorms: small scale processes, *J. Geophys. Res. 90*, 6026-6032.

Illingworth, A. J., and P. R. Krehbiel (1981). Thunderstorm electricity, *Phys. Technol. 12*, 122-128, 139.

Illingworth, A. J., and J. Latham (1977). Calculations of electric field growth, field structure and charge distributions in thunderstorms, *Q. J. R. Meteorol. Soc. 103*, 281-295.

Jacobson, E. A., and E. P. Krider (1976). Electrostatic field changes produced by Florida lightning, *J. Atmos. Sci. 33*, 103-117.

Jayaratne, E. R., C. P. R. Saunders, and J. Hallett (1983). Laboratory studies of the charging of soft-hail during ice crystal interactions, *Q. J. R. Meteorol. Soc. 109*, 609-630.

Krehbiel, P. R. (1981). An analysis of the electric field change produced by lightning, Ph.D. thesis, Univ. of Manchester Institute of Science and Technology.

Krehbiel, P. R. (1984). Corona electrification: A possible means for sustaining or enhancing the electrification of storms, in *Proceedings VII International Conference on Atmospheric Electricity*, American Meteorological Society, Boston, Mass., pp. 188-189.

Krehbiel, P. R., M. Brook, and R. A. McCrory (1979). An analysis of the charge structure of lightning discharges to ground, *J. Geophys. Res. 84*, 2432-2456.

Krehbiel, P. R., M. Brook, R. L. Lhermitte, and C. L. Lennon (1983). Lightning charge structure in thunderstorms, in *Proceedings in Atmospheric Electricity*, L. H. Ruhnke, and J. Lathan, eds., Deepak Publ., Hampton, Va., pp. 408-410.

Krehbiel, P. R., R. Tennis, M. Brook, E. W. Holmes, and R. Comes (1984a). A comparative study of the initial sequence of lightning in a small Florida thunderstorm, in *Proceedings VII International Conference on Atmospheric Electricity*, American Meteorological Society, Boston, Mass., pp. 279-285.

Krehbiel, P. R., M. Brook, S. Khanna-Gupta, C. L. Lennon, and R. Lhermitte (1984b). Some results concerning VHF lightning radiation from the real-time LDAR system at KSC, Florida, in *Proceedings VII International Conference on Atmospheric Electricity*, American Meteorological Society, Boston, Mass., pp. 388-393.

Krider, E. P., and R. J. Blakeslee (1985). The electric currents produced by thunderclouds, *J. Electrostatics 16*, 369-378.

Krider, E. P., and J. A. Musser (1982). Maxwell currents under thunderstorms, *J. Geophys. Res. 87*, 11171-11176.

Larsen, H. R., and E. J. Stansbury (1974). Association of lightning flashes with precipitation cores extending to height 7 km, *J. Atmos. Terr. Phys. 36*, 1547-1553.

Latham, J. (1981). The electrification of thunderstorms, *Q. J. R. Meteorol. Soc. 107*, 277-298.

Lhermitte, R., and P. R. Krehbiel (1979). Doppler radar and radio observations of thunderstorms, *IEEE Trans. Geosci. Electr. GE-17*, 162-171.

Lhermitte, R., and E. Williams (1983). Cloud electrification, *Rev. Geophys. Space Phys. 21*, 984-992.

Lhermitte, R., and E. Williams (1985a). Doppler radar and electric field signatures of downbursts in convective storms, *EOS 66*, 839.

Lhermitte, R., and E. Williams (1985b). Thunderstorm electrification: A case study, *J. Geophys. Res. 90*, 6071-6078.

Ligda, M. G. H. (1956). The radar observation of lightning, *J. Atmos. Terr. Phys. 9*, 329-346.

Livingston, J. M., and E. P. Krider (1978). Electric fields produced by Florida thunderstorms, *J. Geophys. Res. 83*, 385-401.

Loeb, L. B. (1953). Experimental contributions to knowledge of thunderstorm electricity, in *Thunderstorm Electricity*, H. R. Byers, ed., Univ. of Chicago Press, Chicago, pp. 150-192.

MacGorman, D. R. (1978). Lightning location in a storm with strong wind shear, Ph.D. dissertation, Rice Univ., Houston, Tex.

MacGorman, D. R., A. A. Few, and T. L. Teer (1981). Layered lightning activity, *J. Geophys. Res. 86*, 9900-9910.

Malan, D. J., and B. F. J. Schonland (1951). The distribution of electricity in thunderclouds, *Proc. R. Soc. Ser. A 209*, 158-177.

Marshall, T. C., and S. J. Marsh (1985). A thunderstorm sounding of charges and electric field, *EOS 66*, 840.

Marshall, T. C., and W. P. Winn (1982). Measurements of charged precipitation in a New Mexico thunderstorm: Lower positive charge centers, *J. Geophys. Res. 87*, 7141-7157.

Mason, B. J. (1976). In reply to a critique of precipitation theories of thunderstorm electrification by C. B. Moore, *Q. J. R. Meteorol. Soc. 102*, 219-225.

Mazur, V., J. C. Gerlach, and W. D. Rust (1984). Lightning flash density versus altitude and storm structure from observations with UHF- and S-band radars, *Geophys. Res. Lett. 11*, 61-64.

Mazur, V., D. S. Zrnic, and W. D. Rust (1985). Lightning channel properties determined with a vertically pointing Doppler radar, *J. Geophys. Res. 90*, 6165-6174.

Moore, C. B. (1963). Charge generation in thunderstorms, in *Problems in Atmospheric and Space Electricity*, S. C. Coroniti, ed., Elsevier, Amsterdam, pp. 255-262.

Moore, C. B. (1976a). Reply to "In reply to a critique of precipitation theories of thunderstorm electrification by C. B. Moore" by B. J. Mason, *Q. J. R. Meteorol. Soc. 102*, 226-240.

Moore, C. B. (1976b). Reply to "Further comments on Moore's criticisms of precipitation theories of thunderstorm electrification," *Q. J. R. Meteorol. Soc. 102*, 935-939.

Moore, C. B. (1977). An assessment of thunderstorm electrification mechanisms, in *Electrical Processes in Atmospheres*, H. Dolezalek and R. Reiter, eds., Steinkopff, Darmstadt, pp. 333-352.

Moore, C. B., and B. Vonnegut (1977). The thundercloud, in *Lightning, Vol. 1*, R. H. Golde, ed., Academic Press, New York, pp. 51-98.

Moore, C. B., B. Vonnegut, and A. T. Botka (1958). Results of an experiment to determine initial precedence of organized electrification and precipitation in thunderstorms, in *Recent Advances in Atmospheric Electricity*, L. G. Smith, ed., Pergamon, New York, pp. 333-360.

Moore, C. B., B. Vonnegut, E. A. Vrablik, and D. A. McCaig (1964). Gushes of rain and hail after lightning, *J. Atmos. Sci. 21*, 646-665.

Moore, C. B., C. P. Migotsky, and B. Vonnegut (1985). Further observations of inverted polarity clouds, *EOS 66*, 841.

Proctor, D. E. (1971). A hyperbolic system for obtaining VHF radio pictures of lightning, *J. Geophys. Res. 76*, 1478-1489.

Proctor, D. E. (1981). VHF radio pictures of cloud flashes, *J. Geophys. Res. 86*, 4041-4071.

Proctor, D. E. (1983). Lightning and precipitation in a small multicellular thunderstorm, *J. Geophys. Res. 88*, 5421-5440.

Rayleigh (Lord) (1879). The influence of electricity on colliding water drops, *Proc. R. Soc. (London) 28*, 406-409.

Reynolds, S. E., and M. Brook (1956). Correlation of the initial electric field and the radar echo in thunderstorms, *J. Meteorol. 13*, 376-380.

Reynolds, S. E., and H. W. Neill (1955). The distribution and discharge of thunderstorm charge centers, *J. Meteorol. 12*, 1-12.

Reynolds, S. E., M. Brook, and M. F. Gourley (1957). Thunderstorm charge separation, *J. Meteorol. 14*, 426-436.

Richard, P., A. Delannoy, G. Labaune, and P. Laroche (1986). Results of spatial and temporal characterization of the VHF-UHF radiation of lightning, *J. Geophys. Res. 91*, 1248-1260.

Richards, C. N., and G. A. Dawson (1971). The hydrodynamic instability of water drops falling at terminal velocity in vertical electric fields, *J. Geophys. Res. 76*, 3445-3455.

Rust, W. D., and C. B. Moore (1974). Electrical conditions near the bases of thunderclouds over New Mexico, *Q. J. R. Meteorol. Soc. 100*, 450-468.

Rust, W. D., W. L. Taylor, D. R. MacGorman, and R. T. Arnold (1981). Research on electrical properties of severe thunderstorms in the Great Plains, *Bull. Am. Meteorol. Soc. 62*, 1286-1293.

Simpson, G. C., and F. J. Scrase (1937). The distribution of electricity in thunderclouds, *Proc. R. Soc. Ser. A 161*, 309-352.

Standler, R. B., and W. P. Winn (1979). Effects of coronae on electric fields beneath thunderstorms, *Q. J. R. Meteorol. Soc. 105*, 285-302.

Szymanski, E. W., S. J. Szymanski, C. R. Holmes, and C. B. Moore (1980). An observation of a precipitation echo intensification associated with lightning, *J. Geophys. Res. 85*, 1951-1953.

Takahashi, T. (1978). Riming electrification as a charge generation mechanism in thunderstorms, *J. Atmos. Sci. 35*, 1536-1548.

Taylor, W. L. (1978). A VHF technique for space-time mapping of lightning discharge processes, *J. Geophys. Res. 83*, 3575-3583.

Taylor, W. L. (1983). Lightning location and progression using VHF space-time mapping technique, in *Proceedings in Atmospheric Electricity*, L. H. Ruhnke and J. Latham, eds., Deepak Publ., Hampton, Va., pp. 381-384.

Taylor, W. L., W. D. Rust, D. R. MacGorman, and E. A. Brandes (1983). Lightning activity observed in upper and lower portions of storms and its relationship to storm structure from VHF mapping and Doppler radar, in *Proceedings, 8th International Aerospace and Ground Conference on Lightning and Static Electricity*, available from National Technical Information Service, Springfield, Va., pp. 4-1 to 4-9.

Taylor, W. L., E. A. Brandes, W. D. Rust, and D. R. MacGorman (1984). Lightning activity and severe storm structure, *Geophys. Res. Lett. 11*, 545-548.

Teer, T. L., and A. A. Few (1974). Horizontal lightning, *J. Geophys. Res. 79*, 3436-3441.

Vonnegut, B. (1953). Possible mechanism for the formation of thunderstorm electricity, *Bull. Am. Meteorol. Soc. 34*, 378-381.

Vonnegut, B. (1982). The physics of thunderclouds, in *CRC Handbook of Atmospherics, Vol. 1*, H. Volland, ed., CRC Press, Boca Raton, Fla., pp. 1-22.

Vonnegut, B. (1983a). Comments on "The electrification of thunderstorms" by J. Latham, *Q. J. R. Meteorol. Soc. 109*, 262-265.

Vonnegut, B. (1983b). Deductions concerning accumulations of electrified particles in thunderclouds based on electric field changes associated with lightning, *J. Geophys. Res. 88*, 3911-3912.

Vonnegut, B., and C. B. Moore (1960). A possible effect of lightning discharge on precipitation formation process, in *Physics of Precipitation*, Monograph No. 5, American Geophysical Union, Washington, D.C., pp. 287-290.

Vonnegut, B., C. B. Moore, T. Rolan, J. Cobb, D. N. Holden, S. McWilliams, and G. Cadwell (1984). Inverted electrification in thunderclouds growing over a source of negative charge, *EOS 65*, 839.

Warwick, J. W., C. O. Hayenga, and J. W. Brosnahan (1979). Interferometric directions of lightning sources at 34 MHz, *J. Geophys. Res. 84*, 2457-2468.

Weber, M. E., H. J. Christian, A. A. Few, and M. F. Stewart (1982). A thunderstorm electric field sounding: Charge distribution and lightning, *J. Geophys. Res. 87*, 7158-7169.

Williams, E. R. (1981). Thunderstorm electrification: Precipitation vs. convection, Ph.D. thesis, Massachusetts Institute of Technology, Cambridge, Mass.

Williams, E. R. (1985). Large-scale charge separation in thunderclouds, *J. Geophys. Res. 90*, 6013-6025.

Williams, E. R., and R. M. Lhermitte (1983). Radar tests of the precipitation hypothesis for thunderstorm electrification, *J. Geophys. Res. 88*, 10984-10992.

Williams, E. R., C. M. Cooke, and K. A. Wright (1985). Electrical discharge propagation in and around space charge clouds, *J. Geophys. Res. 90*, 6059-6070.

Wilson, C. T. R. (1920). Investigations on lightning discharges and on the electric field of thunderstorms, *Phil. Trans. R. Soc. Ser. A 221*, 73-115.

Wilson, C. T. R. (1929). Some thundercloud problems, *J. Franklin Inst. 208*, 1-12.

Winn, W. P., and L. G. Byerley (1975). Electric field growth in thunderclouds, *Q. J. R. Meteorol. Soc. 101*, 979-994.

Winn, W. P., G. W. Schwede, and C. B. Moore (1974). Measurements of electric fields in thunderclouds, *J. Geophys. Res. 79*, 1761-1767.

Winn, W. P., C. B. Moore, C. R. Holmes, and L. G. Byerley (1978). Thunderstorm on July 16, 1975, over Langmuir Laboratory: A case study, *J. Geophys. Res. 83*, 3079-3092.

Winn, W. P., C. B. Moore, and C. R. Holmes (1981). Electric field structure in an active part of a small, isolated thundercloud, *J. Geophys. Res. 86*, 1187-1193.

Workman, E. J., and S. E. Reynolds (1949). Electrical activity as related to thunderstorm cell growth, *Bull. Am. Meteorol. Soc. 30*, 142-144..

Workman, E. J., R. E. Holzer, and G. T. Pelsor (1942). The electrical structure of thunderstorms, *Aeronaut. Tech. Note 864*, National Advisory Committee on Aeronautics, Washington, D.C., pp. 1-47.

Zrnic, D. S., W. D. Rust, and W. L. Taylor (1982). Doppler radar echoes of lightning and precipitation at vertical incidence, *J. Geophys. Res. 87*, 7179-7191.

Charging Mechanisms in Clouds and Thunderstorms

9

KENNETH V. K. BEARD
University of Illinois at Urbana-Champaign

HARRY T. OCHS
Illinois State Water Survey

INTRODUCTION

Since the time of Benjamin Franklin, a major difficulty in identifying the causes of cloud electricity has been our inability to obtain adequate measurements within clouds. This observational problem is now being remedied by modern electronics and instrumented aircraft. More quantitative theories of charging have become available since the 1940s along with our improved understanding of the atmosphere. In addition, better laboratory simulations of cloud physics in recent decades have led to improved measurements of microscale charge separation. With all these advances we should not be surprised to find that the number of possible charging mechanisms has proliferated. Thus, a modern task has been to sort through the possible mechanisms in trying to identify their relative contribution to cloud electrification (Mason, 1972; Latham, 1981; Takahashi, 1982). A major purpose of this chapter is to describe the mechanisms that charge cloud and precipitation particles. We also evaluate their relative role in cloud-scale electrification and assess our state of knowledge. A broader evaluation of cloud electrification is found in Chapter 10 of this volume.

There are two major categories of charging mechanisms: the microscale separators, which ultimately lead to charged cloud and precipitation particles, and the cloud-scale separators, which can result in field intensification and lightning. The first category includes the creation of ion pairs in the air and charge separation on individual cloud and precipitation particles. These mechanisms are coupled with other microscale separators to produce net charges on cloud and precipitation particles, for example the attachment of ions by diffusion to cloud drops and the charging that results from particle collisions. Once the cloud and precipitation particles become appreciably charged, a larger scale separator such as differential sedimentation is needed to create electrification on the cloud scale. Convection can also act as a cloud-scale separator by redistributing ions and particles. Much of the emphasis in this chapter will be on describing the microscale separators that produce the charged cloud and precipitation particles.

In discussing the charging mechanisms we consider the electrification of convective clouds. These clouds can produce spectacular displays of lightning and are the most important cloud link in the global electric circuit. We start the discussion of charge separation in the simple environment of a small cumulus cloud. This nonprecipitating cloud stage is followed by sections on the rain stage and hail stage. An abbreviated discussion of the charging mechanisms associated with these three stages is found in the evaluation section of this chapter.

CLOUD STAGE

As the first clouds form on a warm summer afternoon, the environment is already set for cloud electrification. The air is filled with ions whose concentrations and mobilities determine the effective conductivity of the atmosphere. In many practical situations the air is an electric insulator, but the conductivity is large enough to permit a relaxation time of less than 7 minutes for discharging the lower atmosphere (Israël, 1971). The discharge current results from the drift of small ions with the mass of a few molecules. Charge separation begins immediately when a field is applied to a mixture of positive and negative ions. Within a cloud the usual result of ion motion is capture by water droplets.

The electric background in which the cloud forms contains vertical gradients in ion concentration with a negative charge at the Earth's surface that is maintained by thunderstorms. The electric field above the ground is reduced by a screening layer of positive ions attracted by the Earth. Positive ions from aloft accumulate near the ground because the field-induced drift is reduced by collisions as the density of the air increases. Capture of ions by aerosols greatly reduces the drift velocity and, during times of heavy pollution, 500 V/m have been measured between the positive space charge and the ground. A field of 130 V/m is more typical of a summer day with a well-mixed boundary layer. This is reduced by about 1 order of magnitude at a 3 km height. Thus, a fresh cumulus cloud forms in an environment of vertical gradients in space-charge density, electric field, ion concentration, and conductivity (see Figure 9.1). The field is oriented toward negative earth with a strong increase in positive space charge below 1 km and a small ion concentration and conductivity that increase with height.

In this electric environment there are several ion capture mechanisms that lead to charged droplets in shallow cumulus clouds. In the following discussion, microscale charge separation is described for diffusion charging, drift charging, and selective ion charging. The cloud-scale separation of charge for these nonprecipitation clouds is discussed below under drift charging.

Diffusion Charging

For the early stage of cloud charging we consider the collection of ions by cloud droplets. The ion-transport equation on the microscale near a cloud droplet gives the charge flux (C/m² sec) or the current density for an ion component as

$$\mathbf{J}_i = \rho_i \mathbf{U} + \rho_i B_i \mathbf{E} - D_i \nabla \rho_i, \qquad (9.1)$$

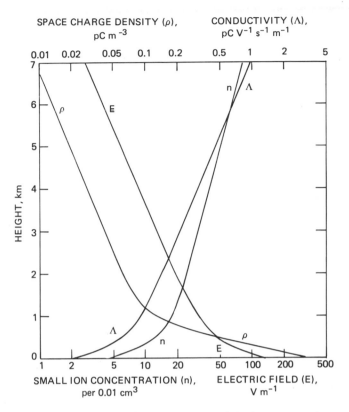

FIGURE 9.1 Average electric properties of the lower atmosphere during fair weather. The variation of the electric field with height is due to Gish (e.g., see Pruppacher and Klett, 1978). The space charge density is a direct result of Gauss's law, whereas the conductivity is obtained by assuming a constant current (2.7 pC m⁻¹ sec⁻¹). The concentration of small ions is proportional to the conductivity and varies inversely with the ion mobility.

where ρ_i is the ion-charge density (C/m³) of a particular species, $\mathbf{U}$ is the air velocity, B_i is the ion mobility (about 2×10^{-4} m²/V sec for small ions in the lowest few kilometers), $\mathbf{E}$ is the electric field, and D_i is the molecular diffusivity (m²/sec). The flux term for the microscale airflow ($\rho_i \mathbf{U}$) is relatively weak because of the low fall speed for cloud droplets. In addition the field is small enough in the cloud stage to ignore the ion drift term ($\rho_i B_i \mathbf{E}$). Thus the charging of small cloud droplets is found by evaluation of the standard diffusion equation.

An important consequence of diffusion charging is a reduction of ion concentration within the cloud by several orders of magnitude. The time constant for depletion can be obtained from the solution to the diffusion equation for a steady-state attachment of ions to a cloud of similar size droplets. The solution is

$$\rho_i = \rho_{i0} \exp(-4\pi R D_i N t), \qquad (9.2)$$

where R is the droplet radius and N is their concentration. For a typical size and concentration in the cloud stage (10 μm, 100 cm^{-3}), the depletion time constant, $t = (4\pi RD_iN)^{-1}$, is 10 sec. Thus equilibrium is achieved rapidly, and the local concentration in clouds can be approximated by a steady-state balance between production by cosmic rays and depletion by recombination and attachment to cloud droplets (Chiu and Klett, 1976).

The amount of charge on a cloud droplet from kinetic theory is a Gaussian-like (Boltzmann) distribution centered on zero charge for equal concentration of positive and negative ions. As is evident by the positive space charge near the ground, the ion mixture is not always neutral. In such cases a net charge is collected by droplets. For the zero-centered distribution the rms Boltzmann charge can provide an estimate for the magnitude attained in diffusion charging:

$$Q = (8\pi\epsilon_0 RkT)^{1/2}, \qquad (9.3)$$

where ϵ_0 is the permittivity of air (8.85×10^{-12} F/m), k is the Boltzmann constant (1.38×10^{-23} J/K), and T is the temperature (Gunn, 1957). This equation can be interpreted as a balance between the stored electric energy on the droplet ($1/2\ Q^2/4\pi\epsilon_0 R$) and the thermal motion energy of the ions (kT). When the Boltzmann charge is evaluated for a typical droplet size in the cloud stage ($R = 10\ \mu$m), the rms charge in number of electrons is $n_e = 6R^{1/2}\ (\mu\text{m}) = 19$. This result is consistent with the spread in droplet charge measured for nonconvective clouds with low electric fields (Gunn, 1957).

Drift Charging

Larger-scale transport of ions is characterized by currents from bulk and eddy transport of ions along with the field-driven drift. The charge-flux equation is the larger-scale version of Eq. (9.1)

$$\mathbf{J}_i = \rho_i\mathbf{U} + \rho_iB_i\mathbf{E} - K\nabla\rho_i, \qquad (9.4)$$

where K is the eddy diffusivity (m^2/sec). If we consider just the drift of ions in the ambient field we find that the vertical drift current ($\rho_iB_i\mathbf{E}$) at cloud top and cloud base must result in an accumulation of positive and negative space charge, respectively (Figure 9.2). An electric balance is achieved fairly rapidly as a screening layer forms with the capture of incoming ions by cloud droplets. The field within the cloud is increased by the charge that accumulates at the boundary.

The amount of charge on droplets in this region can be estimated by considering diffusion capture by ions of only one sign. The maximum charge captured is found from the amount needed to neutralize the induced

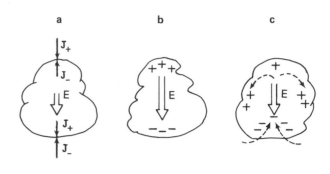

FIGURE 9.2   Electrification of a model cumulus cloud (after Chiu and Klett, 1976): a, vertical drift currents reflecting the ion deficits with the cloud; b, resultant charge accumulation and field enhancement from ion drift; c, effect of convective transport on charges and field.

charge (of opposite sign) from polarization in the electric field:

$$Q = 12\pi\epsilon_0 R^2 E. \qquad (9.5)$$

The capture of ions from the drift current during the cloud stage, for example at cloud base (Figure 9.2b), results in $n_e = 0.002\ R^2\ (\mu\text{m})\ E\ (\text{V/m}) = 20$ for $R = 10\ \mu$m and $E = 100$ V/m. This charge is comparable with the Boltzmann charge given by Eq. (9.3). However charge generation from drift into cloud edges increases with R^2 in contrast to the dependence of Eq. (9.3). Thus charges of several hundred electrons are readily attained for somewhat larger cloud droplets by diffusion of ions of one sign to polarized drops. Somewhat later in this section we shall find out that the size and field dependence given by Eq. (9.5) also applies within clouds for ion capture by polarized drops and for the breakup of these drops.

In addition, for the cloud stage we must also consider the role of the bulk and eddy transport terms in Eq. (9.4). For diffusion charging and the simple convective pattern, illustrated in Figure 9.2c, convection transports negative charge upward within the core and carries positive charge downward along the edges (Chiu and Klett, 1976). The effect of eddy diffusion in this simple model is to smooth the charge distribution produced by the bulk transport and ion drift. When all three terms in Eq. (9.4) are included, the field is enhanced within the cloud but is not so strong as the pure drift case.

For this early stage of cloud electrification, drift charging of droplets with negative charge at cloud base and positive charge at cloud top is apparently the dominant mechanism. The current into cloud base from drift is $\rho_- B_- E$, and the current from convection is U, where $\rho = \rho_+ - \rho_-$ is the space charge. (The value of ρ_- is approximately one half the small ion concentration, beneath cloud base, times a unit charge.) The ratio of drift to convection is 7.5 (using the values for ρ, n, and E on Figure 9.1 at 1 km and $B_- = 2.2 \times 10^{-4}$ m²/V sec and $U = 1$ m/sec) clearly showing the dominance in the drift of negative ions into cloud base over the upward convection of positive space charge. The ratio decreases as the cloud base is lowered and is less than unity for bases below about 300 m. The calculations of Chiu and Klett for a cloud base at 10 m show the dominance of convective transport of positive space charge over drift into cloud base. They found a positive core, but drift still dominated the charging process in the upper cloud with positive charges at cloud edges similar to Figure 9.2c. Thus, for the majority of cases in the cloud stage, drift charging is the most significant electrification mechanism. Convection and eddy diffusion in the single-cell pattern investigated by Chiu and Klett generally weaken electrification by redistributing and mixing the drift-generated charge.

The charge acquired by droplets in the cloud stage has been from diffusion of ions within the cloud and drift charge at cloud edge.

Selective Ion Charging

When equal numbers of positive and negative ions are present there can be a preferred attachment of one sign if the droplet is polarized (Wilson effect). The governing equation for the microscale transport [given by Eq. (9.1)] has two classes of solutions (Whipple and Chalmers, 1944). In the case of "fast ions" the downward drift of positive ions exceeds the fall speed of the droplet ($B_+ E > U$) and ions of both signs are captured at nearly the same rate (Figure 9.3a). The droplet size where $B_+ E = U$ in the lowest few kilometers is $R = 1$ μm for $E = 10$ V/m. Since most droplets in shallow cumulus clouds are larger than 4 μm, ions are captured selectively by the Wilson effect. Larger droplets will acquire a negative charge by the preferential attraction of negative ions, as shown in Figure 9.3b for the "slow ion" case ($B_+ E < U$).

The maximum charge acquired by droplets for the Wilson effect is

$$Q = 2\pi\epsilon_0 R^2 E. \qquad (9.6)$$

This is only one sixth the diffusion charge from the drift current given by Eq. (9.5) and yields a negative charge equivalent to 36 electrons for the largest cloud droplets ($R = 100$ μm) and a downward directed field of 10 V/m.

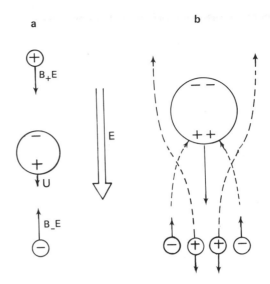

FIGURE 9.3 Selective ion capture from droplet polarization in a downward-directed field (Wilson effect): a, fast-ion case $B_+ E > U$); b, slow-ion case ($B_+ E < U$) with trajectories given by dashed curves.

Thus the largest cloud droplets are charged for the Wilson effect to a magnitude of about the rms Boltzmann charge.

In our cumulus scenario the cloud is only about 1 km deep with a central updraft speed of 1 to 2 m/sec; therefore the small drops that we are considering are carried upward. When the cloud depth increases to about 3 km, drops become large enough to be detected by radar. This is a common circumstance for summer cumulus clouds in mid-latitudes. The cloud top would lie below the level where droplets readily freeze except over elevated terrain or in a more northern climate. The drops associated with the initial radar echo are still quite small and unable to fall out of the cloud. However, we consider the time of the first radar echo as the beginning of the rain stage. In the following section we examine the charging mechanisms associated with drizzle drops and raindrops: selective ion charging, breakup charging, and induction charging.

RAIN STAGE

Selective Ion Charging

When we make the transition to the new stage, microscale separation of charge becomes more powerful because of the R^2 dependence of charge captured by polarized drops [Eqs. (9.5) and (9.6)]. As the drizzle drops begin moving downward in the cloud a larger scale separation of charge can result as drizzle drops capture negative ions (Wilson effect) and the excess positive ions become attached to cloud droplets. The enhanced electric field within the cloud would provide a positive feedback to the Wilson effect by increasing the polarization on the drops. If the field should reach about 10 kV/m the ion drift velocity would increase to a few meters per second. This is the situation where the velocity of positive ions is about the same as the fall speed of small raindrops. Consequently generation by selective ion charging and the simple feedback mechanism for the Wilson effect is limited. [The same limit does not apply to diffusion charging at cloud edges, Eq. (9.5)].

In the rain stage, microscale separation of charge from ion capture continues while new mechanisms make their appearance. The role of convection remains central to cloud development as well as the motion of charged droplets within the updraft. Transport and drift are important factors in ion movement within the cloud and to droplets at the boundaries. The Wilson effect appears to be responsible for some of the field enhancement but must be considered along with the additional mechanisms of breakup charging and induction charging.

Breakup Charging

The collisions between drizzle ($R = 100\text{-}1000\ \mu m$) and cloud droplets ($R = 10\text{-}100\ \mu m$) usually result in coalescence growth and the production of rain. In contrast the collisions between raindrops ($R = 1\text{-}6$ mm) and drizzle often result in only transient coalescence followed by fragmentation. Such events can result in charged drops in the presence of an electric field. The polarization charge of one sign on a spherical drop is

$$Q = 3\pi\epsilon_0 R^2 E. \qquad (9.7)$$

This equation gives the net positive or negative charge that can be separated by "slicing" a polarized drop in half. To apply Eq. (9.7) to the rain stage, consideration of some details of drop collisions and the role of the electric field follows.

Four general kinds of breakup phenomena are illustrated in Figure 9.4 (neck, sheet, disk, and bag) based on the laboratory study of McTaggart-Cowan and List (1975). The amount of charge separated in bag breakup is given approximately by Eq. (9.7) (Matthews and Mason, 1964), but the charge has not been determined for the other cases shown in Figure 9.4. The most frequent kinds of breakup result after a vertical elongation of the coalesced drop pair followed by a neck or sheet that tears into numerous droplets. This is not the ideal "slicing" required for Eq. (9.7). For example, an elongation increases the polarization charge by a factor of 4 if the

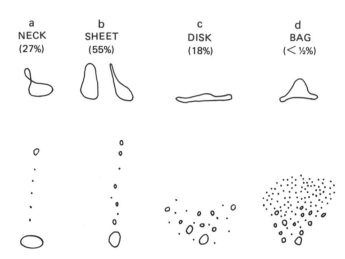

FIGURE 9.4   The four observed breakup types with percentages of occurrence: a, neck; b, sheet (two views taken perpendicular to each other); c, disk; d, bag (from McTaggart-Cowan and List, 1975).

distorted drop is modeled as a prolate spheroid with a major axis of 5 times the spherical diameter. Thus Eq. (9.7) gives a rather conservative estimate of the microscale charge separation in breakup.

An important feature of breakup collisions is that they occur slowly compared with the charge-relaxation time (i.e., the time required to redistribute the charge). The breakup time is given roughly by the raindrop diameter divided by the velocity difference between the colliding drops (about 0.5 msec). In contrast, the charge relaxation time for pure water is about 100 times faster and for rainwater with impurities, 10^4 to 10^6 times faster. Thus, the distribution of polarization charge on a distorting raindrop is in approximate equilibrium with the electric field.

In breakup charging, the electric field separates charge on individual drops by polarization, breakup separates charge between colliding drops, and gravity separates charge on a large scale. The sheet breakup shown in Figure 9.4b, with a downward directed field, will result in a positive charge on the large fragment ($R > 1$ mm) given approximately by Eq. (9.7) and a negative charge of the same magnitude distributed over the small fragments ($R \simeq 100\ \mu m$). The difference in fall speed between these sizes (6 to 10 m/sec) gives the cloud-scale separation rate.

Although breakup charging contains the microscale and cloud-scale mechanisms of charge separation necessary for cloud electrification, it does not reinforce the existing field. However, it may contribute significantly to drop charging found in both the rain and hail stages.

Induction Charging

In addition to coalescence and breakup, the collision between drops can result in interactions where the two drops bounce apart. The amount of charge transferred between drops that are polarized depends on the contact angle relative to the field, the contact time, the charge relaxation time, and the net charge on the drops as well as the magnitude of the polarization. Induction charging was first considered by Elster and Geitel (see Chalmers, 1967) for contact between a large and small sphere along a line parallel to the field. Later studies included the effects of image charges, contact angle, and net charge with extensions of theory to ice particles. In the rain stage we will consider only induction charging for drops while reserving the ice aspects of this mechanism for the hail stage.

The importance of contact angle is immediately obvious from the induced surface charge on a conducting sphere in a uniform electric field given by $3\pi\epsilon_0 E \cos\theta$ and shown schematically on Figure 9.5a. However, the contact angle is a hydrodynamic problem of two deformable bodies in a gaseous medium with its own set of governing parameters. Even in the absence of electric effects, such interactions are understood only in terms of broad categories in a manner similar to the breakup phenomena. For example, contact angles of 50-90° are associated with bouncing drops, and angles of 60-80° with partial coalescence. These phenomena occur over sizes intermediate to the ranges for coalescence and breakup. Laboratory experiments indicate bouncing between large and small drizzle drops and partial coalescence between drizzle and large cloud drops. We will emphasize the charge transfer between dissimilar sizes, as the above ranges suggest. Interactions between similar size drops are relatively unimportant because their similar fall speeds lead to infrequent collisions.

The maximum charge on a large drop (R) acquired by collision with a small drop (r) for dissimilar sizes is given approximately by

$$Q = 12\cos\theta\,\pi\epsilon_0 r^2 E. \qquad (9.8)$$

If we assign an average contact angle of 70 (whereby $12\cos\theta = 4$), the result is similar to Eqs. (9.6) and (9.7) except for the scaling by r^2 instead of R^2. It should be obvious from the r^2 dependency that the induction mechanism is not a powerful means of direct charge separation. However, as Figure 9.5 demonstrates, the microscale charge separation followed by differential sedimentation reinforces the existing field. Therefore the induction mechanism may be capable of significant charge separation on the cloud scale through positive feedback to Eq. (9.8).

There are several possible limitations on induction charging between drops. First, charge transfer must occur on a time scale compatible with contact. When we consider the interaction between drizzle drops and large cloud droplets, appropriate for partial coalescence or bouncing, the contact time ranges from 1 to 50 μsec. This is several orders of magnitude longer than the charge-relaxation time for cloud and rain water. Therefore the contact time is not a limitation for charge transfer between drops.

A second possible limitation occurs because an electric field can transform partial coalescence (or bounce) into complete coalescence. This effect has not been studied in detail; however, laboratory simulations of induction charging (Jennings, 1975) showed that the separation probability is reduced by an order of magnitude when the field is increased from 10 kV/m to 30 kV/m. Other studies with charged drops of similar sizes show suppression of bounce at charges comparable with those induced by the above fields. Hence, there is evidence suggesting a limit on induction charging but at fields well above those found in the rain stage (typically less than 1 kV/m).

A third possible limitation, one that applies to bouncing drops but not to partial coalescence, is that charge transfer must occur across an air gap. Transfer mechanisms such as field emission or corona, in a small air gap, usually require very high fields (greater than 10^7 V/m). Since the field between drizzle drops is enhanced by induced charges by a factor of only about 50 to 500, thunderstorm fields appear to be required for charge transfer across the air gap between bouncing drops. However, neither charge transfer between bouncing drops nor the limitation of field-induced coalescence have been adequately investigated.

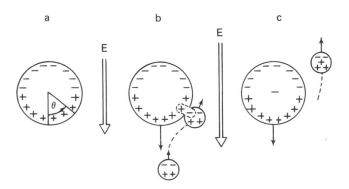

FIGURE 9.5   Charge transfer by the induction mechanism for colliding drops in a downward-directed field: a, charge distribution on a polarized drop; b, contact at moderate angle; c, charge generation after separation.

Drop charging occurs in the rain stage from drift charging at cloud edges and selective ion capture within the cloud. The latter mechanism enhances the electric field by gravitational separation of negatively charged precipitation drops and positively charged cloud droplets. In addition, breakup charging of colliding drops may result in significant charges on raindrops and drizzle drops. Since fields are weak in the clouds we have considered, induction charging is ineffective. However, under the special circumstance of deep (warm) convection, as discussed in the evaluation of charging mechanisms, induction may lead to higher fields through a positive feedback.

As the cloud top rises above the freezing level the newly formed cloud droplets, as well as drizzle drops carried in the updraft, remain in the liquid state. Soon some of the larger drops freeze until, at about the $-15°C$ level, the cloud top takes on a fuzzy outline indicating a substantial number of ice particles. Such a glaciated cloud usually undergoes a growth spurt from the release of latent heat. If the air above is not too warm, convection may continue up to the base of the stratosphere, resulting in an intense thunderstorm.

Typically on a day that has isolated thunderstorms, the first cumulus clouds in the early afternoon reach only the cloud stage. Somewhat later cloud tops are higher and reach the rain stage. It is often not until middle afternoon that cloud tops are high enough to glaciate. This is the onset of the hail stage, since what follows is the beginning of hail-like precipitation as droplets collide and freeze onto larger ice particles.

HAIL STAGE

The glaciated portion of the cloud contains ice crystals in a saturated vapor environment maintained by the presence of more numerous cloud droplets. Since the saturation vapor pressure for ice is less than water, the ice crystals grow rapidly by vapor diffusion. As the ice crystals grow larger than about $100\ \mu m$, they begin to collect cloud droplets. This riming process continues within the upper portion of the cloud until the particles are transformed into soft hail (also termed "snow pellets" or "graupel"). These ice particles are not nearly so dense or large as typical hailstones. When the size and liquid water concentration are large enough, the accretion of water occurs too rapidly for immediate freezing. Water will then infiltrate the rime structure and increase the particle mass, fall speed, and growth rate. In a strong updraft the water in soft hail may refreeze higher in the cloud, resulting in ice pellets (i.e., small hailstones). Further growth by riming may be followed by a descent to a region where wet growth can again

occur. A cycle of wet and dry growth may be repeated several times to produce the multiple layering found in larger hailstones (see also Chapter 7, this volume).

The above description for the growth of soft hail and hailstones indicates the complexity of particle interactions in the hail stage. Although the growth of ice precipitation is governed by the collection of cloud droplets, the charging of ice precipitation appears to be linked to collision with smaller ice particles (Gaskell and Illingworth, 1980; Latham, 1981; Jayaratne *et al.*, 1983). Therefore, we will consider the separation of charge for collisions of precipitation, such as soft hail and hailstones, with frozen drops and ice particles. First, the induction charging discussed for the rain stage will be extended to drops and ice particles rebounding from ice precipitation with dry or wet surfaces. Then we discuss thermoelectric charging and interface charging.

Induction Charging

The concept of induction described for the rain stage can be applied in the hail stage after considering a few alterations. Of primary importance is the charge relaxation time for ice that is a factor of 1000 slower than for liquid water. Theoretical estimates by Gaskell (1981) for charge transfer between a 100-μm ice sphere and a much larger one, including the effect of a surface conductivity, yield a relaxation time constant of $\tau = 100$ μsec, which is much longer than an estimated contact time of less than $1\ \mu sec$. Since the amount of charge transferred during contact is proportional to $1 - e^{-t/\tau}$, an ice sphere charges at less than one hundredth the rate of a comparable water drop. This would increase the induction-charging time from a few minutes for drizzle drops to several hours for ice pellets. However, induction charging of wet hail would proceed to the maximum amount in Eq. (9.8) about as rapidly as in the water-drop interactions.

Another aspect of induction charging is the effective contact angle found from the average over the range of bouncing interactions. In the rain stage the average is about $70°$ for collisions between large and small drizzle drops. Experimental measurements of charge separation for larger precipitation particles (both water drops and ice spheres) colliding with cloud droplets indicate that the average contact angle is greater than 85 (e.g., see Jennings, 1975; Gaskill, 1981). In contrast, ice particles almost always separate after colliding with hail, resulting in an average contact angle of $45°$.

When we consider both the effects of contact angle and charge relaxation, we can compare the strength of induction charging for various particle interactions in the hail stage. For collisions between dry hail and ice particles the relaxation time is too long to permit charging to the maximum value given by Eq. (9.8). As stated above, the charge attained by an ice pellet is less than one hundredth that of drizzle in a comparable time. In addition, the average contact angle of 45° yields only a factor-of-2 increase over the 70° angle assumed for drizzle. Thus, induction charging between dry hail and ice particles is severely limited by the long relaxation time for ice.

In collisions between wet hail and cloud drops or ice crystals, charge relaxation should be controlled by liquid water and the charging rate comparable to induction in the rain stage. The maximum charge attained would be governed by the average contact angle in Eq. (9.8) of over 85° for wet hail and cloud droplets (or small drizzle drops) and 45° for wet hail and rebounding ice crystals. Thus, the most powerful interaction for induction charging in the hail stage would appear to be collisions between wet hail and ice crystals.

Thermoelectric Charging

Up to this point, we have discussed mechanisms that depend on the ambient electric field. We now turn to charge transfer between cloud and precipitation particles where an external field is unnecessary. This class of microscale separation mechanism originates from intrinsic charge carriers and their relationship to bulk properties. Thermoelectric charging is the result of a thermally induced gradient in the concentration of carriers that transport positive and negative charges. For a linear gradient in temperature, the steady-state balance between carrier diffusion and drift produces a field of $E = k\,dT/dx$ with an empirically determined coefficient of $k = 2$ mV/°C (Latham and Mason, 1961). The corresponding surface charge density from Gauss's law is about (10^{-15} C/cm °C) dT/dx (where the gradient is in degrees per centimeter). We can estimate the steady-state charge (in coulombs) for a short ice cylinder of radius r by

$$Q = 10^{-15}\,r\Delta T \qquad (9.9)$$

for a temperature difference ΔT (°C) across a length of πr (cm) and a surface charge on an area of πr^2.

In the hail stage a temperature gradient would occur during the contact between a precipitation particle, warmed by the freezing of rime (e.g., soft hail) and a smaller particle. A negative charge would be transferred to the precipitation particle with $Q = 10^{-16}$ C for a small particle using $r = 100$ μm and with a rather large temperature difference of $\Delta T = 10$°C. Thus, the charging in a single collision is rather insubstantial when com-

pared with values of greater than 1 pC measured for soft hail in thunderstorms.

These estimates of thermoelectric charging are further reduced when we account for the limitation imposed by transient contact. Both theory and experiment show that about 10 msec are required to reach a maximum charge comparable to Eq. (9.9) (Latham and Stow, 1967). Since the contact time between precipitation and cloud particles is many orders of magnitude smaller we can reasonably expect that our estimate of 10^{-16} C for a single collision would be reduced to well below 10^{-18} C. Even under the most favorable conditions (i.e., 10^4 collisions within 20 minutes for high ice crystal concentrations at 100 per liter), the accumulated charge would be less than 10^{-14} C.

In contrast to the estimate of considerably less than 10^{-18} C per event for thermoelectric charging, recent laboratory studies have yielded up to 0.3 pC per collision between a small ice particle and a simulated hailstone (Gaskell and Illingworth, 1980). Thus, there is evidence to demonstrate that mechanisms far more powerful than thermoelectric charging are at work in the hail stage.

Interface Charging

Two types of interface charging will be discussed: freezing potentials involving impurities and contact potentials.

Charge can be transferred across a freezing interface by selective incorporation of ions, originating from dissolved salts and gases, into the advancing ice. In the steady state, a balance is reached between the selection process and the relaxation of charge in the ice. A transient in potential is observed when a plane interface advances past an electric probe. Early workers measured large potentials in the freezing of aqueous solutions containing naturally occurring salts over the range of concentrations found in precipitation (Workman and Reynolds, 1948). Subsequent researchers have made more refined measurements and developed a theoretical description of freezing potentials (e.g., see Caranti and Illingworth, 1983a). Others have investigated charge transfer between solution drops and simulated hailstones (e.g., Latham and Warwicker, 1980). As a result of these later studies and related ones (Gaskell and Illingworth, 1980; Jayaratne et al., 1983; Caranti and Illingworth, 1983b), the role of interface charging in hailstage electrification is being clarified.

As indicated above, two methods are used to study interface charging: (1) potentials are measured as a function of solute impurities and supercooling, with differing growth rates and interface areas; and (2) transfer of charge is measured for collisions between particles and a much larger "target" electrode coated with ice as a function of solute impurity, supercooling, and speed of impacting drops. Other conditions have also been varied, such as the target temperature and the riming rate of the target for a mixture of supercooled drops and ice crystals. Investigation of these various parameters covers many of the conditions found in the hail stage and helps to sort out contributions of freezing potentials from contact potentials and the thermoelectric effect.

Recent studies have shown that interface potentials for bulk solutions near 0°C are substantially reduced by supercooling, apparently from the effects of the dendritic interface (Caranti and Illingworth, 1983a). Potentials could not be measured for 100-μm-diameter droplets, in the range -1 to -20°C, impacting on an ice substrate because the potential was either too small (less than 100 mV) or it decayed too rapidly (in less than 5 msec). A reduction in charge transfer was also found by Latham and Warwicker (1980) for millimeter-size drops splashing from ice targets in comparison with earlier findings for drops not completely cooled by the air (Workman, 1969). In fact, the charging was of the wrong sign for the freezing of sodium chloride solutions and independent of concentration, indicating that the freezing potential was not the dominating mechanism. A more likely cause was a common form of interface charging associated with the disruption of an air-water interface (i.e., spray electrification). The charges on the air-water interface are readily overwhelmed by polarization charges. For example, Latham and Warwicker (1980) found that charge transfer was substantially increased by applying a field of only 100 V/m.

As the above comparisons demonstrate, the relative importance of charge transfer during the freezing of aqueous solutions is greatly diminished by supercooling. The effect of dissolved ions on charging seems to be negligible in the splashing of supercooled solution drops from ice targets. However, the above considerations do not rule out the freezing potential as a factor in the transfer of charge for a target collecting solution drops and also colliding with ice crystals.

Before examining collisions involving ice crystals during rime formation, we will consider the transfer of charge between an ice-coated target and rebounding ice particles in the absence of droplets. Experiments have shown that a target of ice accumulates either negative or positive charge in collision with ice particles depending on whether the target surface has undergone sublimation or deposition (Buser and Aufdermaur, 1977). It was concluded from an additional experiment that the condition of the surface was controlling the charge transfer rather than a thermal gradient as would have been expected for the thermoelectric effect. Buser and Aufder-

maur (1977) also found that the transfer of charge between ice particles and targets of various metals was proportional to the contact potential. Thus variations in the surface state and, in particular, the free energy of the charge carriers is a major factor in this type of charge separation. The differing signs in the ice-target experiments can be attributed to differing surface characteristics of the target. The surface exposed during sublimation was composed of ice originally formed near 0°C, whereas a frosted surface was produced by deposition at the experimental condition of -45°C.

More recently Gaskell and Illingworth (1980) studied interface charging in the temperature range from -5 to -25°C. They also found negative charging for a subliming target and positive charging during deposition without any direct evidence of the thermoelectric effect. Frozen droplets of 100-μm diameter transferred charges of about -0.015 pC for sublimation and $+0.10$ pC for deposition. Little variation in charging was found over the temperature range or when the ice target contained ion impurities. These results are consistent with interface charging by a contact potential mechanism whereby the surface states of the charge carriers differ between the smooth surface formed near 0°C, exposed during sublimation, and the frosted surface formed by deposition. Additional support for the contact potential hypothesis comes from measurements of the effects of impact velocity and droplet size on charging, since the transfer of charge was found to increase with both of these parameters in a manner consistent with an increase in contact area (Gaskell and Illingworth, 1980).

Charging was also examined by Gaskell and Illingworth (1980) for a target undergoing simultaneous collisions with ice particles of 100 μm diameter and supercooled droplets at low to moderate liquid water contents (0.05 to 0.85 g/m^3). The charge transferred to the target was positive at -5°C and negative at -15°C with the transition near -10°C. The sign reversal was possibly caused by changes in the contact potential with rime structure at different temperatures (Caranti and Illingworth, 1980).

An estimate of interface charging in collisions between various sizes of cloud and precipitation particles can be obtained from the work of Gaskell and Illingworth (1980) in which the charge was found to be related to the contact area through the impact speed and ice particle size. In the following formula, we have combined their relation for impact velocity ($Q \propto U^{1.6}$) with an expression for the velocity of hail ($U \propto R^{0.8}$, e.g., see Pruppacher and Klett, 1978) and have included their scaling for ice particle size ($Q \propto r^{1.7}$):

$$Q = FR^{1.3} r^{1.7}. \qquad (9.10)$$

The factor F is a function of the interface potential, depends on the nature of the contact surfaces, and may be evaluated from laboratory data. For example, in the riming experiment of Gaskell and Illingworth (1980) the collisional charge was $Q \simeq -0.04$ pC in the range -15 to $-20°C$ for $r = 50 \mu m$ and $R = 0.43$ cm (a size with a terminal velocity at the laboratory impact speed, 8 m/sec). Thus, for the interface conditions corresponding to these experiments the factor for the interface potential is $F = -970$ (with Q in picocoulombs and the radii in centimeters).

The latest investigation of ice crystals rebounding from riming targets provides additional evidence for interface charging (Jayaratne *et al.*, 1983). In this study the target electrode was moved through a cloud of supercooled droplets that was seeded to produce ice crystals. Charging of the target began shortly after seeding and ended after about 4 minutes when the ice crystals settled out of the cloud. At low liquid-water contents the maximum current was positive at a temperature below about $-10°C$ and negative at temperatures above, indicating a reversal in sign of the charge in a manner similar to that of Gaskell and Illingworth (1980). The results of Jayaratne *et al.* are difficult to interpret in terms of charge transfer because of transients in the ice crystal size, the liquid-water content, and the current measured at the target. However, the charge was estimated for single events by the investigators from the target current and ice crystal concentration. In one case they estimated $Q \simeq 0.01$ pC for an ice crystal size of 125 μm diameter, where the target speed was 2.9 m/sec, the temperature $-6°C$, and the liquid-water content 2 g/m^3. The factor for the interface potential that corresponds to these experimental conditions is $F = 870$ (with Q in picocoulombs and the radii in centimeters).

The variation in the sign-reversal temperature with liquid-water content was attributed to changes in the freezing process or to variations in the structure and density of the rime. This view is consistent with interface charging by the contact potential mechanism. However, sufficient data are unavailable to express the factor for the interface potential as a function of the riming rate or fundamental parameters such as temperature and liquid-water content.

Another aspect investigated by Jayaratne *et al.* (1983) was the variation in charging with solute impurities in the cloud droplets. For cloud water containing natural amounts of sodium chloride they determined that about -0.003 pC was transferred by ice crystals of 50-μm diameter at $-10°C$ and 1 g/m^3. A charge of the same magnitude but of the reverse sign was found for ammonium sulfate under the same conditions. At $-20°C$ the charge was -0.08 pC for sodium chloride and $+0.08$ pC for ammonium sulfate. It should be noted that with

these impurities there was no sign reversal in charging over the temperature range investigated (-4 to $-20°C$). The factor for the interface potential based on the sodium chloride measurements is $F \simeq -1100$ for $-10°C$ and $F \simeq -3100$ at $-20°C$. The corresponding factors for the ammonium sulfate measurements are the same but have a positive value.

In contrast to the above study, Caranti and Illingworth (1983b) found that solute impurities at natural concentrations did not have a measurable effect on the contact potential of a riming substrate. Thus there seems to be contradictory evidence on the role of solutes in contact charging. A direct comparison between these two studies may be inappropriate because the rime structure and the resultant factor for the interface potential may have been affected by differing experimental conditions. If the solute influenced the rime structure in the experiments of Jayaratne *et al.* (1983) either directly during the freezing or indirectly by alternating the cloud properties, then their results would have been affected by the contact potential mechanism.

In summary, interface charging occurs between riming precipitation particles and ice crystals with a negative charge acquired by the precipitation for temperatures below about $-15°C$ or $-20°C$ depending on the liquid-water content. The separation of charge appears to result from an interface potential with an amount given by Eq. (9.10) that incorporates the variation in contact area through the ice crystal size and hail size (i.e., impact speed). The interface potential enters Eq. (9.10) through an empirical factor, F. Natural amounts of solute also affect the magnitude and sign of charging. However, it is unclear whether the resultant change in interface potential occurs as a freezing potential or indirectly as a contact potential through an altered rime structure.

Before applying the results of these laboratory findings to the hail stage we first consider the appropriate value of the factor for the interface potential. In the evaluations presented here F was found to be in the range -3100 to $+3100$. Thus the charge obtained by a precipitation particle of $R = 1$ mm in a collision with an ice particle of $r = 50 \mu m$ has a range, according to Eq. (9.10), of -0.02 to $+0.02$ pC. This range is increased to $+0.15$ pC for $R = 2$ mm and $r = 100 \mu m$. These calculated values also correspond to the measured range presented in this section because the particle sizes correspond approximately to the experimental sizes (r) and impact speeds. For a particular application to the hail stage we consider a situation similar to the experiment of Gaskell and Illingworth (1980) at low to moderate liquid-water contents and in the temperature range -15 to $-20°C$. For this set of conditions the type of rime is appropriate for soft hail. The estimated charge transferred to a soft hail particle (R = 1 mm) by an ice particle (r = 50 μm) is −0.006 pC using F = −970 (determined previously for this experiment).

Since contact charging happens much more rapidly than discharging of net charge between particles by conduction, a buildup of charge can readily occur in multiple collisions. A charge of 20 pC could accumulate under favorable conditions with ice crystal concentrations of 100 per liter after 3000 collisions (or about 6 minutes). This amount is comparable with negative charges found on ice precipitation in highly electrified regions of thunderstorms (e.g., see Latham, 1981). Thus, interface charging appears to be capable of microscale charge separation in amounts that can account for a major aspect of thunderstorm electrification. However, the wide range and sign variation for the interface factor (and charge) found in laboratory studies seems to be at odds with observations of the predominantly negative charge center for thunderstorms. It is apparent that additional laboratory measurements along with more detailed field observations are required to sort out the discrepancy. The relation of the interface factor (F) to temperature, liquid-water content, and impurities must be known before interface charging can be reliably applied to the variety of conditions in the hail stage.

This completes our detailed discussion of microscale charge separation. We have gone from diffusion charging in the cloud stage to the more complex mechanisms involving precipitation in the rain and hail stages. In the following section we consider the relative importance of these mechanisms, and in particular, we evaluate their contributions by comparison with observations of clouds and thunderstorms.

EVALUATION

The requirements for a satisfactory explanation of charge separation in clouds and thunderstorms are fairly well known (e.g., Mason, 1972; Latham, 1981; Takahashi, 1982). Any theory must be capable of explaining microscale and cloud-scale charge separation on a suitable time scale. For a fairly complete assessment of charge-separation mechanisms, we need to take into account the evolution of charges and fields, and their interactions, in at least two dimensions. Such an evaluation is well beyond the scope of this paper (see Chapter 10, this volume, on cloud modeling.) What we can do here is reiterate some of the pronounced strengths and weaknesses for the various charging mechanisms. We can also indicate where model studies would be helpful in quantifying our gross conclusions and point to areas in need of further laboratory or field research. In making our evaluation we consider the major requirements

for an adequate theory of charge separation in three stages: (1) the cloud stage for small cumulus clouds that contain only cloud droplets and drizzle drops; (2) the rain stage for larger cumulus clouds that contain raindrops formed by accretion of cloud droplets; and (3) the hail stage for the upper regions of large cumulus clouds where precipitation (notably, soft hail) is formed by accretion of supercooled droplets.

Cloud Stage

Electrification is generally weak in the cloud stage. Ion mechanisms dominate because of the absence of precipitation and their associated charge-separation mechanisms. Charges have been observed to range from about 1 to 20 electrons for small cloud droplets with a normal distribution centered near zero charge. Values of the average charge magnitude are indicated in Figure 9.6 by region C_1 from measurements in stratocumulus clouds (Phillips and Kinzer, 1958). The observations agree with a Boltzmann equilibrium (line c) after the

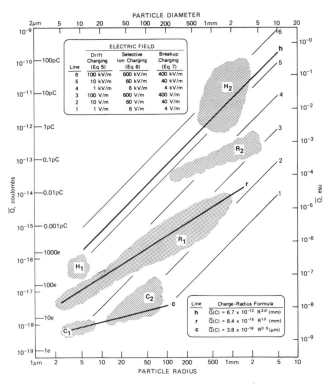

FIGURE 9.6   Average charge magnitude for cloud and precipitation particles. Regions C_1 and C_2 show cloud stage (after Phillips and Kinzer, 1958; Gunn, 1957), regions R_1 and R_2 show rain stage for shallow and deep convection (after Takahashi, 1973a, 1978), and regions H_1 and H_2 show hail stage (after Takahashi, 1973a; Latham, 1981). Lines labeled c, r, and h are from equations given by lower inset. Lines 1 through 6 have R^2E dependence on charge (see upper inset for corresponding mechanisms and fields).

rms charge given by Eq. (9.3) is converted to a mean deviation (see lower inset on Figure 9.6 for the corresponding $\bar{Q}$ equation). We can expect that the charge distributions for large cloud drops would be skewed (no longer centered on zero charge) with higher averages as indicated by region C_2 from the influence of the electric field by drift charging at cloud edges and selective ion charging. Note that charging under the influence of the electric field is depicted on Figure 9.6 by the lines labeled 1 through 6. The corresponding values of electric field for three mechanisms are found in the upper inset. For example, line 2 shows the charge magnitude for drift charging at cloud base (or top) in a field of 10 V/m and selective ion charging in a field of 60 V/m.

For a small cumulus, charge separation in the cloud stage is consistent with microscale ion capture and cloud-scale convective transport of charge. This combination can account for such features in the cloud stage as the negative core and positive edges. It can also explain a positive charge in the base of a cloud very near to the ground. Charging by ion capture appears to be limited by cosmic-ray production within the cloud (Wormell, 1953) and transport from outside, and therefore additional mechanisms are required to produce the fields and charges found in the rain stage and hail stage.

Rain Stage

Electrification in convective clouds of less than about 3 km deep is characterized by the drop charges for the rain stage indicated in Figure 9.6 by region R_1 (Takahashi, 1973a). The mean value ($\bar{Q}$) is approximated by the straight line proportional to $R^{1.3}$ (i.e., line r with equation shown on lower inset from Pruppacher and Klett, 1978). Since the electric field associated with these clouds is often 10 to 100 V/m, it is apparent from Figure 9.6 that drift charging (lines 2 and 3) and also selective ion charging and breakup charging can produce charges of the observed magnitude (Q) for drizzle and raindrops ($R > 100~\mu m$). What is not so apparent is how cloud drops and small drizzle drops acquire their relatively high charge in the rain stage. One possibility is by evaporation of drops with higher charge. Other explanations involve selective ion capture from the effects of surface potentials (Takahashi, 1973b; Wahlin, 1977). However, the details of these mechanisms are poorly understood, and consequently their role in drop electrification remains uncertain. Additional research is needed to clarify the microscale-separation mechanisms responsible for charging cloud drops and drizzle drops.

Another aspect of electrification in the rain stage is the predominant sign of charge for cloud drops, drizzle, and raindrops. We can consider charge separation for a convective cloud of about 3 km deep (after Takahashi, 1982). The trajectories of drizzle and raindrops are depicted in Figure 9.7 to indicate differences for the preferred sign of charge. The drizzle drop is in a region of lower updraft speed (dashed arrow), which results in a shorter growth time within the cloud. Negative charging occurs by the Wilson effect (for a downward-directed field) and by drift current at cloud base. In addition, drizzle may be produced by breakup of raindrops, resulting in negatively charged drizzle for the field near and below cloud base. Raindrops become electrified positively by breakup charging. At an earlier time, raindrops near cloud top may also acquire a positive charge from the capture of droplets.

Although this picture of drop trajectories in Figure 9.7 is greatly simplified it does illustrate some essential differences that can occur in growth histories and in the resultant charge-separation mechanisms for cloud droplets, drizzle drops, and raindrops. Our conclusions about the predominant sign of charge, based on trajectories and separation mechanisms, are consistent with extensive observation of tropical cumulus clouds (e.g., see Takahashi, 1982). These observations show a predominance of positive droplets near cloud top and negative drizzle drops and positive raindrops within and below the cloud.

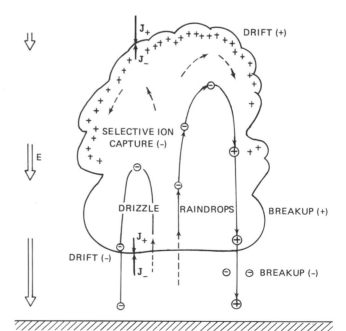

FIGURE 9.7 Rain-stage electrification based on simplified growth histories for drizzle and raindrops (modified from Takahashi, 1982). Air currents are shown by dashed arrows and ion drift currents by heavy arrows.

A fascinating aspect of electrification in the rain stage is the reported occurrence of lightning for clouds warmer than 0°C (e.g., see Moore *et al.*, 1960). The evidence is incomplete because this phenomenon has not been verified by in situ measurements of cloud temperature. Such lightning is apparently restricted to the tropics and probably occurs only in clouds that are deeper than discussed above. For warm clouds of about 5 km depth average charges are typified by region R₂ shown on Figure 9.6 (Takahashi, 1978). These charges were measured at the ground with associated fields of less than 1 kV/m. Drift charging, selective ion capture, and breakup charging can probably account for such high charges providing the fields in or around the cloud reach about 1 kV/m. (Fields as large as 3 kV/m were measured from an aircraft in the vicinity of warm cloud lightning by Moore *et al.*, 1960.) There is also the possibility that induction charging could contribute to the field intensification within the cloud, but the enhancement of drop coalescence in fields of 30 kV/m suggests that lightning cannot be achieved by induction alone. We should keep in mind, however, that induction charging has been investigated for only a narrow range of drop sizes and that lightning in deep (warm) convection has not been studied in detail. Therefore, additional research on charge separation mechanisms is required to understand the strong electrification that occurs in deep warm clouds.

Hail Stage

There are three aspects of charging in the hail stage that must be explained: (1) the observed region of negative charge, (2) the buildup of fields, and (3) the average values of charge. Regions of negative charge lie gener-

ally between the -10 and $-25°C$ level even in winter thunderstorms, as illustrated by the location of lightning sources shown on Figure 9.8 (from Krehbiel *et al.*, 1983). The space-charge densities associated with the region of negative charge average about 1 C/km³ from estimates based on lightning currents and particle charges (Latham, 1981). The second aspect of charging in the hail stage is the development of large electric fields. The maximum fields measured in thunderstorms are consistent with estimates of the requirement for lightning initiation (about 400 kV/m).

The average magnitude of charge on individual particles ($\bar{Q}$) in the hail stage is shown on Figure 9.6 by region H₁ for cloud droplets and region H₂ for precipitation particles, with line *h* giving an approximate $\bar{Q}$ from Grover and Beard (1975). The charges on raindrops are in the lower portion of H₂ (after Takahashi, 1973a), whereas charges on solid precipitation reside in the upper portion of H₂ (Latham, 1981). In both cases the average for the negative charges usually exceeded the positive charges by a significant amount, with charges on individual particles sometimes in excess of 100 pC. As a result of the high fields found in the hail stage, breakup charging would be highly efficient (Figure 9.6, line 6), with the sign depending on the orientation of the local field. (Note that a variety of field orientations must occur within thunderstorms around charge centers, see Figure 9.8.) Drift charging would also be efficient in strong fields if ion concentrations are enhanced by corona from ice crystals (Griffiths and Latham, 1974). Thus the generally high charges found on cloud and precipitation particles are probably an indication of field-driven mechanisms that separate charge on the microscale. For the causes of high fields we will consider the

FIGURE 9.8 Schematic diagram illustrating the levels and distribution of charge sources for ground-flash lightning observed for summertime in Florida and New Mexico and for wintertime in Japan (from Krehbiel *et al.*, 1983).

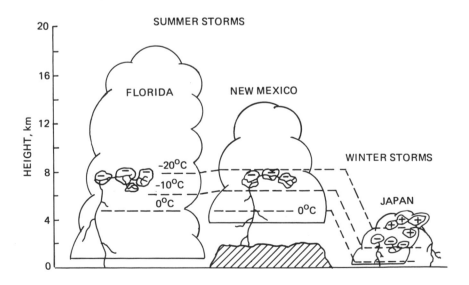

microscale and cloud-scale mechanisms that lead to the negative-charge region observed to occur between − 10 and − 25°C.

We first evaluate convection charging, which has held a controversial position among theories of thunderstorm electrification (e.g., see arguments in Mason, 1976; Moore, 1977). These theories, proposed by Grenet (see Chalmers, 1967) and Vonnegut (1955), rely on the transport by updrafts and downdrafts of space charges and screening layers. In the hail stage, with a highly electrified cloud, the charge from the positive corona at the ground is carried into the cloud base and by the updraft, to the cloud top where it attracts a negative screening layer. The downdrafts carry the negative charge back toward the cloud base to strengthen the negative field between the cloud base and the ground, thus enhancing the positive corona. The notion of negative charge in downdrafts has been reintroduced in the form of nonrandom mixing from the cloud top by Telford and Wagner (1979) to provide a qualitative explanation for a negative-charge region near the − 10°C level.

This picture of descending negative charges appears to be at odds with the early stages of convective electrification modeled by Chiu and Klett (1976). They found a positive screening layer at a cloud top and positive descending charges. Convection and mixing were found to weaken the field within the cloud. Since Chiu and Klett did not expect their model outcome to change appreciably for clouds deeper than 5 km, it is difficult to envision how convective (single-cell) transport could be the source of strong electrification. However, in the hail stage with multicell convection and with corona from precipitation, the ion concentrations would differ considerably from the model of Chiu and Klett. Thus, a better model is required to evaluate the importance of convection in highly electrified clouds.

Another criticism of convection charging is that updrafts and downdrafts may disorganize their associated charges through mixing (e.g., see Chalmers, 1967). The study of Chiu and Klett clearly shows that single-cell convection with eddy diffusion diminishes the electric field within the cloud. If we picture cloud turrets as convective cells (similar to Figure 9.2) with interspersed updrafts and downdrafts, then the possibility for disorganization by mixing between adjacent charge regions becomes rapidly apparent. Questions regarding the importance of mixing in cloud electrification probably will not have a satisfactory answer until we have more quantitative models of turret scale motions and entrainment. In addition, the common occurrence of a negative charge center near − 15°C suggests that transport of ions and charged particles by convection is not so impor-

tant as microscale charge separation involving ice particles.

In evaluating induction charging in the hail stage, we consider that the most efficient microscale interactions are collisions between wet hail and ice particles. The maximum charge according to Eq. (9.8) is about 1 pC in the limiting field of 400 kV/m for a particle of 100-μm radius and 10 pC for a 300-μm particle. Since the wet growth of hail requires sizes larger than about 10 mm diameter, the induction limit of 1 to 10 pC even for R = 10 mm is well below the charge expected from ion capture in high fields (see line 6, Figure 9.6). Thus, induction charging cannot be directly responsible for the average charges (1 to 100 pC) found on precipitation particles (0.5 to 2 mm radius) in the hail stage.

Another aspect of the induction charging of wet hail is its importance to the region of negative charge near the − 15°C level. For hailstones with a maximum charge of 10 pC at a maximum concentration of 1 m^{-3}, the resulting space-charge density is 0.01 C/km^3. This estimate of the maximum charge density is several orders of magnitude smaller than a charge of about 1 C/km^3 found from measurements of particles and from estimates based on lightning currents. Although the induction mechanism provides negative charge on precipitation particles and field intensification through feedback, it appears to be too weak to account for the charge densities associated with the hail stage. Its shortcomings are twofold: the maximum charge is limited by the effects of size and contact angle given in Eq. (9.8), and the charge density is limited by the instrinsically low concentration of hailstones. The microscale separation mechanism in the negative-charge region is probably associated with smaller precipitation particles (e.g., soft hail), because their higher concentration could lead more easily to a sufficient charge density. For example, particles at a concentration of 100 m^{-3} carrying 10 pC would result in a more realistic 1 C/km^3.

The charging of soft hail has been simulated in the laboratory by collisions between ice particles and an ice electrode in the process of riming (Gaskill and Illingworth, 1980; Jayaratne et al., 1983). Although our understanding of the separation mechanism is incomplete, the evidence points to interface charging from contact potentials with freezing potentials having a secondary role. Investigations of temperature effects in the above studies have ruled out thermoelectric charging. The formula applicable to these results scales with contact area and is the same as line r on Figure 9.6 when evaluated for an ice particle of r = 60 μm. Thus charge transfer for individual collisions between soft hail and small ice particles is around 0.01 pC. Since contact time is relatively short compared with the time required to discharge colliding particles by conduction, an accumulation of charge can occur in multiple collisions. In this manner an increase of 3 orders of magnitude, well into region H$_2$, may be attained with high concentrations of ice crystals in several minutes.

A key result of the above experiments is a reversal in the sign of charging at temperatures of -10 to $-25°C$ depending on liquid-water content. Riming particles acquire negative charges if colder than the reversal temperature and positive if warmer. Thus, soft hail would be charged negatively above the reversal level in the cloud and positively below this level.

Jayaratne *et al.* (1983) postulated that descending precipitation particles should have their maximum negative charge near the reversal level and that rebounding ice crystals carrying negative charge from below would also contribute to the negative region. The fields directed toward the reversal level would intensify during the process of charge separation by differential sedimentation.

There are many features of interface charging that need clarification before we can feel comfortable with the above description of charging in the hail stage. First, the roles of temperature, liquid-water content, and solutes are poorly understood. These appear to influence contact potential through changes in rime structure.

(Solutes may also affect interface charging by transient freezing potentials.) Second, the details of charge transfer have not been explained, although the charge carriers are probably associated with the contact surfaces. This concept is consistent with a rapid transfer of charge that scales with contact area. Finally, our recently acquired understanding of interface charging, even though somewhat limited, cannot be adequately assessed until it is placed within the framework of a multidimensional cloud model.

CONCLUSIONS

The charging mechanisms in clouds and thunderstorms are varied and numerous. Some are simple and readily appreciated, whereas others are complex. Several important mechanisms are poorly understood. Feedback readily occurs through changes in ion concentration and the electric field making it difficult to identify the primary causes for electrification. However, some simplification can result by considering the charging mechanisms in three stages of cumulus cloud development: the cloud, rain, and hail stages.

The charging mechanisms discussed in this chapter are summarized in Table 9.1. Each mechanism is listed with the microscale and cloud-scale separators (with

TABLE 9.1   Charge Separation in Clouds and Thunderstorms

Mechanism	Microscale	Cloud Scale	Major Roles
1. Diffusion charging[a]	Ion capture by diffusion		Removes ions within cloud
2. Drift charging[a,b]	Ion capture in drift currents	Drift currents Convection (Sedimentation)	Charges particles Enhances field
3. Selective ion charging[c,d]	Ion capture by polarized drops	Sedimentation (Convection)	Charges particles Enhances fields
4. Breakup charging[e]	Collisional breakup of polarized drops	Sedimentation (Convection)	Charges drops
5. Induction charging[d,f]	Charge transfer between polarized particles	Sedimentation (Convection)	Charges particles Enhances field
6. Convection charging[d,g,h]	Space-charge production Ion capture in drift currents	Convection	Enhances field (Charges particles)
7. Thermoelectric charging[i]	Charge transfer between particles of differing temperatures	Sedimentation (Convection)	(Charges particles)
8. Interface charging[j,k]	Charge transfer between particles involving contact potentials (freezing potentials)	Sedimentation (Convection)	Charges particles Enhances field

[a]Gunn (1957); [b]Chiu and Klett (1976); [c]Wilson (1929); [d]Chalmers, (1967); [e]Matthews and Mason (1964); [f]Elster and Geitel (1913); [g]Grenet (1947); [h]Vonnegut (1955); [i]Latham and Mason (1961); [j]Workman and Reynolds (1948); [k]Buser and Aufdermaur (1977).

items of secondary importance shown in parentheses). Charging appears to be well described by diffusion, drift, and selective ion capture for the nonprecipitating cloud stage (mechanisms 1-3). The situation in the rain stage is complicated by the addition of breakup and induction (mechanisms 4 and 5). We suspect that drift, selective ion capture, breakup, and induction are responsible for charges and fields in shallow clouds. However, it is difficult to find an explanation for the stronger electrification in convective clouds over a few kilometers deep. The basis of lightning from clouds with tops warmer than freezing remains a mystery. A major problem in the rain stage is that our knowledge of the suspected mechanisms is still rather rudimentary. There is clearly a need for additional research on charging by ions, breakup, induction, and convection to understand the electrification of warm clouds.

In the hail stage we add thermoelectric and interface charging (mechanisms 7 and 8). Recent laboratory studies of charge transfer involving ice particles rebounding from simulated hailstones in the process of riming have shown that interface charging is the dominant mechanism. The roles of temperature, liquid-water content, and solutes are most likely important in altering the rime structure and thereby the contact potential and contact area. More research is required to understand these effects and the details of charge transfer.

The electrification process becomes more complex as a cloud develops. The cloud stage involves mechanisms 1-3, whereas the rain stage includes 1-6. All the mechanisms listed in Table 9.1 may occur in the hail stage. We might ask, as many have before us, which separation mechanisms are essential to cloud electrification. The answers, if we had them, would depend on which aspect of cloud electrification we consider. For example, the essential mechanism for lightning depends on whether we are looking at the field development in the rain or hail stages or whether we are concerned with the charge centers associated with cloud-to-ground, in-cloud, or cloud-to-cloud lightning. Yet another aspect of lightning is the mechanism that initiates the stroke. Clearly the idea of an "essential" mechanism is an oversimplification. A more useful approach is to examine the interdependencies. We should be asking how the various charge-separation mechanisms are related. Some answers should be forthcoming as we incorporate the knowledge gained from recent laboratory studies of individual mechanisms into models of cloud electrification and compare the findings to field observations. With continued progress in laboratory, field, and modeling research we should achieve, in the next decade, a much improved perspective of the charging mechanisms in clouds and thunderstorms.

ACKNOWLEDGMENTS

We appreciate the helpful comments of Bernice Ackerman, David Johnson, Anthony Illingworth, and an anonymous reviewer. This review was supported in part by a grant from the National Science Foundation under ATM-83-14072.

REFERENCES

Buser, O., and A. N. Aufdermaur (1977). Electrification by collision of ice particles on ice or metal targets, in *Electrical Processes in Atmospheres*, N. Dolezalek and R. Reiter, eds., Steinkopff, Darmstadt, p. 294.

Caranti, J. M., and A. J. Illingworth (1980). Surface potentials of ice and thunderstorm charge separation, *Nature 284*, 44.

Caranti, J. M., and A. J. Illingworth (1983a). Transient Workman-Reynolds freezing potentials, *J. Geophys. Res. 88*, 8483.

Caranti, J. M., and A. J. Illingworth (1983b). The contact potential of rimed ice, *J. Phys. Chem. 87*, 4125.

Chalmers, J. A. (1967). *Atmospheric Electricity*, Pergamon Press, New York, 515 pp.

Chiu, C. S., and J. D. Klett (1976). Convective electrification of clouds, *J. Geophys. Res. 81*, 1111.

Elster, J., and H. Geitel (1913). Zur Influenztheorie der Niederschlags- elektrizität, *Phys. Z. 14*, 1287.

Gaskell, W. (1981). A laboratory study of the inductive theory of thunderstorm electrification, *Q. J. R. Meteorol. Soc. 107*, 955.

Gaskell, W., and A. J. Illingworth (1980). Charge transfer accompanying individual collisions between ice particles and its role in thunderstorm electrification, *Q. J. R. Meteorol. Soc. 106*, 841.

Grenet, G. (1947). Essai d'explication de la charge électrique des nuages d'orages, *Ann. Geophys. 3*, 306.

Griffiths, R. F., and J. Latham (1974). Electrical corona from ice hydrometeors, *Q. J. R. Meteorol. Soc. 100*, 163.

Grover, S. N., and K. V. Beard (1975). A numerical determination of the efficiency with which electrically charged cloud drops and small raindrops collide with electrically charged spherical particles of various densities, *J. Atmos. Sci. 11*, 2156.

Gunn, R. (1957). The electrification of precipitation and thunderstorms, *Proc. IRE 45*, 1331.

Israël, H. (1971). *Atmospheric Electricity*, Israel Program for Scientific Translations, Ltd., 317 pp.

Jayaratne, E. R., C. P. R. Saunders, and J. Hallett (1983). Laboratory studies of the charging of soft-hail during ice crystal interactions, *Q. J. R. Meteorol. Soc. 109*, 609.

Jennings, S. G. (1975). Electrical charging of water drops in polarizing electric fields, *J. Electrostatics 1*, 15.

Krehbiel, P. R., M. Brook, R. L. Lhermitte, and C. L. Lennon (1983). Lightning charge structure in thunderstorms, in *Proceedings in Atmospheric Electricity*, L. H. Ruhnke and J. Latham, eds., A. Deepak Publ., Hampton, Va., pp. 408-410.

Latham, J. (1981). The electrification of thunderstorms, *Q. J. R. Meteorol. Soc. 107*, 277.

Latham, J., and B. J. Mason (1961). Electric charge transfer associated with temperature gradients in ice, *Proc. R. Soc. Lond. A 260*, 523.

Latham, J., and C. D. Stow (1967). The distribution of charge within ice specimens subjected to linear and non-linear temperature gradients, *Q. J. R. Meteorol. Soc. 93*, 122.

Latham, J., and R. Warwicker (1980). Charge transfer accompanying the splashing of supercooled raindrops on hailstones, *Q. J. R. Meteorol. Soc. 106*, 559.

Mason, B. J. (1972). The physics of the thunderstorm, *Proc. R. Soc. London A 327*, 433.

Mason, B. J. (1976). In reply to a critique of precipitation theories of thunderstorm electrification by C. B. Moore (Moore, 1977), *Q. J. R. Meteorol. Soc. 102*, 219.

Matthews, J. B., and B. J. Mason (1964). Electrification produced by the rupture of large water drops in an electric field, *Q. J. R. Meteorol. Soc. 90*. 275.

McTaggart-Cowan, J. D., and R. List (1975). Collision and breakup of water drops at terminal velocity, *J. Atmos. Sci. 32*, 1401.

Moore, C. B. (1976). Reply (to comments by B. J. Mason, 1976), *Q. J. R. Meteorol. Soc. 102*, 225.

Moore, C. B. (1977). An assessment of thunderstorm electrification mechanisms, in *Electrical Processes in Atmospheres*, N. Dolezalek and R. Reiter, eds., Steinkopff, Darmstadt, p. 333.

Moore, C. B., B. Vonnegut, B. A. Stein, and H. J. Survilas (1960). Observations of electrification and lightning in warm clouds, *J. Geophys. Res. 65*, 1907.

Phillips, B. B., and G. D. Kinzer (1958). Measurements of the size and electrification of droplets in cumuliform clouds, *J. Meteorol. 15*, 369.

Pruppacher, H. R., and J. D. Klett (1978). *Microphysics of Clouds and Precipitation*, D. Reidel Publishing Co., Dordrecht, 714 pp.

Takahashi, T. (1973a). Measurement of electric charge of cloud droplets, drizzle and raindrops, *Rev. Geophys. Space Phys. 11*, 903.

Takahashi, T. (1973b). Electrification of condensing and evaporating liquid drops, *J. Atmos. Sci. 30*, 249.

Takahashi, T. (1978). Electrical properties of oceanic tropical clouds at Ponape, Micronesia, *Mon. Weather Rev. 106*, 1598.

Takahashi, T. (1982). Electrification and precipitation mechanisms of maritime shallow warm clouds in the tropics, *J. Meteorol. Soc. Japan 60*, 508.

Telford, J. W., and P. B. Wagner (1979). Electric charge separation in severe storms, *Pure Appl. Geophys. 117*, 891.

Vonnegut, B. (1955). Possible mechanism for the formation of thunderstorm electricity, in *Proceedings International Conference on Atmospheric Electricity*, Geophys. Res. Paper No. 42, Air Force Cambridge Research Center, Bedford, Mass., p. 169.

Wahlin, L. (1977). Electrochemical charge separation in clouds, in *Electrical Processes in Atmospheres*, N. Dolezalek and R. Reiter, eds., Steinkopff, Darmstadt, p. 384.

Whipple, F. J. W., and J. A. Chalmers (1944). On Wilson's theory of the collection of charge by falling drops, *Q. J. R. Meteorol. Soc. 70*, 103.

Wilson, C. T. R. (1929). Some thunderstorm problems, *J. Franklin Inst. 208*, 1.

Workman, E. J. (1969). Atmospheric electrical effects resulting from the collision of supercooled water drops and hail, in *The Physics of Ice*, N. Riehl, ed., Plenum Press, New York, p. 594.

Workman, E. J., and S. E. Reynolds (1948). A suggested mechanism for the generation of thunderstorm electricity, *Phys. Rev. 74*, 526.

Wormell, T. W. (1953). Atmospheric electricity: Some recent trends and problems, *Q. J. R. Meteorol. Soc. 79*, 3.

Models of the Development of the Electrical Structure of Clouds

ZEV LEVIN
Tel Aviv University

ISRAEL TZUR
National Center for Atmospheric Research

INTRODUCTION

Thunderstorms are highly variable in their intensity, dimensions, composition, and electrical structure. Some generalizations can be made about them, however. The lightning activity follows strong vertical air currents and precipitation. As a consequence of this correlation, lightning is most frequently observed in cumulus clouds, rarely in stratus clouds, and never in isolated cirrus clouds. Both satellite and ground observations reveal lightning activity at all latitudes between 60° N and 60° S with the most frequent occurrence at low latitudes and over land. The high occurrence rate over land is believed to be related to the more convectively unstable conditions normally present over land. In high latitudes, the lightning frequency decreases because of the reduced convection from colder surfaces and the reduced absolute humidity.

Most thunderstorms contain both water drops and ice crystals, they usually have mass contents (water + ice) greater than 3 g/m^3, and they have precipitation rates (involving particles larger than 100 μm) in excess of 20 mm/h. Although lightning has been observed most often in clouds containing both ice and water, there have been a few observations of lightning from all-water clouds (e.g., Lane-Smith, 1971). Lightning has been observed in clouds that are completely at temperatures below

0°C, but these clouds usually contain both supercooled water droplets and ice.

The complexity of the processes leading to the development of both the precipitation and electrical structure in the clouds makes it impossible to construct or validate theories of cloud electrification from simple field experiments. It is only through the complementary efforts of laboratory experimentation, field observations, and mathematical simulations that we can hope to understand the physical processes involved in thunderstorms. A recent review by Latham (1981) summarized some of the main observations of thunderstorm electrification in a coherent fashion, and we refer the interested reader to it.

Improved understanding of the major processes leading to the buildup of strong electrical fields and their mutual interaction with precipitation can lead to better forecasting of thunderstorm activity for use in aviation and protection of forested areas, to the development of methods for artificially modifying lightning activity, and even to the development of more efficient rain-enhancement operations.

As an ultimate test of the various theories of how electrical charge separates in thunderclouds, it would be necessary to design a model that simulates as accurately as possible the three-dimensional and time-dependent nature of the cloud and its environment, including the

131

microphysical development of the liquid and solid phases in the cloud and all the possible electrical processes that are operating. These requirements are not yet attaintable with our current state of knowledge and available computers. The number of processes involved and the large range of scales (from the molecular level to the dynamic scale—1000 m) cannot all be included in a single model. Therefore, attempts have been made to deal with the problem of the development of the electrical structure of clouds by emphasizing some processes and ignoring others. A few models, for instance, simulate the microphysical processes only, neglecting the macroscale dynamics altogether, whereas others have gone to the other extreme and simulate the dynamics in detail while simplifying the microphysics dramatically. To fill in the gap, some modelers have tried to deal with both the microphysics and the dynamics with sacrifices at both ends of the scale.

We will review some of these models, try to establish a common denominator from their results and conclusions, and draw attention to some unanswered points that need further work.

GENERAL REQUIREMENTS FROM ELECTRICAL MODELS OF THUNDERCLOUDS

The validity of any thunderstorm model is determined by its ability to simulate observed features. Owing to the large natural variability of the various processes in thunderclouds, it is difficult to find a "typical" storm with which all models could be compared. It is possible, however, to list some common observed features to use as general criteria for such comparisons.

The following summary by Mason (1971) of the basic thunderstorm observations still appears to be valid:

1. The average duration of precipitation and electrical activity from a single cell is about 30 min.
2. The average electric moment destroyed in a lightning flash is about 100 C km, the corresponding charge being 20 to 30 C.
3. In a large, extensive cumulonimbus, this charge is separated in a volume bounded approximately by the $-5°C$ and $-40°C$ levels and has an average radius of perhaps 2 to 3 km.
4. The negative charge resides at altitudes just above the $-5°C$ isotherm. Krehbiel *et al.* (1979) observed that the negative charge transferred by lightning originates from regions between $-10°C$ and $-17°C$, independent of the height above ground and regardless of the geographical location of the thunderstorm. The main positive charge is situated several kilometers higher. An-

other subsidiary small positive charge may also exist near cloud base, centered at or below the 0°C level.
5. The charge-separation processes are closely associated with the development of precipitation, probably in the form of soft hail (particles containing both liquid water and ice).
6. Sufficient charge must be separated to supply the first lightning flash within 12 to 20 min of the appearance of precipitation particles of radar-detectable size ($d \geq 200 \, \mu m$).

MECHANISMS OF CHARGE SEPARATION

For space-charge centers to build up in clouds, charge must be separated first in the microscale, and then larger-scale processes can act to separate the opposite charges in space. When accomplished, this dual-scale process leads to the buildup of a space-charge distribution similar to that in Figure 10.1.

In thunderclouds the charge separated on a microscale by particle interactions is subsequently separated on a macroscale with the help of convection and gravitational settling. Convection plays a role in cloud particle growth by forcing the condensation of water vapor until the particles are large enough to coalesce. Some of the interactions between cloud particles, particularly those followed by rebounding, may result in charge separation (as will be discussed later). These charges are then separated by differential terminal settling velocities. The larger particles, which carry predominantly one

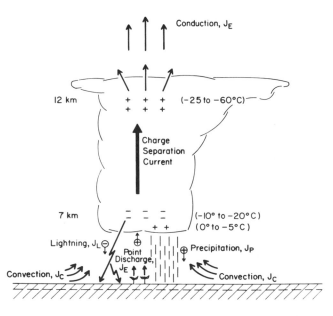

FIGURE 10.1 A schematic of the main space-charge distribution and currents in a thundercloud.

sign of charge, fall faster and farther down with respect to the convected air, leaving the oppositely charged smaller particles above.

As the charge centers build up, discharging mechanisms become more effective, and these should be included in any complete description of cloud electrification. Two kinds of discharge are possible: (1) discharge by collisions between drops and ions of opposite polarity and (2) discharge by collision and coalescence with cloud particles of opposite charge (see Colgate et al., 1977). The first mechanism depends on the electrical conductivity of the small ions ($\lambda = nek$, where n is the ion number density, e is the electronic charge, and k is the electrical mobility) and on the ion diffusivity. Often ions of different polarities have different diffusivities; those with higher diffusivity attach preferentially to cloud and free aerosol particles. However, after some charge is built up on a cloud particle, further ion diffusion to it will be limited.

Another factor affecting ion attachment to cloud particles is the electric field in the cloud. Strong fields will move free ions from one region to another. In so doing, they also increase the conduction currents and, hence, the discharge of cloud particles. In addition, when the electric field exceeds about 50 kV/m, corona discharge begins near the corners of ice crystals and high concentrations of ions are generated. These ions increase the electrical conductivity and help to prevent or slow down the further buildup of the space charge.

The other discharge mechanism (collection of oppositely charged cloud particles) takes place at all stages of particle growth. This mechanism is enhanced if the interacting particles are highly charged, have opposite polarity, and when the ambient field is strong. In this case the collection efficiency increases by increasing the collision efficiency (Coulomb attraction changes the trajectories of particles relative to each other) or by increasing coalescence efficiency (not allowing bouncing, and hence no charge separation, to occur) or by both.

Therefore, for a charge mechanism to be effective, it has to separate sufficient charge at a rate sufficiently high to overcome these discharge processes. The many charge-separation mechanisms and their complexity require a detailed discussion that is beyond the scope of this chapter. We recommend that the interested reader refer to Chapter 9 (this volume) by Beard and Ochs and to Mason (1971).

The various charging mechanisms that have been proposed as possible major contributors to electrification of thunderstorms can be divided into two major classifications: (1) precipitation mechanisms requiring particle interactions with subsequent space-charge separation by gravitational sedimentation and (2) ion attachment to cloud or precipitation particles and then charge separation by either gravitational settling or by atmospheric convection (updrafts or downdrafts). Mechanisms from group (1) above are divided into two major subprocesses—inductive and noninductive. Most models to date treat these mechanisms with various degrees of detail. On the other hand, only a few models are available that treat the ion convective process. Consequently, since the intention here is to review the present state of knowledge in modeling electrical development in clouds, most of the emphasis is placed on the models dealing with the precipitation mechanisms. As discussed later, there are still a great many questions that these models cannot answer.

Inductive Process

Charge can be separated by the inductive process during rebounding collisions of particles embedded in an electric field. This mechanism, which is relatively simple to formulate, was treated intensively in cloud electrification models. According to Sartor (1967) and Scott and Levin (1975) the amount of charge that is separated per collision by this process is

$$\Delta Q = [-4\pi\epsilon_0 E\Phi r^2\cos\theta + (\omega + 1)Q + \omega q]$$
$$[1 - \exp(-t_c/\tau)]. \quad (10.1)$$

In this equation ΔQ represents the charge transfer to the large particle as a smaller particle of radius r collides and rebounds in an electric field E (defined as positive when a positive charge is overhead), making an angle θ between the field and the line connecting the centers of the particles at the point of separation; Φ and ω are constants that depend on the ratio of sizes of the colliding particles (Ziv and Levin, 1974); Q and q represent the initial charge on the particles before the interaction; t_c is the contact time of the colliding particles and τ the relaxation time of the charge carrier ($= \epsilon\epsilon_0/K$, where ϵ and K are the dielectric constant and the electrical conductivity, respectively; and ϵ_0 is the permittivity of free space).

The first term on the right-hand side of Eq. (10.1) represents the charge that is transferred from the small to the large particle because of the inductive polarization effect. One can see that the stronger the field or the larger the size of the smaller particle, the larger is the charge separated. The constant Φ represents the enhancement of the electric field around the colliding particles, as compared with the the ambient field. Particles may collide at the head-on position but will skid or roll on each other and finally separate at the angle θ. For water drops, θ may vary between 50° and 90° (Levin and Machnes, 1977). Large liquid particles sometimes may also be separated at $\theta \geq 90°$, leading to their discharge (Al-Saed and Saunders, 1976). On the other hand, nonspherical solid or liquid particles can separate larger charges by the inductive process as a result of the much enhanced electric fields near them (Censor and Levin, 1974).

The second and third terms on the right side of Eq. (10.1) represent the limitation of charge transfer due to the initial charge on the large (Q) and small (q) particles, respectively. The constant ω then is a geometrical factor that represents the effect of the capacitance of the two on the charge transfer.

The last term on the right represents the limitation to charge transfer due to the electrical conductivity of the materials that compose the particles (Sartor, 1970; Caranti and Illingworth, 1983; Illingworth and Caranti 1984). Ice particles at low temperatures, for example, have low bulk and surface electrical conductivities that lead to longer relaxation times τ. This means that in any given collision there is the possibility that not all the available charge will be transferred, since the contact time t_c might be shorter than τ. Indeed a recent laboratory study by Illingworth and Caranti (1984) on the dependence of charge transfer during ice-ice collisions on the surface conductivity of ice, suggests that for ice-ice interactions the inductive mechanism is not efficient. Interactions of two particles can result in either collection or rebound. To describe the probability of these two end results a collision efficiency, E_1, and a coalescence efficiency, E_2, are defined. E_1 represents the probability of two cloud particles to interact, and E_2 represents the probability of the interacting particles to coalesce. Therefore, the collection probability is $E_1 \cdot E_2$, and the rebound probability is $(1 - E_2)E_1$. To separate charge an electrical contact among rebounding particles must be achieved. Only a fraction, E_3, of the particles that collide and rebound make such electrical contact. We will refer to E_3 as the electrical contact probability. Therefore, the probability for separating charge between two cloud particles is $P = E_1(1 - E_2)E_3$.

The rate of charge buildup on the large particles per unit volume as a result of collisions of particles can be expressed as

$$dQ/dt = \pi(R + r)^2(V - v)Nn(P\Delta Q - E_1E_2q),$$

$$(10.2)$$

where R and r are the radii of the large and small particles, respectively, V and v are their fall speeds, and N and n are their concentrations. The term $P\Delta Q$ represents the charge separated per interaction, while E_1E_2q accounts for the discharge of the large particles resulting from collection of oppositely charged particles (Scott and Levin, 1975).

Noninductive Process

Many noninductive mechanisms have been proposed to explain the formation of electricity in thunderstorms. Among the most powerful are the thermoelectric effect (Reynolds *et al.*, 1957; Latham and Mason, 1961), freezing potentials (Workman and Reynolds, 1948), and contact potentials (Buser and Aufdermaur, 1977; Caranti and Illingworth, 1980). All of these rely on the electrochemical nature of water or ice for charge separation.

Thermoelectric Effect   Charge separation results from interactions of ice particles of different surface temperatures. On contact the temperature gradient across the surface causes the H^+ ions to migrate from the warmer particle to the cold one, leaving OH^- ions on the warmer ice particle. Subsequent rebound of these particles will result in charge separation. The amount of charge separated in this process depends on the temperature and the temperature gradient. In most models, the value of the charge separation per interaction is taken as a constant, regardless of the temperature or temperature gradient.

Freezing Potentials   Workman and Reynolds (1948) and Pruppacher *et al.* (1968) observed that high electrical potentials develop across an ice-water interface when the water contains small amounts of impurities ($\sim 10^{-5}$ molar). These potentials develop as a result of preferential incorporation of certain ions from the solution into the ice lattice, leaving the ice and the liquid solution oppositely charged. In clouds, if such a situation occurs, fragments of the solution can be thrown off as a result of the impact of other particles. These fragments carry away charge of one sign, leaving the ice particle with the opposite charge. Gravitational settling can then separate the two charges in space.

These early works suggested that the magnitude and sign of the separated charge depend critically on the amount and type of impurity used. Various laboratory experiments conducted to simulate this charging mechanism have resulted in a surprisingly wide range of charge transfer. Most investigators (e.g., Weickmann and Aufm Kampe, 1950; Latham and Mason, 1961) measured charging rates that correspond to roughly 3×10^{-16} to 3×10^{-15} C per collision. Schewchuk and Iribarne (1971) observed about 10^{-11} C per collision for very large water drops ($R = 2.9$ mm), a value that decreased as the drop size and impact velocity decreased. On the other hand, they observed very little dependence on impurities but much stronger dependence on temperature. In most of these experiments the rebounding droplets received positive charge, leaving the target ice negatively charged.

In experiments by Latham and Warwicker (1980), these general findings were confirmed, but a maximum charge of only 10^{-14} C per collision was observed with slightly smaller drops, in conformity with most other investigations. It is also clear from these experiments that the charge separation is more a function of drop size than of impurities.

Contact Potentials Buser and Aufdermaur (1977) and more recently Caranti and Illingworth (1980) observed that a surface potential develops during riming of supercooled water droplets on ice. This surface potential increases steadily with decreasing temperature down to $-10°C$ and remains constant to $-25°C$. Caranti and Illingworth (1980) also observed that impurities, such as NH_4OH, NaCl, or HF, made no detectable difference in the surface potentials. In clouds, charge could be transferred by collisions and subsequent rebounds of small unrimed ice crystals with a surface potential near zero from the surface of a rimed crystal with negative surface potential.

The electric charge buildup by each of the noninductive processes can be expressed as in Eq. (10.1), except that ΔQ has the form

$$\Delta Q = [-A + (\omega + 1)Q$$
$$+ \omega q][1 - \exp(-t_c/\tau)], \quad (10.3)$$

where A is the charge transfer per collision resulting from one of the above mechanisms. The value of A for small cloud particles varies from 10^{-15} to about 10^{-14} C per collision, depending on the type and size of the particles and on temperature (Takahashi, 1978). Owing to the lack of comprehensive data about the charge transfer as a function of size and temperature, all available numerical models take this value as a constant.

One should keep in mind that the charge buildup by both the inductive and the noninductive processes depends on the interaction probabilities E_1, E_2, and E_3 and on the ratio t_c/τ. In the models, the values of E_1, E_2, and E_3 have to be specified. A detailed discussion of the probabilities and the way they are calculated and measured is beyond the scope of this paper. The interested reader is referred to Pruppacher and Klett (1978, Chapter 14) for details.

The collision efficiency E_1 of water drops used in the numerical models is based on calculations of particle trajectories (e.g., Davis and Sartor, 1967). A few calculations are available for collisions of ice particles with water drops and with other crystals. These are limited owing to the complex geometrical shapes of the ice crystals and their dependence on temperature.

The coalescence efficiency of water drops or ice crystals has not been theoretically evaluated and is determined by experimental measurements. For water drops, the coalescence efficiencies of Whelpdale and List (1971) and Levin and Machnes (1977) are often used. These values vary from almost zero for interactions of large drops among themselves to a value close to unity for interactions of very small drops with large ones. The experiments on interaction of ice particles with water drops did not differentiate between collision and coalescence and only measured the end result such as collection (E_1E_2) or rebound [$E_1(1 - E_2)$] (e.g., Aufdermaur and Johnson, 1972, and some other works summarized by Pruppacher and Klett, 1978). Aufdermaur and Johnson (1972) observed that rebound occurred on only about 1 percent of the impacting drops; this implied about a 99 percent collection. However, this experiment was conducted with a limited range of drop sizes and temperatures. Unfortunately, not enough information is available on this parameter.

The values of E_3 are the least known, and a large range of values is usually tested in the models.

To simplify things, some models do not use the detailed formulation of E_1, E_2, and E_3, but rather combine them into one parameter ($P = E_1(1 - E_2)E_3$. The relaxation time for charge transfer between the interacting particles, τ, depends on their electrical conductivity. This conductivity, either surface (electrons) or bulk (ions), is temperature dependent. The relaxation time of ionic charge transfer of pure ice decreases from 6.8×10^{-3} sec at $-10°C$ to 2.8×10^{-2} sec at $-19°C$ (Sartor, 1970). However, for slightly impure ice (doped with 3×10^{-6} M chloride, for example) this relaxation time will be shortened by two orders of magnitude but become more temperature dependent (Gross, 1982). The relaxation time of charge transfer by surface electrons on the other hand is believed to be about 30 times shorter than bulk ions. It is therefore the surface electrons that are probably responsible for the transfer of charge during interactions of ice particles (Gross, 1982).

The contact time t_c has been estimated to vary between 10^{-4} to 10^{-6} sec (Sartor, 1970; Caranti and Illingworth, 1980). Therefore, the ratio t_c/τ will vary with temperature by a few orders of magnitude. As the temperature decreases, the factor $[1 - \exp(-t_c/\tau)]$ in Eqs. (10.1) and (10.3) inhibits the charge transfer. For water drops, this factor is almost unity, because water has a higher conductivity than ice.

CHARGING BY ION ATTACHMENT

Attachment of ions to cloud particles can also charge them. Three kinds of mechanisms should be considered: ion diffusion, ion conduction, and ion convection. Diffusion of ions through air is a function of the temperature and the sizes of the ion. At altitudes typical of thunderstorms, negative ions have a diffusivity about 25-40 percent larger than that of positive ions. This would suggest that at the early stages of cloud development, when all other charge mechanisms are not effective, charge separation by ions would dominate.

At later stages when the strength of the electric field increases, ions can be conducted to the cloud particles because of the electrical forces (ion conduction). At the same time ions can be transported toward the particles because of the relative velocities between them. Wilson (1929) pointed out that ions, which move because of the presence of the electrical forces and the air flow, selectively interact with cloud particles moving under the action of gravity and air flow (ion convection). This selective ion current depends on the fall speed of the particle, its charge, and the magnitude and direction of the external electric field.

The attachment of ions to cloud particles reduces their concentration and the electrical conductivity. Phillips (1967) calculated the electrical conductivity existing in electrified clouds under a quasi-static situation. His calculations were based on the balance between the ion production from cosmic-ray ionization, the rate of ion loss from ion recombination, and ionic diffusion and conduction to cloud particles. Similar formulation was used by Griffithes *et al.* (1974) for calculating the electrical conductivity for three different cloud types—cumulus congestus, strato-cumulus, and fog. They concluded that a decrease in conductivity of about 3 orders of magnitude occurred under highly electrified conditions. This decrease was found to be sensitive to variations in the liquid-water content and the electrical field but only slightly affected by changes in altitude, particle charge, and the manner in which the charge is distributed over the size spectrum. When a secondary source of ion production, resulting from corona currents emitted from ice particles under the influence of a strong electric field, was introduced into the calculations, a large increase in conductivity was predicted.

The process of ion attachment to cloud particles continues until enough charge is accumulated, at which point any additional charge can be quickly neutralized by attachment of ions of opposite charge. Some charge on the cloud particles often exceeds this saturation threshold value owing to charging by other mechanisms, so that within the main charge centers of the cloud ion attachments will generally act as discharge processes.

When the cloud is electrified the conducting environment reacts. Atmospheric ions that have the same polarity as the charge center within the cloud will be re-

pelled, while those with an opposite polarity will be conducted from the surroundings toward the cloud. The ions that enter through cloud boundaries are attached to cloud particles and generate a charged screening layer. This process was first recognized by Grenet (1947) and independently by Vonnegut (1955). Brown *et al.* (1971) and Klett (1972) presented detailed calculations of the charge distribution and accumulation process in the screening layer.

Recently it has been suggested (Wahlin, 1977) that negative ions not only have higher mobility than positive ones but also have higher electrochemical affinity to surfaces and will rapidly attach to cloud particles. Therefore, negative ions that are brought, along with positive ions, into the cloud by an updraft, will preferentially attach to cloud particles near cloud base, leaving the free positive ions to be carried to cloud top. This mechanism also relies on falling precipitation particles and updraft for charge separation to occur. Without large precipitation particles falling with respect to the updraft, all the charges (negatively charged particles and positive free ions) would occupy the same volume and mask each other completely. However, this mechanism is probably too weak to produce strong fields during cloud development, since the ion concentration produced by cosmic rays below the cloud base is too low to produce extensive charge separation (Wormell, 1953).

A mechanism that also relies on atmospheric ions for the charging but does not require gravitational settling of precipitation particles for charge separation, is the convective model proposed by Vonnegut (1955). It depends on air currents to bring abundant positive ions from the ground up to cloud level. Cloud droplets that collect these ions at the cloud base carry them to the cloud top in the updrafts. The resulting region of positive charge, according to Vonnegut, will preferentially attract negative ions from the free atmosphere above to form a screening layer at the cloud top. Downdrafts, produced by the vortex circulation of the air in the cloud, which is enhanced by the negatively buoyant air created by overshooting the thermal equilibrium point and by the evaporation of cloud particles at the cloud top, will transport the negative ions down to the cloud base, forming a vertical electrical dipole.

Latham (1981) suggested that the convective mechanism plays only a minor role in the charging of thunderstorms because rates of ion production by cosmic rays are far too small to produce enough charge that can be separated and produce lightning. On the other hand, calculations by Martell (1984) suggested that the ion pair production over continental surfaces is greater than that produced by cosmic rays by more than an order of magnitude because of the decay of radioisotopes. If these calculations are confirmed experimentally, the relative contribution of the convective mechanism will have to be re-examined.

SURVEY OF THEORETICAL MODELS

Parallel-Plate Models

Parallel-plate models are the simplest models of cloud electrification. They completely ignore the contributions of the air motions and focus on the microphysics. However, even in this area they consider only a small fraction of the microphysical processes that take place. To simplify things, they assume that any charge separated in the charging volume is accumulated on two parallel plates, simulating the centers of the space charges in the cloud. Therefore, the models cannot predict the vertical structure of the charges in the cloud. The cloud is assumed to be composed of water drops alone, ice crystals with hail pellets, or a combination of them. The simplest of these models allows the particles to grow with time at preassigned rates (Mason, 1972), whereas the more detailed models allow the growth to proceed by semicontinuous (Ziv and Levin, 1974) or stochastic interactions (Scott and Levin, 1975). These models do not explicitly consider the effect of ions on the charging but assume discharge of particles (owing to attachment of ions of opposite signs) that exponentially depends on the field. All these models tested the effectiveness of the inductive process only.

Mason (1972) and Sartor (1967) assumed that charge is separated by collisions of ice crystals and hail pellets. They concluded that the inductive process is a very powerful one and is capable of separating enough charge for the field to reach a few kilovolts per centimeter in about 500-600 sec.

Scott and Levin (1975), who treated the particle growth in more detail, concluded that the inductive process could account for the first lightning of a thundercloud provided the electrical contact probability, E_3, is greater than 0.1 (see Figure 10.2). That is, of the cloud particles that do make contact and then rebound, about 10 percent need to separate charge in order for the process to be effective. For water drops, the value of the charge separation probability, which contains E_3 in it, is thought actually to be lower than 0.1, thus making this process ineffective in producing enough charge separation. For ice-ice collisions, on the other hand, this efficiency could be as high as 0.9. There is still great uncertainty as to its value for water drops colliding with ice pellets.

As mentioned before, the charge transferred per collision of ice particles should decrease with decreasing

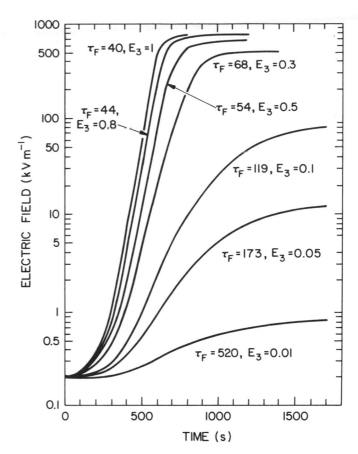

FIGURE 10.2 The growth of the electric field as a function of time under the inductive process with water drops only and calculated by the infinite cloud model of Scott and Levin (1975). The different curves represent different values of E_3, the electrical contact efficiency. The values of τ_F correspond to the time constants during the time of the maximum growth rate of the electric field.

temperature. Ziv and Levin (1974) simulated this feature for ice-ice collisions and found greatly diminished charge and field buildup.

Other important factors determining the electrical development in clouds are the relative sizes of the colliding particles and the number of concentrations of the cloud elements. The first factor affects the charge that is separated per collision, since the charge transferred increases with increasing size of the rebounding particle. The second factor affects the number of collisions and, hence, the rate of charge (and field) buildup. When intense precipitation occurs (rates > 30 mm/h) the field can develop to large values with the inductive process only. However, for smaller precipitation rates (smaller particles and lower concentrations) it takes longer than the times set by the criteria above for the field to buildup.

One-Dimensional Models

Illingworth and Latham (1975) correctly pointed out that horizontally infinite parallel-plate models overestimate the electric-field development because they lack a finite horizontal extent for the cloud. They constructed a simple one-dimensional model in which precipitation ice particles descended from the cloud top downward and interacted with smaller ice crystals (Illingworth and Latham, 1977). During these interactions, charge was separated by either inductive or noninductive processes. The linear dependence of the charge separation in the noninductive process [Eq. (10.3)], and its independence of the ambient field, caused the field to grow early in a linear fashion (see Figure 10.3). The inductive process, on the other hand, started later since it relies on the magnitude of the ambient field. Superposition of the two processes led to both a rapid linear field development in the early stages as a result of the noninductive process and a subsequent enhancement of the field owing to the inductive process. One of the important results of this

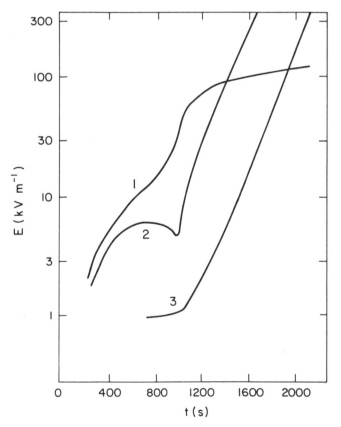

FIGURE 10.3 The variation of the maximum field E_m with time for the ice-ice noninductive charging mechanism (curve 1), the ice-ice inductive charging mechanism (curve 3), and the combined ice-ice mechanisms (curve 2). From Illingworth and Latham (1977).

simple model is its ability to predict the vertical dipole in the cloud and even the small positive pocket at cloud base.

Tzur and Levin (1981) developed a much more detailed model that included a macroscale dynamical framework in one-and-a-half dimensions (height as an independent variable and a finite cloud radius with lateral mixing) and fully interactive microphysics of the precipitation development. Electrically the model treated in great detail free ions and their attachment to cloud particles and inductive and some noninductive processes with both ice and water, all in a time-dependent frame. From the results of the model Tzur and Levin concluded that charge separation in the liquid section of the cloud is not likely to be effective since the efficiency of bouncing and charge separation by water-water interaction is probably low. Similarly, collisions between ice particles and ice pellets in the absence of water droplets, either by the inductive or thermoelectric effects, namely, near the cloud top (temperatures $< -25°C$), are not likely to contribute greatly to cloud electrification either [see Figures 10.4(a) and 10.4(b)]. This is because of the small value of t_c/τ at these temperatures (low surface and bulk electrical conductivities in ice). Also, at these altitudes the number of ice particles is relatively low, reducing the collision frequency and the charge separation.

At higher temperatures (about $-10°C$ or warmer) ice particles interact with both ice crystals and water droplets. From the model results, Tzur and Levin concluded that the collisions of the ice crystals with water drops, by a mechanism such as the Workman-Reynolds effect, are very effective charge separators [see Figure 10.4(c)] owing to the large concentration of water droplets as compared with that of ice.

Comparison of Figures 10.4(a) and 10.4(c) shows that the charging rate by the inductive process changes rapidly with time once charging starts. On the other hand, the charging rate by the noninductive mechanism is almost constant with time, in agreement with the recent measurements by Krider and Musser (1982). These measurements show that the total currents (Maxwell currents) below electrified clouds remained fairly constant with time while at the same time the electric field in the cloud increased by a few orders of magnitude.

Testing the inductive process revealed that very high fields and large charges can be produced only after 3000 sec from cloud initiation [see Figure 10.5(a)] or about 20 min after precipitation particles appeared in reasonable concentration for radar detection. As in the case of the simpler model of Illingworth and Latham (1977), the noninductive process produced linear field development. However, as opposed to Illingworth and Latham

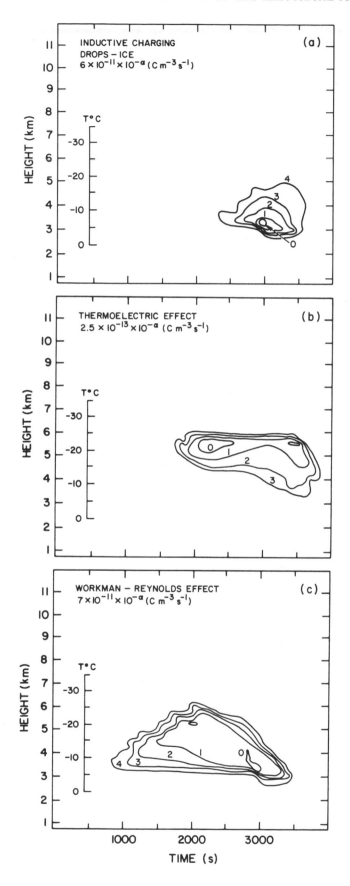

(1977), who assumed that only ice-ice collisions separate charge, Tzur and Levin (1981) assumed that charge separation by ice-ice collision is temperature dependent, and hence less effective than interactions of ice and water, which occur at warmer temperatures. Although the noninductive mechanism that they considered is the Workman-Reynolds process, any electrochemical process in which charge is separated by interaction of ice pellets and water droplets during riming is applicable to these calculations.

Ions contributed only slightly to charge buildup, either by diffusion to charged particles or by conduction. Their contribution can be pronounced, on the other hand, in the early stages of the cloud buildup, when droplets are very small, and during rain below the cloud base. This latter charging of raindrops becomes significant when the field near the ground passes the threshold for corona discharge. During this stage the charge of raindrops can be greatly modified by attaching of oppositely charged ions to them.

A closer look at the charge structure produced by the noninductive process [Figure 10.5(a)] reveals that a "classical" dipole develops with a negative charge center at about $-8°C$ and with the main positive charge center at higher altitudes (about $-18°C$) (see Figure 10.5 at $t = 2400$ sec). Large fields are already formed by 2500 sec after cloud initiation (about 10 min after precipitation particles appear), and with precipitation rates less than 20 mm/h. A positive charge pocket develops near the cloud base at temperatures warmer than $0°C$.

On the other hand, the inductive process alone delays the field buildup for about 3000 sec. It produces the negative charge center between about -10 and $-20°C$ and the positive charge center still higher up at temperatures lower than $-20°C$ [see Figure 10.5(b)]. A positive pocket extending from the $-5°C$ isotherm to the cloud base is also found. This means that the noninductive mechanism produces space-charge centers at slightly lower altitudes, at earlier times, and with lower precipitation rates than does the inductive process.

FIGURE 10.4   (a) The charging rate in coulombs per cubic meter per second by the ice-water inductive process as a function of height and time, from Tzur and Levin (1981). The value of each contour is $6 \times 10^{-11} \times 10^{-\alpha}$ with α displayed near each one. Note that the charging rate rapidly varies with time and reaches a maximum value at about 3000 sec. (b) The charging rate by the ice-ice thermoelectric (noninductive) effect. Note that the values of the contour are two orders of magnitude smaller than in (a). (c) The charging rate by the ice-water (noninductive) Workman-Reynolds effect. Note that the values of the contours are similar to those of (a). Also note that charging starts early and tends to remain fairly constant with time for most of the lifetime of electrical production.

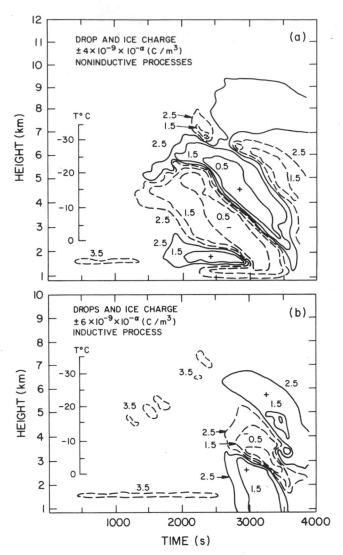

FIGURE 10.5 (a) The net charges in coulombs per cubic meter on cloud and precipitation particles (ice and water) resulting from the ice-water noninductive process, as a function of height and time, from Tzur and Levin (1981). Solid lines represent net positive charges, and dashed lines represent net negative charges. The value of each contour is given by $\pm 4 \times 10^{-9} \times 10^{-\alpha}$ with α displayed next to each one. Note that the maximum charges are produced around $t = 2500$ sec with negative charge near $-10°C$ and positive charge around $-25°C$. Another small positive charge appears just below the $0°C$ level. (b) As in (a) except for the inductive (ice-water) process. Note the delay in the development of the space charges as compared with (a).

A combination of the two processes produced strong field and space-charge distributions, which are almost a linear superposition of the two individual cases. Specifically, a strong field develops early ($t \sim 2500$ sec) but is enhanced later ($t \sim 3000$ sec). Since the inductive process begins to operate when the field is stronger, a new space-charge center (negative charge) is produced at the

cloud top. This charge center, as with all other charge centers, then descends as precipitation falls. At a particular height it seems as if the charges switch signs with time. This implies that at this stage the inductive process is so effective that charged particles falling below a certain space-charge volume are rapidly charged oppositely owing to the reversal of the field direction below the charge center. Had the effectiveness of the inductive process been reduced, the charge centers might have spread out over a greater cloud depth and would have prevented the field reversal.

One of the limitations, of course, of the one-dimensional, time-dependent models is their poor simulation of the air circulation within the cloud and the entrainment of air from the environment on the sides and top. Since in this model any mixing in of drier air, or detrainment of cloudy air, is immediately averaged over the entire layer of the cloud, it actually affects the whole cloud development in the model, as compared with nature, where relatively smaller effects are produced by mixing at cloud edges only. For a better simulation of these effects, two- or three-dimensional models are needed.

Two-Dimensional Models

Two-dimensional models have been developed to improve the simulation of the macroscale dynamics and its effect on the charge distribution and electrical development.

Chiu (1978) simulated a vortex-type thunderstorm in a two-dimensional, time-dependent axisymmetric model. In his model, only water drops were considered, and charge was allowed to develop via the inductive process and ion attachment. Cloud microphysics was not dealt with in detail, and cloud water was converted to precipitation particles, of a preassigned distribution, by a parameterized formulation. In each time step in the model, the number of possible particle interactions was calculated based on known collision efficiencies. From it a net charge separation was derived. Simultaneously, ion attachment by diffusion and conduction was permitted to take place, and the total net charge at each Δt was found. Chiu's results also indicate that the inductive process could be a very effective charge separation mechanism, provided that large precipitation rates are present. The results also indicate the development of a vertical dipole of a proper "classical" polarity with an additional small positive space charge near the cloud base. As in the one-and-a-half-dimensional model of Tzur and Levin (1981) large charges and strong fields developed only after rain formed. The evolution of the horizontal electric field, with a maximum at 30 min, is

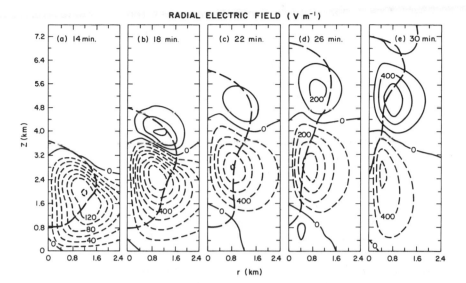

RADIAL ELECTRIC FIELD (V m⁻¹)

FIGURE 10.6 Evaluation of the radial electric field, from Chiu (1978). (a) At 14 min. Contour interval is 2 V/m, and the range is − 14 to 0 V/m. (b) At 18 min. Contour interval is 10 V/m, and the range is − 70 to 30 V/m. (c) At 22 min. Contour interval is 200 V/m, and the range is − 1000 to 200 V/m. (d) At 26 min. Contour interval is 10^3 V/m, and the range is $(-4-2) \times 10^3$ V/m. (e) At 30 min. Contour interval 2×10^3 V/m, and the range is − 6 to 6×10^3 V/m.

shown in Figure 10.6. The effect of ions, either by diffusion in the early stages or by conduction at the later stages, was found to be relatively small and did not significantly alter the charging of the cloud. The entrainment of ions from cloud sides and tops did not greatly modify the electrical development.

Heldson (1980) used the same model for a two-dimensional slab cloud and simulated the effect of artificial chaff seeding for the prevention of lightning. The introduction of chaff into the cloud creates centers for ion production by corona discharge as the electric-field strength approaches that needed for lightning. The results of the model suggest that the presence of excessive ions at this stage increases the cloud electrical conductivity and enhances the discharge of the cloud particles. This in turn prevents the further buildup of the electric field and charges. The results of this model demonstrate one practical use for modeling of electrical processes in thunderclouds.

Thunderstorms usually contain both water and ice. The models of both Chiu and Heldson are therefore limited since no ice formation was simulated even though the clouds in their models reached heights where ice is usually found.

Kuettner *et al*. (1981) developed another two-dimensional model. Their model superposes a kinematic flow model, including cloud particle growth, on an electrical charge separation model. The cloud model uses either vortex or shear flow to simulate a steady-state flow configuration. Precipitation ice particles were introduced about 1 km above the cloud base and allowed to be moved with the airflow. During their ascent and descent they grew by collecting cloud droplets and separated charge through rebounding collisions of either wa-

ter droplets or small ice crystals. The model did not consider ion attachment or particle growth by condensation. Particle growth by collection was calculated with a constant probability of charge separation. The model also did not address the problem of entrainment or turbulent diffusion. However, the merit of this model is its relative simplicity and the capability of testing the electrical development under different airflow conditions such as those observed in the field. The results of this model point out that the noninductive process, incorporating ice-water and to some extent ice-ice interactions with an average value of observed charge separation per collision, can produce an electrical dipole at realistic altitudes but cannot enhance the field to a value comparable with the breakdown value.

On the other hand, the inductive process, involving charge separation by ice-water interactions, produced very high fields but generated a very complex space-charge configuration. The complex field and space-charge structure arises as a consequence of the high efficiency with which the inductive process operates when large precipitation particles appear. As these large particles descend through a space-charge center and become exposed to an electric field of opposite direction, their charge polarity reverses in response to the reversal of the electric field.

Combination of the inductive with the noninductive mechanisms produced both a proper charge distribution and a rapid growth of the field. The results of Kuettner *et al*. (1981) suggest that charge separation processes involving ice-ice collisions are not very powerful, being limited by both the long relaxation time of the charge carriers and by the relatively low concentration of ice crystals (resulting in low collision rates). In addition, in agreement with the other two-dimensional models, this model demonstrates that both strong horizontal and vertical fields can be produced by charge separation mechanisms that depend on precipitation. The horizontal fields are generated by horizontal displacement of the charged particles by the air circulation. Even under very weak shear conditions the space-charge centers were found to be displaced horizontally and produce very strong horizontal fields even close to the cloud base. The presence of the shear was found to smooth the development of the charge centers by limiting mixing of precipitation particles of opposite charges.

Takahashi (1979) developed a two-dimensional, time-dependent model of a small warm cloud, which treats the microphysics and the macroscale dynamics in detail. Electrification due to the inductive mechanism and to ion attachment by diffusion and conduction is considered in a way that seems to explain the weak electrification of warm maritime clouds. The most important mechanism responsible for charging in such clouds, according to this model, is the attachment of ions to cloud and precipitation drops. This attachment is significantly enhanced during condensation and evaporation. During the former, positive ions are preferentially incorporated into the growing drops, whereas during evaporation negative ions are preferentially attached.

In an attempt to evaluate the effectiveness of the convection electrification process, Ruhnke (1972) and Chiu and Klett (1976) developed two-dimensional, axisymmetric steady-state models. Ruhnke calculated the electric fields and charges that arise from ion attachment to cloud water in solenoidal flow, intended to represent the flow in an isolated convective cloud. The actual cloud volume (where liquid water exists) was assumed to be spherical and to be entirely within the updraft. Space charge arises owing to local differences in electrical conductivity. These differences stem from attachment of ions to cloud particles (assumed to depend only on liquid-water content), which form ion currents consisting of both conduction and convection currents. By assigning a specific liquid-water content and a relation between it and ion conductivity, Ruhnke avoided dealing with interactions between ions and cloud droplets. The steady-state assumption precludes any detail of the initial development of the convective electrification. His results show that only very small fields can be developed by this process.

Chiu and Klett (1976) improved on this model by using a more realistic convective circulation in which the updraft was within the cloud and the downdraft was at its edges. They also considered the effect of turbulent diffusion in addition to conduction and convection currents on the transport of ions. Attachment of ions to cloud drops was affected by the liquid-water content and by the ambient electric field. Chiu and Klett's results show that convective electrification by itself cannot explain the strong electrification in thunderclouds. One should bear in mind that the terms that are highly variable with time such as the rate of charge buildup, especially at the later stages of thunderstorm development, are ignored in these steady-state models. In a fully developed time-dependent model, such terms may modify the above conclusions. In addition, the dynamics used in the convective models is parameterized and may not be realistic enough to simulate the real convective charging process that is highly dependent on cloud dynamics.

Three-Dimensional Models

To date, only one three-dimensional, time-dependent model of an electrical cloud has been developed (Rawlins, 1982). This model uses pressure as the vertical coordinate with grid spacings of 50 mbar vertically and 1 km horizontally. The microphysical parameterization of Kessler (1969) is used to describe the growth of cloud particles into precipitation size. Ice, initiated by ice nuclei that freeze the supercooled water drops, is represented by three size classes: 0-100 μm, 100-200 μm, and 200-300 μm in radius. Hail is designated as ice greater than 300 μm in radius, and it is forced to be distributed exponentially in size (Marshall and Palmer, 1948).

With this model Rawlins tested the effectiveness of various electrical processes, such as the inductive and the contact surface potential (noninductive) mechanisms. He assumed that ion attachment to cloud particles can be ignored altogether. He concluded that the noninductive process is able to produce fields of high enough intensity to initiate lightning within about 20 min after precipitation begins. This process produced a "classical" dipole [see Figure 10.7(a)] but without the small positive charge center closer to the cloud base.

The inductive process, involving ice-ice collisions, was capable of producing strong fields only in the presence of high concentrations of ice particles and only 30 min after precipitation particles appeared. With this process, a very complex space-charge structure emerged [see Figure 10.7(b)] as was also found by Kuettner *et al.* (1981). Allowing ice crystals to rebound more than once from ice pellets reduced the calculated maximum field to below that needed for lightning [compare the values of E_z in Figures 10.7(a) and 10.7(b)].

However, one should note that the same restriction of multiple collisions was not applied to Rawlins's calculations of the noninductive process. Multiple collisions, if

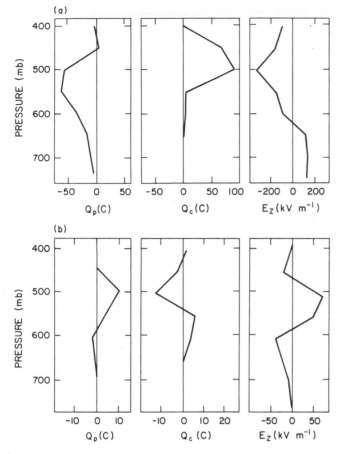

FIGURE 10.7 The vertical distribution of space charges on precipitation, Q_p, and cloud, Q_c, particles and the vertical electric field, from Rawlins (1982). (a) Ice-ice noninductive charging mechanisms after 36 min of cloud growth with charge separation per collision of $Q = 10\,fC$. (b) Ice-ice inductive process after 44 min of cloud growth. Note the simple dipole structure and the intense field in (a) as compared with the more complex structure and weaker field in (b), implying the ineffectiveness of the inductive process under the assumptions of this model.

allowed for, will restrict charge transfer during particle collisions regardless of the process considered.

DISCUSSION

From this survey, it is clear that present models can describe both the electrical development and the growth of precipitation in some detail. An interesting common conclusion of all the models is the insensitivity of the results to small changes in free ion concentration or conductivity. These parameters do become important during the early stages of cloud development and below the cloud base during rain. They are probably also important just at the onset of lightning or immediately after-

ward, but none of the models described here has dealt with this complex problem.

The emergence of the more complex models of two or three dimensions provides a clear visualization of the ability of the precipitation charging mechanisms to produce strong horizontal displacement of charges. These are often found in clouds and frequently lead to horizontal lightning strokes. It was shown that such numerical models could also be used to test the feasibility of preventing lightning by limiting the electric field growth. Multidimensional models can also greatly aid interpretation of the results of field experiments such as chaff dispersal in real clouds because they incorporate more realistic air circulations than do one-dimensional models.

One of the main purposes of all the models discussed here is to test the various proposed mechanisms of charge separation in clouds. It seems that now that the models are capable of simulating the main features of the electrical charge separation in the cloud in a framework that combines air circulation and precipitation growth, however, reliable values for some of the various parameters are desparately needed. In particular, the electrical contact probabilities of the various particles (primarily ice with water and ice with ice), the coalescence probabilities, the relaxation time of the charge carriers on ice as a function of temperature, and the length of time the particles actually make contact before rebounding are all essential, and not yet known, for evaluating the effectiveness of the various mechanisms. Such parameters can only be obtained by careful laboratory experiments.

Despite the uncertainties in the values of the main parameters involved in the precipitation processes of cloud electrification, it is still impressive to see that virtually all of the models appearing within the past 10 years, regardless of their complexity, agree that precipitation mechanisms can explain the main features observed in thunderclouds. They explain the presence of the space-charge centers at the proper altitudes and temperatures. They show that strong fields can be developed within 20 to 30 min of the appearance of precipitation in the cloud. Some show that noninductive charge separation processes [either ice-ice (Rawlins, 1982) or ice-water (Tzur and Levin, 1981)] can produce very strong fields with low precipitation rates as is sometimes observed in nature (Gaskell et al., 1978). In addition results with noninductive processes show that the electric field grows linearly with time, as observed by Winn and Byerly (1975). These results also agree with the recent measurements of Krider and Musser (1982), which suggest that the charging rates in thunderclouds are independent of the field and fairly constant with time [Figure 10.4(c)]. However, the observations of Williams and Lhermitte (1983) pointed out that the Musser and Krider results can also be explained by the convective charge transport. Their observations showed that falling precipitation may not be the only cause for the electrification of thunderstorms. All the models agree that the inductive process requires higher precipitation rates in order to operate effectively. Some models show that the most effective method to produce strong fields is to let both inductive and noninductive mechanisms operate simultaneously. While noninductive mechanisms can be powerful, particularly early in the development of the electric field, it is difficult to see how one can ignore the inductive process altogether. This process should operate in general whenever an ambient electric field is present. In some cases, it may discharge the particles, while in others it will charge them, but it should always operate. If, on the other hand, its effectiveness is very low, as reported by Illingworth and Caranti (1984), it will not be felt in the cloud. Thus, if a charge greater than that predicted by Eq. (10.1) is found on some of the particles (Christian *et al.*, 1980), the inductive process should have discharged them. Since such charges were observed, it must be concluded that in these cases the inductive process did not effectively operate.

Most investigators seem to feel that charge separation through interactions among water drops only is not effective since most collisions result in coalescence, thus limiting the possibilities for charge separation. Nevertheless, laboratory experiments (Levin and Machnes, 1977; Beard *et al.*, 1979) suggest that the coalescence efficiency is far from being understood, so the role of water-drop interactions should not yet be ignored completely.

Laboratory measurements of the surface potentials of ice under various growth conditions (Buser and Aufdermaur, 1977; Caranti and Illingworth, 1980) reveal the complexity of the charge-transfer problem. Again, additional experiments are needed to resolve the dependence of charge separation by this process on temperature and on the strength of an external electric field.

In spite of the fact that the numerical models thus far rule out convective electrification as an effective mechanism for producing strong fields by itself, it must be emphasized that these models are only quasi-static and contain parameterized dynamics. To simulate this mechanism effectively, more detailed cloud dynamics, ion convection and conduction, and precipitation processes must be included. Thus far, no such model has been developed. Such a detailed model is urgently needed, especially following the recent experiments by Vonnegut *et al.* (1984) that reversed the polarity of a

thundercloud by emitting negative ions from a long cable electrified to 100 kV and suspended below the cloud. Their observations suggest that the negative ions penetrated the cloud, ascended to the cloud top, and attracted positive ions from the free atmosphere above and were carried down by the air currents to the cloud base—thus reversing the previous polarity of the cloud. If the ion concentration was too small to produce this effect, it is still possible that the additional ions changed the initial conditions of the cloud electrification, which led to the reversal in the cloud polarity.

With the newly available data and faster computers we can look forward to a new generation of models incorporating cloud microphysics and dynamics together with the convection and precipitation electrification mechanisms.

REFERENCES

Al-Saed, S. M., and C. P. R. Saunders (1976). Electric charge transfer between colliding water drops, *J. Geophys. Res. 81*, 2650-2654.

Aufdermaur, A. N., and D. A. Johnson (1972). Charge separation due to riming in an electric field, *Q. J. R. Meteorol. Soc. 98*, 369-382.

Beard, K. V., H. T. Ochs III, and T. S. Tung (1979). A measurement of the efficiency for collection between cloud drops, *J. Atmos. Sci. 36*, 2479-2483.

Brown, K. A., P. R. Krehbiel, C. B. Moore, and G. N. Sargent (1971). Electrical screening layers around charged clouds, *J. Geophys. Res. 76*, 2825-2835.

Buser, O., and A. N. Aufdermaur (1977). Electrification by collisions of ice particles on ice or metal targets, in *Electrical Processes in Atmospheres*, N. Dolezalek and R. Reiter, eds., Steinkopff, Darmstadt, p. 294.

Caranti, J. M., and A. J. Illingworth (1980). Surface potentials of ice and thunderstorm charge separation, *Nature 284*, 44-46.

Caranti, J. M., and A. J. Illingworth (1983). Frequency dependence of the surface conductivity of ice, *J. Phys. Chem. 87*, 4078-4083.

Censor, D., and Z. Levin (1974). Electrostatic interaction of axisymmetric liquid and solid aerosols, *Atmos. Environ. 8*, 905-914.

Chiu, C. S. (1978). Numerical study of cloud electrification in an axisymmetric liquid and solid aerosols, *J. Geophys. Res. 83*, 5025-5049.

Chiu, C. S., and J. N. Klett (1976). Convective electrification of clouds, *J. Geophys. Res. 81*, 1111-1124.

Christian, H., C. R. Holmes, J. W. Bullock, W. Gaskell, J. Illingworth, and J. Latham (1980). Airborne and ground-based studies of thunderstorms in the vicinity of Langmuir Laboratory, *Q. J. R. Meteorol. Soc. 106*, 159-174.

Colgate, S. A., Z. Levin, and A. G. Petschek (1977). Interpretation of thunderstorm charging by polarization induction mechanisms, *J. Atmos. Sci. 34*, 1433-1443.

Davis, M. H., and J. D. Sartor (1967). Theoretical collision efficiencies for small cloud droplets in Stokes flow, *Nature 215*, 1371-1372.

Gaskell, W., A. J. Illingworth, J. Latham, and C. B. Moore (1978). Airborne studies of electric fields and the charge and size of precipitation elements in thunderstorms, *Q. J. R. Meteorol. Soc. 104*, 447-460.

Grenet, G. (1947). Essai d'explication de la charge électrique des nuages d'orages, *Ann. Geophys. 3*, 306-310.

Griffithes, R. F., J. Latham, and V. Myers (1974). The ionic conductivity of electrified clouds, *Q. J. R. Meteorol. Soc. 100*, 181-190.

Gross, G. W. (1982). Role of relaxation and contact times in charge separation during collision of precipitation particles with ice targets, *J. Geophys. Res. 87*, 7170-7178.

Heldson, J. H., Jr. (1980). Chaff seeding effects in a dynamical-electrical cloud model, *J. Appl. Meteorol. 19*, 1101-1183.

Illingworth, A. J., and C. M. Caranti (1984). Ice conductivity restraints on the inductive theory of thunderstorm electrification, in *Conference Proceedings, VII International Conference on Atmospheric Electricity*, American Meteorological Society, Boston, Mass., pp. 196-201.

Illingworth, A. J. and J. Latham (1975). Calculations of electric field growth within a cloud of finite dimensions, *J. Atmos. Sci. 32*, 2206-2209.

Illingworth, A. J., and J. Latham (1977). Calculations of electric field growth, field structure and charge distributions in thunderstorms, *Q. J. R. Meteorol. Soc. 103*, 231-295.

Kessler, E. (1969). *On the Distribution and Continuity of Water Substance in Atmospheric Circulation*, Meteor. Monogr. Vol. 10, No. 32, American Meteorological Soc., Boston, Mass., 84 pp.

Klett, J. D. (1972). Charge screening layers around electrified clouds, *J. Geophys. Res. 77*, 3187-3195.

Krehbiel, P. R., M. Brook, and R. A. McCrory (1979). An analysis of the charge structure of lightning discharges to ground, *J. Geophys. Res. 84*, 2432-2456.

Krider, E. P., and J. A. Musser (1982). Maxwell currents under thunderstorms, *J. Geophys. Res. 87*, 11171-11176.

Kuettner, J., Z. Levin, and J. D. Sartor (1981). Inductive or noninductive thunderstorms electrification, *J. Atmos. Sci. 38*, 2470-2484.

Lane-Smith, D. R. (1971). A warm thunderstorm, *Q. J. R. Meteorol. Soc. 97*, 577-578.

Latham, J. (1981). The electrification of thunderstorms, *Q. J. R. Meteorol. Soc. 107*, 277-298.

Latham, J., and B. J. Mason (1961). Generation of electric charge associated with the formation of soft hail in thunderclouds, *Proc. R. Soc. London A260*, 537-549.

Latham, J., and R. Warwicker (1980). Charge transfer accompanying the splashing of supercooled raindrops on hailstones, *Q. J. R. Meteorol. Soc. 106*, 559-568.

Levin, Z., and B. Machnes (1977). Experimental evaluation of the coalescence efficiencies of colliding water drops, *Pure Appl. Geophys. 115*, 845-867.

Marshall, J. S., and W. M. K. Palmer (1948). The distribution of raindrops with size, *J. Meteorol. 5*, 165-166.

Martell, E. A. (1984). Ion pair production in convective storms by radon and its radioactive decay products, in *Conference Proceedings, VII International Conference on Atmospheric Electricity*, American Meteorological Society, Boston, Mass., pp. 67-71.

Mason, B. J. (1971). *The Physics of Clouds*, Oxford Univ. Press, Cambridge, 671 pp.

Mason, B. J. (1972). The physics of thunderstorms, *Proc. R. Soc. London A327*, 433-466.

Phillips, B. B. (1967). Ionic equilibrium and the electrical conductivity in thunderclouds, *Mon. Weather Rev. 95*, 854-862.

Pruppacher, H. R., and J. D. Klett (1978). *Microphysics of Clouds and Precipitation*, Reidel, Dordrecht, 714 pp.

Pruppacher, H. R., E. H. Steinberger, and T. L. Want (1968). On the electrical effects that accompany the spontaneous growth of ice in supercooled aqueous solutions, *J. Geophys. Res. 73*, 571-584.

Rawlins, F. (1982). A numerical study of thunderstorm electrification using a three dimensional model incorporating the ice phase, *Q. J. R. Meteorol. Soc. 108*, 778-880.

Reynolds, S. E., M. Brook, and M. F. Gourley (1957). Thunderstorm charge separation, *J. Meteorol. 14*, 426-436.

Ruhnke, L. H. (1972). Atmospheric electron cloud modeling, *Meteorol. Res. 25*, 38-41.

Sartor, J. D. (1967). The role of particle interactions in the distribution of electricity in thunderstorms, *J. Atmos. Sci. 24*, 601-615.

Sartor, J. D. (1970). *General Thunderstorm Electrification*, National Center for Atmospheric Research, Boulder, Colo., p. 99.

Schewchuk, S. R., and J. V. Iribarne (1971). Charge separation during splashing of large drops on ice, *Q. J. R. Meteorol. Soc. 97*, 272-282.

Scott, W. D., and Z. Levin (1975). Stochastic electrical model of an infinite cloud charge generation and precipitation development, *J. Atmos. Sci. 32*, 1814-1828.

Takahashi, T. (1978). Riming electrification as a charge generation mechanism in thunderstorms, *J. Atmos. Sci. 35*, 1536-1548.

Takahashi, T. (1979). Warm cloud electricity in a shallow axisymmetric cloud model, *J. Atmos. Sci. 31*, 2236-2258.

Tzur, I., and Z. Levin (1981). Ions and precipitation charging in warm and cold clouds as simulated in one dimensional time-dependent models, *J. Atmos. Sci. 38*, 2444-2461.

Vonnegut, B. (1955). Possible mechanism for the formation of thunderstorms electricity, in *Proceedings International Conference Atmospheric Electricity*, Geophys. Res. Paper No. 42, Air Force Cambridge Research Center, Bedford, Mass., p. 169.

Vonnegut, B., C. B. Moore, T. Rolan, J. Cobb, D. N. Holden, S. McWilliams, and G. Cadwell (1984). Inverted electrification in thunderclouds growing over a source of negative charge, *EOS 65*, 839.

Wahlin, L. (1977). Electrochemical charge separation in clouds, in *Electrical Processes in Atmospheres*, H. Dolezalek, and R. Reiter, eds., Steinkopff, Darmstadt, p. 384.

Weickmann, H. K., and J. J. Aufm Kampe (1950). Preliminary experimental results concerning charge generation in thunderstorms concurrent with the formation of hailstones, *J. Meteorol. 7*, 404-405.

Whelpdale, D. M., and R. List (1971). The coalescence process in rain drop growth, *J. Geophys. Res. 76*, 2836-2856.

Williams, E. R., and R. M. Lhermitte (1983). Radar tests of the precipitation hypothesis for thunderstorm electrification, *J. Geophys. Res. 88*, 10984-10992.

Wilson, C. T. R. (1929). Some thundercloud problems, *J. Franklin Inst. 208*, 1-12.

Winn, W. P., and L. G. Byerly III (1975). Electric field growth in thunderclouds, *Q. J. R. Meteorol. Soc. 101*, 979-994.

Workman, E. J., and S. E. Reynolds (1948). Suggested mechanism for the generation of thunderstorm electricity, *Phys. Rev. 74*, 709.

Wormell, T. W. (1953). Atmospheric Electricity: Some recent trends and problems, *Q. J. R. Meteorol. Soc. 79*, 3.

Ziv, A., and Z. Levin (1974). Thundercloud electrification cloud growth and electrical development, *J. Atmos. Sci. 31*, 1652-1661.

III
GLOBAL AND REGIONAL ELECTRICAL PROCESSES

Atmospheric Electricity in the Planetary Boundary Layer

11

WILLIAM A. HOPPEL, R. V. ANDERSON, and JOHN C. WILLETT
Naval Research Laboratory

INTRODUCTION

The planetary boundary layer (PBL) is that region of the lower atmosphere in which the influences of the Earth's surface are directly felt. The primary influences of the surface are drag, heating (or cooling), and evaporation (or condensation). These processes cause vertical fluxes of momentum, sensible heat, and moisture, which penetrate into the lower atmosphere to a finite height. These fluxes, in turn, generate turbulence, ultimately controlling the mean profiles of wind speed, temperature, and water vapor in the PBL. Since the height of penetration depends on the direction, magnitude, and persistence of the surface fluxes and on the large-scale meteorological conditions, the PBL can range in thickness from tens of meters to a few kilometers.

Because of its position next to the Earth's surface, the PBL has been the site of the vast majority of atmospheric-electrical measurements. If the history of atmospheric electricity begins with Franklin, Lemonnier, and Coulomb, as is customary, there have been more than two centuries of effort in this region. It was observed early that measured quantities such as electric field respond strongly to meteorological processes. This led Lord Kelvin to suggest that the day would come when weather forecasting would be done with an electrometer (Dolezalek, 1978). Unfortunately, this optimistic prediction has not materialized, largely because of the complexity of the dependence of electrical variables on a bewildering variety of other phenomena.

Most atmospheric processes are interrelated and cannot be studied in isolation, but it is usually possible to identify one or two dominant influences. In the case of atmospheric electricity in the PBL, however, separating the various causes and their effects can be extremely difficult. In fact, this field may be unique with respect to its sensitivity to many disparate phenomena spanning a tremendous range of scales in both space and time. For example, locally produced turbulent fluctuations in space-charge density have an effect roughly comparable in magnitude to that of changes in global thunderstorm activity on electric-field variations within the PBL.

Over the years this responsiveness of atmospheric electricity has led to its exploitation for many different purposes. Electrical measurements have been made in the PBL to observe large-scale processes such as the global circuit, to study local phenomena like boundary-layer turbulence, or simply to examine unusual electrical signatures in their own right. In each type of investigation it has been found necessary to minimize the effects of unwanted processes on the data. This filtering is never entirely successful, however, and investigators must always be aware of the whole range of influences and alert for contamination.

While atmospheric-electrical variables respond to many processes, they usually have little influence on the phenomena to which they respond. Thus the electrical state of the PBL is irrelevant to the fields of environmental radioactivity, air pollution, boundary-layer turbulence, and global meteorology, for example. The inverse is not true, however. Atmospheric electricity in the PBL is a truly interdisciplinary study requiring a knowledge of all these areas in addition to ionic conduction in gases, aerosol physics, and electrostatics.

Recent advances in the disciplines cited above make it a propitious time to re-examine the relationships between atmospheric electricity and PBL processes. Such a re-examination should ultimately lead to a better understanding not only of atmospheric-electrical phenomena but of the related disciplines as well. With this in mind, the remainder of this chapter presents a brief overview of atmospheric electricity in the PBL. First the primary physical mechanisms influencing the electrical phenomena are discussed. The gross phenomenology is described next, including spatial and temporal variability of the important electrical parameters. Then the most important aspects of modeling and theory are summarized in an effort to relate the physical causes and their electrical effects. Finally, the chapter is closed with a discussion of principal applications and areas of needed research.

PHYSICAL MECHANISMS THAT AFFECT ATMOSPHERIC ELECTRICITY IN THE PLANETARY BOUNDARY LAYER

Ionization

The electrical conductivity of the air is due to ions produced primarily by ionizing radiation. The early investigations of the sources of atmospheric ions led to the discovery of cosmic rays. Cosmic radiation is the primary source of ions over the oceans and above a couple of kilometers over land. In the PBL the cosmic-ray contribution to the ionization rate is about 1 to 2 ion pairs per cubic centimeter per second. It is quite constant in time, and the latitudinal dependence caused by the Earth's magnetic field is well understood.

The primary source of ions in the PBL over land is natural radioactivity originating from the ground. This ionization source can be divided into two parts: (i) αs, βs, and γs radiating directly from the Earth's surface and (ii) radiation from radioactive gases and their radioactive daughter products exhaled from the ground. The

gases originate in both the uranium and the thorium decay series where one of the daughters is the noble gas, radon. In the uranium decay series, the daughter is ^{222}Rn with a half-life of 3.8 days; and in the thorium series, ^{220}Rn (thoron) with a half-life of 54 sec. During the radon part of the decay the radioactivity can diffuse from the ground into the atmosphere and contribute to the volume ionization. The amount of radon that escapes depends on the amounts of ^{226}Ra and ^{232}Th in the ground; the type of ground cover; and porosity, dampness, and temperature of the soil. The height distribution in the atmosphere depends on atmospheric mixing in the boundary layer and the half-life of the isotope.

It is obvious that radiation directly from the ground will vary greatly depending on the geographical variations in ground radioactivity. Ground radiation intensity also decreases with height; α ionization is confined to the first few centimeters, β to the first few meters, and γ to the first few hundred meters. The amount of ionization in the first few centimeters resulting from αs is largely unknown. Values of ionization due to βs in the first few meters are typically in the range of 0.1 to 10, and those due to γs in the lowest hundred meters are in the range of 1 to 6 ion pairs $cm^{-3} sec^{-1}$.

Ionization due to radioactive gases in the air is even more variable and depends not only on the amount exhaled from ground but also on atmospheric dispersion. Direct measurements of ionization due to radioactive gases in the atmosphere are difficult and have not generally been satisfactory. Estimation of the ionization rate is therefore based on measurements of radioactive products in the air. The height distribution of Rn in the atmosphere as a function of turbulent diffusion has been the subject of a number of investigations and is often used to determine the turbulent diffusion coefficient. On cool nights with nocturnal temperature inversions the radioactive gases can be trapped in a concentrated layer close to the ground, whereas during unstable convective periods, the gases can be dispersed over an altitude of several kilometers. Ionization at 1 to 2 m due to radioactive gases and their short-lived daughter products is typically in the range of 1 to 20 ion pairs $cm^{-3} sec^{-1}$ and is predominantly caused by α particles.

Figure 11.1 illustrates the vertical variation of ionization in the PBL. Ionization due to cosmic rays is nearly constant in the first kilometers. The ionization from ground β and γ radiation will vary geographically depending on the abundance of radioactivity in the local soil. The curves shown are typical, with β radiation being predominant below 1 m and the effect of γ radiation extending to several hundred meters. The curves labeled Q_{max} and Q_{min} represent the sum of ionization due to cosmic rays, γs and βs from the ground, and the decay of

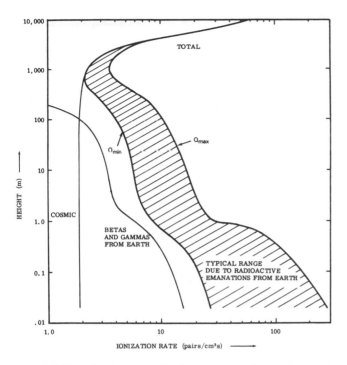

FIGURE 11.1   Ionization profiles showing range of values that might be expected over land. Q_{max} would represent high values; Q_{min} low values.

^{220}Rn and ^{222}Rn gases and their daughter products in the atmosphere. The area between the two curves illustrates the variability expected depending on atmospheric and soil conditions. The high ionization rate below 1 m indicated by Q_{max} is due to an accumulation of ^{220}Rn in the lowest meter and occurs only under strong surface-temperature inversions and low winds. During convective (daytime) periods with significant wind shear the ionization profile below 10 m would be expected to follow Q_{min} values closely. The data used to construct Figure 11.1 were taken largely from Ikebe (1970), Ikebe and Shimo (1972), Crozier and Biles (1966), and Moses *et al.* (1960). There is a limited amount of data on the vertical distribution of ^{220}Rn (thoron), and therefore Q_{max} is the expected maximum ionization based on measurements at only three sites. Geographical areas with high ground radioactivity may exhibit even larger ionization rates than shown in Figure 11.1.

Ionization within the first few centimeters of the Earth's surface due to α particles has never been adequately investigated. There is the possibility that α-emitting daughter products of radon that attach to aerosol particles are deposited on the Earth's surface, enhancing surface activity. This activity is usually neglected in the study of environmental radiation because it represents a small fraction of the total environmental

activity. Yet this source of ionization is likely to be important in determining the ion concentration at the surface. Surface ion concentration is an important boundary condition influencing the atmospheric-electrical structure in the interior of the PBL.

Our intention here is not to review the field of environmental radioactivity but rather to point out that over land the ionization in the PBL depends primarily on ground radiation and radioactive gases and to emphasize the necessity of studying the geographical variation of ground radioactivity and the dynamics of the dispersion of radioactive gases if we are to understand atmospheric electricity in the PBL.

Over the oceans far from land the ionization rate is determined solely by cosmic rays. The 3.8-day half-life of ^{222}Rn permits it to advect over oceans for hundreds of kilometers before its ionization is negligible compared with that of the cosmic-ray background.

Compared with the ionization due to natural sources, ionization from nuclear power plants and weapons is negligible on a global scale. This was true even during the active period of nuclear weapons testing in the 1950s and 1960s (Israël, 1973). There can, of course, be locally significant effects (Huzita, 1969).

In addition to ionizing radiation, electrical discharges can also form ions in the atmosphere. This requires high electric fields that generally occur only in disturbed weather near thunderstorms and in regions of blowing dust or snow. The field is greatly augmented at points on electrically grounded, elevated objects such as vegetation and antennas. As the electric field increases, the field in very small regions near such points reachs breakdown values and a small ionic current is discharged into the atmosphere. A large number of unipolar ions is injected locally into the atmosphere, and the ionic space charge thus formed tends to reduce the high electric field.

Ions can also be produced by the bursting of water films. In nature this occurs in waterfalls, falling rain, and breaking waves. Ions generated by this mechanism are not formed in pairs, and a net charge is introduced into the atmosphere. In most cases the residue remaining after evaporation of the water is much larger than a small ion and is more appropriately identified as a charged aerosol particle.

Properties of Ions

The radioactive ionization process separates an electron from a molecule of nitrogen or oxygen. The electron attaches rapidly to a neutral molecule to form a negative ion. During the next few milliseconds both positive and negative ions undergo a series of chemical, charge-exchange, and clustering reactions with the molecular species present in the air. Much progress has recently been achieved in understanding this chain of ionic reactions in well-defined laboratory gases (Huertas *et al.*, 1978) and in the upper atmosphere (Arnold and Ferguson, 1983). In the troposphere, where trace gases are numerous and variable, the ion chemistry is complicated, and the composition of the terminal ion must be regarded as uncertain at best. If mass spectroscopic measurements can be extended to identify ambient ions, it is possible that this identification can be used as a measure of certain elusive trace gases in the troposphere.

Fortunately, the ion nature enters the equations that govern atmospheric electricity in the PBL only as it affects the ionic mobility and recombination coefficient. The mobility is defined as the mean drift velocity of the ion per unit electric field. The mobilities of ambient ions are more easily determined than their mass or chemical composition. (There is no unique relationship between mobility and mass, and a rather large uncertainty in mass translates into a much smaller uncertainty in mobility.) Values for aged positive and negative ion mobilities at STP are about 1.15 and 1.25 cm^2 V^{-1} sec^{-1}, respectively (Mohnen, 1977), are inversely proportional to air density, and are independent of electric-field strength.

If ions are formed in pairs by ionizing radiation and annihilated in pairs by ion-ion recombination, then the positive and negative ion concentrations are equal and, in the absence of particulates, are given by

$$n_{\pm} = (q/\alpha)^{1/2}, \qquad (11.1)$$

where q is the ionization rate and is the recombination coefficient. If there are no aerosol particles in the atmosphere the ion lifetime is given by

$$\tau = n_{\pm}/q = (1/q\alpha)^{1/2}. \qquad (11.2)$$

Recombination of ions in the troposphere is a three-body process. The present theoretical treatment of three-body recombination is inadequate for calculation of absolute values of the recombination coefficient. This is due in part to our inability to identify the ion chemistry. Measurements of the recombination coefficient yield a value of about 1.4×10^{-6} cm^3/sec for aged atmospheric ions at STP (Nolan, 1943).

The movement of ions in an electric field gives rise to the electrical conductivity of the atmosphere. The conductivity is defined as

$$\lambda = e(n_+ k_+ + n_- k_-), \qquad (11.3)$$

where $k_{\pm}$ is the mobility and e is the elementary charge.

The conductivity is non-ohmic (field dependent) if the ion densities depend on electric field, as is the case near a boundary (electrode effect).

Equations (11.1)-(11.3) predict concentrations of about 3000 ions/cm^3, lifetimes of about 5 min, and a conductivity of about 1×10^{-13} mho/m for an ionization rate of 10 ion pair cm^{-3} sec^{-1}. In reality the values of these variables are considerably smaller because ions are destroyed not only by recombination but also by attachment to aerosol particles, as described in the next section.

Attachment of Ions to Aerosol Particles

Suspended in the atmosphere are many small particles mostly in the size range between 0.01 to 0.5 μm radius. The concentration of aerosol particles varies from a few hundred per cubic centimeter over remote regions of the oceans to a hundred thousand per cubic centimeter in polluted urban environments. Ions diffuse to these particles and, on contact, transfer their charge to them; thus the particles act as centers of recombination. In most continental areas the loss of ions by attachment to aerosol particles is greater than the loss by ion-ion recombination. The attachment process also establishes a size-dependent statistical charge distribution on the aerosol particles. (Some are negatively, some positively, and some neutrally charged.)

To treat the problem of ion-aerosol attachment in its totality requires the solution of a system of balance equations for ions and particles with various numbers of elementary charges. Since we are interested here only in the effect of aerosols on ion depletion, we can write a simplified ion balance equation as

$$\frac{dn}{dt} = q - \alpha n^2 - \beta nZ, \qquad (11.4)$$

where Z is the total particle concentration irrespective of the charge state of the individual particles and β is the effective ion-aerosol attachment coefficient. For steady-state conditions and $q = 10$ ion pairs cm^{-3} sec^{-1}, the dependence of ion density on the particle concentration is shown in Figure 11.2 for an effective attachment coefficient of $\beta = 2 \times 10^{-6}$ cm^3 sec^{-1}. It is readily seen that when the particle concentration is greater than about 1000 cm^{-3}, the concentration of ions is more dependent on the aerosol attachment than on ion-ion recombination. Only in remote oceanic and Arctic regions can the effect of particles on the ion density sometimes be ignored.

Atmospheric aerosols are hygroscopic, and their equilibrium radius increases as the relative humidity in-

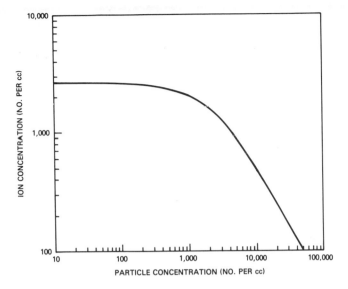

FIGURE 11.2 Ion depletion as a function of increasing aerosol concentration for an ionization rate of 10 ion pairs $cm^{-3} sec^{-1}$ and a typical value of ion-aerosol attachment coefficient.

creases. This growth is especially pronounced at relative humidities above about 90 percent. An increase in the size of aerosol particles makes them more efficient scavengers of ions, resulting in a lower conductivity. When the relative humidity exceeds 100 percent, some of the particles are activated and grow rapidly (by about 2 orders of magnitude) to radii larger than 1 μm, forming fog or cloud droplets. These droplets are effective scavengers of ions and are responsible for the low conductivities found in fogs and clouds. The atmospheric-electric fog effect (see section below on Phenomenology of Atmospheric Electricity in the Planetary Boundary Layer), where conductivity decreases and the electric field increases before the formation of fog, is probably related to aerosol growth with increasing humidity.

Recent advances in aerosol measurements have resulted in accurate aerosol size distributions that could now be used to evaluate more thoroughly the ion-aerosol equilibrium. An accurate treatment would include not only a realistic aerosol size distribution but also the statistics of diffusional charging of particles by ions.

Effect of the Global Circuit

Electric fields and currents in the PBL arise primarily from the voltage impressed across it by the global circuit discussed in Chapter 15 (this volume). The weak conductivity of the PBL causes it to act as a resistive element in the global circuit, conducting a fair-weather current density of about 1 to 3 pA/m^2 to ground where the fair-

weather electric field is about 100-200 V/m. Because the PBL is the region of greatest electrical resistance, it largely controls the discharge rate of the global circuit.

If the conductivity of the atmosphere were uniform, it would be a passive ohmic medium with no accumulation of space charge to alter the electric field. Space charge is generated internally in the unperturbed atmosphere in two ways: (1) by conduction down a conductivity gradient and (2) by the imbalance of ion flow near a boundary. The first mechanism can be understood in terms of Ohm's law, $J = \lambda E$, where J is the conduction-current density. For steady-state conditions and in the absence of any convective charge transport, J is uniform, and, therefore, the electric field is inversely proportional to the conductivity. Since conductivity changes with altitude, there must be an inverse altitude dependence of the electric field. Poisson's equation requires this change in field to be accompanied by a space charge given by

$$\rho = \epsilon_0 \nabla \cdot E = \frac{\epsilon_0}{\lambda} \nabla \cdot J - \frac{\epsilon_0}{\lambda} E \cdot \nabla \lambda, \quad (11.5)$$

where ϵ_0 is the dielectric permeability of air. By the steady-state assumption, the first term on the right-hand side vanishes and the space charge is proportional to the electric field and conductivity gradient and inversely proportional to the conductivity. This process can be thought of as a pileup of space charge due to conduction down a conductivity gradient.

The second mechanism for producing space charge operates only near a boundary. Across any horizontal area in the atmosphere stressed by a vertical electric field, positive and negative ions will flow in opposite directions. However, at a boundary, ions of one sign can flow to that boundary, but there will be no compensating flow of the opposite sign away from it. This imbalance in ion flow gives rise to a space charge in the vicinity of the boundary. This second mechanism for generating space charge is aptly referred to as the electrode effect. For the case of uniform volume ionization in laminar airflow with no aerosols and bounded on the bottom by a conducting surface, the ion-balance equations can be solved together with Poisson's equation to give the solution shown in Figure 11.3. The ionization rate was taken to be 10 ion pairs $cm^{-3} sec^{-1}$, and the solution illustrates that the effect of the electrode, in this idealized case, would extend to about 3 m (Hoppel, 1967). In the turbulent atmosphere, the electrode effect extends to much higher altitudes and the space charge formed by these two mechanisms is dispersed by turbulent mixing, causing a convective flux of charge in addition to the conduction current.

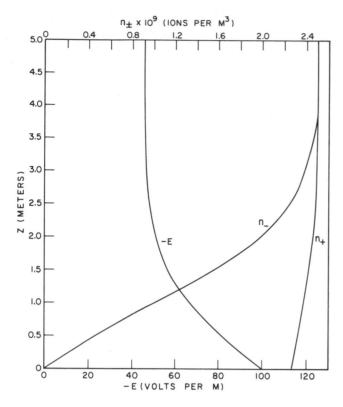

FIGURE 11.3 Simple electrode effect in nonturbulent air with constant volume ionization rate of 10 ion pair cm^{-3} sec^{-1} over a plain surface (from Hoppel, 1967).

Turbulent Transport of Electrical Properties

Most aspects of the structure of the PBL are dominated by the effects of turbulence (Haugen, 1973; Wyngaard, 1980), and electrical processes are no exception. Turbulent mixing prevents the buildup of radioactive emanations in shallow layers near the ground except under very stable conditions, disperses aerosols over a greater depth increasing the columnar resistance, and redistributes space charge, producing convection currents. The almost continual state of turbulent motion in the atmosphere is caused by the combined influences of drag, heating, and evaporation from the underlying surface. It is only in cases of extremely low wind speed and strong surface cooling that laminar flow may be found, and even then only for short periods and over limited areas.

Drag generates turbulence through shear instability, which transfers kinetic energy from the mean flow to the turbulence. The energy goes into eddies on the scale of the mean velocity gradient, which is strongest near the surface, and tends to produce turbulence with a lo-

cal integral scale comparable to the height above the surface. Both heating and evaporation from the surface tend to produce an unstable density gradient. The structure of the turbulence produced by buoyant instability can be quite different from that of shear-generated turbulence because warm, moist parcels starting near the surface accelerate as they rise through an unstable environment. The resulting eddies tend to be elongated vertically and to have a size scale determined by the thickness of the entire PBL.

The turbulent kinetic energy injected into the flow by these two mechanisms is not dissipated at the scales where it is produced. Instead, energy cascades down to ever smaller eddy sizes and is dissipated primarily at scales smaller than a centimeter and results in a wide range of eddy sizes. These eddies efficiently mix passive contaminants like ions, aerosols, space charge, and radioactive gases. The turbulent transport of space charge is equivalent to an electrical convection current through the atmosphere. It is evident from Figure 11.3 that turbulent diffusion would disperse the space charge resulting from the imbalance of positive and negative ions in the lower layers, producing a net upward flow of charge (opposing the downward conduction current).

Electric field fluctuations are also caused by turbulent movement of this space charge. To specify the average electric field, the instantaneous field must therefore be averaged over an interval of time longer than the period of the largest eddy. Figure 11.4 shows an electric-field profile over the ocean in the tradewinds off the coast of Barbados and illustrates the increase in vertical extent caused by turbulence. Each circle represents an average of 50 to 300 measurements of 10-sec duration, and the bars are the standard deviations. The large deviations illustrate the variability in the instantaneous field and the necessity of averaging to obtain a meaningful profile in the PBL. The variations are greatest where the gradient is steepest and are much less at a height of 160 m above the surface. The solid and dotted lines are the result of numerical modeling where χ is a parameter related to the strength of turbulent mixing. It is obvious from this discussion that any treatment of atmospheric electricity in the PBL must include the effects of turbulence.

This variability with time and position at low altitudes demonstrates the danger inherent in balloon soundings used to obtain integrated ionospheric potential. Neither spatial nor temporal averaging is possible during the rapid ascent and only a coarse vertical resolution is available where fields are largest and most variable.

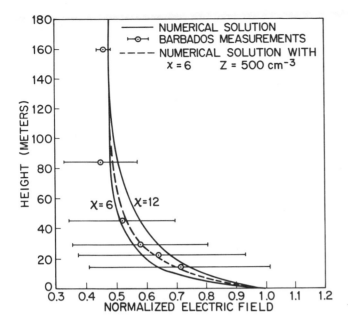

FIGURE 11.4 Variation of the electric field with height (electrode effect) found over the tropical ocean illustrating the observed large fluctuations of the instantaneous field from the average field caused by atmospheric turbulence. The curves are numerical solutions for the governing equations using gradient diffusion model for ion transport (from Hoppel and Gathman, 1972).

PHENOMENOLOGY OF ATMOSPHERIC ELECTRICITY IN THE PLANETARY BOUNDARY LAYER

Although typical average values are often cited for atmospheric-electrical observables, the greatest interest and significance by far has been attached to variations with time over a broad range of time scales. There have been studies of atmospheric-electrical variations with annual and even 11-yr periodicities; short-period fluctuations have also received some attention in the past, and they are currently a subject of renewed interest, as will be discussed later. By far the greatest effort has been focused on diurnal variations with respect to either universal or local time and, to a lesser extent, on seasonal changes in diurnal patterns.

One of the first, and certainly the most famous, demonstrations of a reproducible variation pattern with universal time is the hourly average potential gradient curves obtained over the world ocean on the *Carnegie* cruises. These curves have long served as *de facto* standards with which to evaluate the viability of attempts to measure universal variations. The result of the 1928-1929 *Carnegie* cruise (Torreson *et al.*, 1946) is plotted in Figure 11.5 along with several of the more successful subsequent attempts to measure this average variation.

These are integrated (and extrapolated) ionospheric potential from 125 balloon ascents (Mühleisen, 1977), 20 aircraft profiles in the Bahamas (Markson, 1977), and 6 aircraft profiles at a selected Arctic location (Clark, 1958). Also shown are quasi-continuous measurements of current density made above the PBL during the Arctic winter, which, when allowance is made for the small (6 percent) measured change in columnar resistance during the measurement period, should accurately mirror variations in ionospheric potential (Anderson, 1967).

Consideration of these curves strongly indicates that there is indeed a well-defined average global diurnal variation but that there are equally important real local deviations with time. This is demonstrated in Figure 11.6 (Israël, 1973) in which the effect of turbulent processes on warmer (March to October) afternoons is seen to dominate the average diurnal curve. For this reason a better knowledge of ionospheric potential variations would play a key role in the identification of effects attributable to PBL processes. However, subtraction of

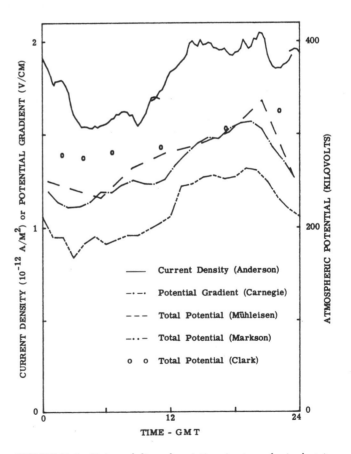

FIGURE 11.5 Universal diurnal variations in atmospheric electricity from different measurement techniques.

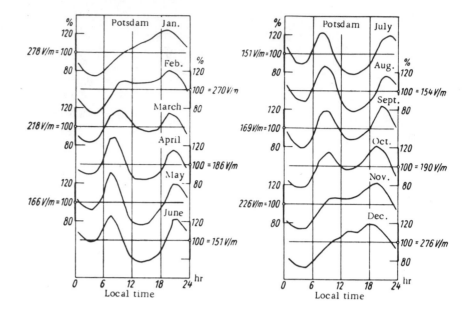

FIGURE 11.6 Average diurnal variations in potential gradient showing effect of afternoon turbulence mixing (from Israël, 1958).

the average diurnal variation does not necessarily isolate a particular local phenomenon.

Electric field and vertical current density are driven by the global circuit and, as shown in Figure 11.5, can exhibit the characteristic variation pattern even within the PBL in certain cases if suitably averaged. There are other observables such as the ionic conductivity, which are only weakly influenced by global processes while depending strongly on local effects. Conductivity has been seen to be a result of ionization, ionic mobility, recombination, and attachment of ions to particulates, all of

which can be influenced by local conditions and variable trace constituents. It is reasonable therefore to expect that large variations in conductivity will be primarily determined within the PBL and have a marked dependence on local time. The columnar resistance is a parameter usually defined as the resistance of a vertical column of unit cross section from ground level to the base of the ionosphere. It is commonly derived from a measured vertical profile of atmospheric conductivity, and it is observed that most of this resistance lies within the PBL. The validity of this is seen in Figure 11.7,

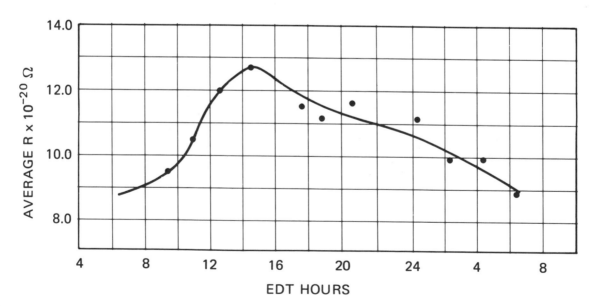

FIGURE 11.7 Average diurnal variation in columnar resistance as a function of local time (from Sagalyn and Faucher, 1956).

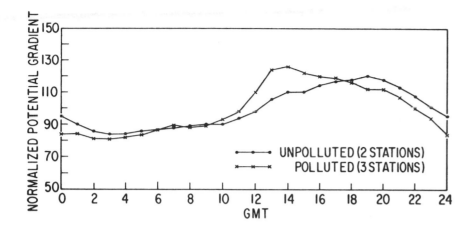

FIGURE 11.8 Normalized potential gradient showing rush-hour effect at polluted urban sites (from Anderson and Trent, 1969).

which shows the resistance R of a 1-cm^2 column from the surface to 4.57 km over land as a function of local time. The strong effect of the midday upward dispersion of aerosols on R is readily apparent, and the shape of the curve and the 40 percent variation are in reasonable accord with our knowledge of the daily variation of turbulent activity.

The strong dependence of the columnar resistance on PBL conditions provides a mechanism for the modification of electrical observables with local time. Such mechanisms can embrace the entire PBL as does fully developed turbulent convection, or they can be confined to a shallow region near the surface. One example of a shallow effect is seen in the observed response to the typical urban morning rush hour (Anderson and Trent, 1969). In the morning there is still little turbulent activity, since solar heating of the surface has just begun, and there is an abrupt injection of combustion products into the stratified atmosphere. Conductivity is reduced by particulates, but this reduction is confined to a shallow layer; so the total columnar resistance and thus the vertical current are largely unaffected. Consequently, the local surface field increases as required by Ohm's law. This increase is seen in Figure 11.8 between 1100 and 1600 GMT (0600 and 1100 EST) at three sites located in urban areas. Because of these local diurnal variations, single-station potential gradient recordings, even when heavily averaged, rarely exhibit a classical *Carnegie*-type diurnal variation pattern (Israël, 1961).

These diurnal variations with both universal and local time are not the only fluctuations observed in atmospheric-electricity recordings. Shorter-period variations are always observed and usually dismissed as noise. Observed fluctuations on atmospheric-electricity recordings made within the PBL are comparable in magnitude to the mean values of those recordings (Takagi and Toriyama, 1978). A typical recording is seen in Figure 11.9. Much of the fluctuation content in the range from roughly 0.1 sec to tens of minutes is produced by local turbulence.

There are two obvious consequences of this coupling. First, all the methodology of turbulence analysis, such as eddy-correlation, profile, and dissipation analyses, can be applied to the electrical observations. The second consequence is the possibility of utilizing electrical observations as a tool in the study of turbulent processes. The relationship between atmospheric electricity and turbulence will be considered in more detail in the section below on Modeling and Theory. There is also a suggestion that fluctuations in the total Maxwell current density in the range of 10 to 1000 sec can correlate at intercontinental distances (Ruhnke *et al.*, 1983). It is clear that short-period fluctuations in atmospheric-electrical recordings strongly influence attempts to observe global scale phenomena, are relatively underexploited, and constitute a fertile area for possible applications.

Although horizontal gradients of atmospheric-electrical variables within the PBL are much smaller than vertical ones (largely a consequence of the geometric scales involved), significant horizontal variability is observed. Significant instances include the effect of organ-

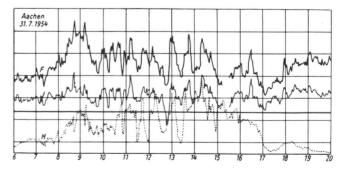

FIGURE 11.9 Records of the potential gradient (F), air-earth current density (I), and brightness (H) at Aachen on July 31, 1954 (from Israël, 1958).

ized convection activity on a large scale as seen in Figure 11.10 (Markson, 1975), terrain effects produced by mountains and coastlines, and the sunrise effect caused by differences in turbulent mixing between the heated and dark regions. To a first approximation such effects are seen to be the result of imposition of a local perturbation on an otherwise uniform situation and are, hence, essentially comparable with local phenomena such as the rush-hour effect previously described.

The spatial variation on which attention has been focused is in the vertical dimension. The interest in global-scale phenomena has led to the use of a vertical profile as a convenient observational unit. Measurements are made of one or more atmospheric-electrical variables, typically field, conductivity, and/or current density, at a variety of altitudes in a relatively short time span (from a few to tens of minutes). The sensors are carried aloft with aircraft, balloons, or rockets, and data are presented both as profiles and as numerically integrated totals. Profiles have been made over land because of convenience and to study specific terrain effects and over water in attempts to eliminate land effects. Profile data have been responsible for the detection of convection currents in the PBL comparable in magnitude with the total current, the classical electrode effect over water under stable conditions, the response of columnar resistance to pollutant buildup, and the classical diurnal variation in ionospheric potential.

Typical vertical profiles of atmospheric potential through the PBL are seen in Figure 11.11. The Greenland profiles are characterized by extremely low levels of particulate contamination, and the vertical variation of conductivity closely approximates that predicted theoretically for an aerosol-free atmosphere. The addition of particulate burdens, whether in a shallow layer as in curve C or in a thick layer as in D, markedly affects the observations within the PBL. It is apparent that, in the presence of atmospheric contaminants, the voltage drop across the PBL is significantly greater than in the Greenland observations.

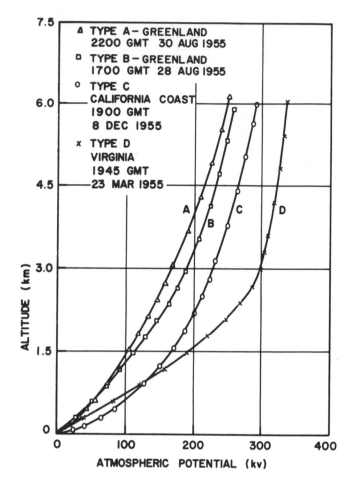

FIGURE 11.11   Typical vertical distributions of atmospheric potential (from Clark, 1958).

The conduction-current density, defined as the product of electric field and conductivity, can easily be computed from airborne measurements. Two typical profiles of conduction-current density are shown in Figure 11.12. Above the PBL the current density is seen to be essentially constant with altitude. This is a direct result of the small space-charge density above the PBL and the greatly reduced turbulent mixing found there. This vertical constancy led to the aforementioned use of current-density measurements above the PBL to follow universal variations. The increases seen at low altitudes were the first unambiguous evidence of the existence and significant magnitude of convective charge transport within the PBL, as discussed in the next section.

In addition to variations that depend on time or height are variations that are associated with a specific phenomenon. The most well known such case is the atmospheric-electric fog effect. It has been observed that the conductivity decreases markedly in fog and that the start of the decrease may precede the actual fog onset.

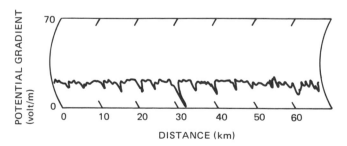

FIGURE 11.10   Horizontal variations in potential gradient showing effect of organized convection (from Markson, 1975).

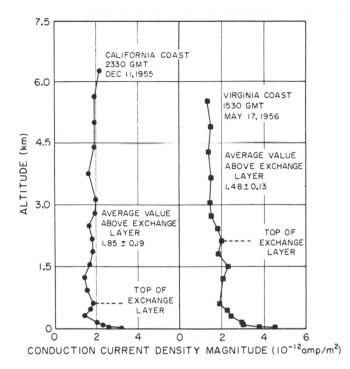

FIGURE 11.12 Two profiles of convection current density computed from aircraft measurements of field and conductivity (from Kraakevik, 1958).

Analogously, the increase in conductivity at the termination of the fog event may also precede actual dissipation. A typical data recording through a fog event is seen in Figure 11.13. This phenomenon has been reported by many observers (Dolezalek, 1963), but it has not as yet received an adequate physical analysis.

Similarly, there are effects associated with surf and waterfalls wherein some charge separation is produced

by mechanical breakup of the water surfaces (Blanchard, 1963). For example, the positive space charge produced by breaking surf can lead to an appreciably larger electric field on shore than outside the surf zone during onshore winds. In contrast to the positive charge produced along ocean coasts, sufficient negative charge has been observed along the shore of a freshwater body (Lake Superior) during heavy surf to reverse the fair-weather electric field (Gathman and Hoppel, 1970). Negative charge is also observed from waterfalls. In some cases there are also strong local effects associated with smoke plumes and with volcanic eruptions. Again there is a separation of charge, which then diffuses away from the source.

MODELING AND THEORY

The complicated dependencies of the local electrical variables on the ionization profile, aerosol concentrations, turbulent structure of the PBL, and temporal variations in the global electrical circuit make it dangerous to trust intuitive notions when interpreting measurements made in the PBL. Increased insight into the meaning of the observations is obtained by modeling various physical mechanisms mathematically. Some of the more important results of these theoretical efforts follow.

The electric field tends to be nearly vertical in fair weather, and the meteorological structure of the PBL usually changes slowly in comparison to the electrical relaxation time. This has led naturally to the assumption of a quasi-steady, horizontally homogeneous mean state and to one-dimensional, time-independent models of the mean electrical structure. These assumptions imply that the conduction- and convection-current densities

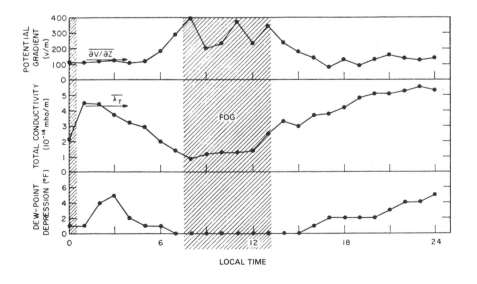

FIGURE 11.13 Atmospheric-electric fog effect—a typical example of successful forecasts of both onset and dissipation (from Serbu and Trent, 1958).

are vertically directed and that their sum, the total current density, is height independent, leading to great simplifications from a modeling point of view. About this mean state, of course, are fluctuations caused by turbulent eddies. (A parallel may be drawn here to the prevalence of one-dimensional models of the mean meteorological structure of the PBL, in which small-scale phenomena such as turbulence enter only in terms of their horizontally averaged effects.)

The primary goal of electrical modeling of the boundary layer to date has been to understand the mean profiles of the electrical variables resulting from currents driven by the global circuit. Thus, theoretical developments have focused on the sources of charge within the PBL and the phenomena produced by the vertical turbulent transport of that charge. After an introduction to the modeling of turbulent mixing, we shall examine the electrode layer and the convection currents resulting from electrode-effect space charge.

Turbulence Modeling

The simplest and most pervasive mathematical description of turbulent mixing is the gradient-diffusion model. Applied to the vertical transport of space-charge density by vertical velocity fluctuations w', it states that the mean turbulent flux (or convection current) is proportional to the local mean gradient:

$$\overline{w'\rho'} = -K(z)\frac{\partial \overline{\rho}}{\partial z}. \qquad (11.6)$$

Here $K(z)$ is the eddy-diffusion coefficient, usually taken to be proportional to height above the surface. Values at 1 m generally lie between 0.01 and 1.0 m²/sec.

This model is a form of first-order closure of the conservation equation for space charge. It is generally believed to provide an acceptable description of mixing near the surface, where shear production predominates over buoyant production of turbulent kinetic energy. Gradient diffusion is often applied throughout the PBL, sometimes with a stability-dependent coefficient and a decrease at the top to represent an inversion, although it is known to provide poor results in the interior of an unstable mixed layer. In atmospheric electricity its use is best restricted to modeling of the electrode effect.

There are many more sophisticated (and more complex) approaches to turbulent transport modeling. A popular one is second-order closure, in which conservation equations are derived for evaluating the second moments, such as $\overline{w'\rho'}$ and $\overline{\rho'\theta'_v}$ (θ_v is the virtual potential temperature), in terms of each other and of mean variables like $\partial\overline{\rho}/\partial z$. Such models allow the mean charge flux

at a given height to depend on the dynamics of the entire PBL, not just on local properties. This capability is essential for the correct modeling of unstable mixed layers, where local flux-gradient relationships are known to break down.

Before leaving this discussion of turbulent transport modeling, some cautionary mention should be made of secondary flows. There are many circumstances when the largest scale of motion in the boundary layer is organized into a nonrandom structure. The most familiar example is the stationary roll vortices, which sometimes form over the tropical ocean. Such organized motions can carry a large fraction of the vertical transports; and since they cannot be adequately represented by turbulence models of the types discussed above, these transports must be described explicitly with a dynamical model of two or more dimensions.

The Electrode Effect

The electrode effect has already been defined, and its simplest manifestation has been described for laminar flow and uniform ionization in aerosol-free air (Figure 11.3). In this case the electrode layer has a thickness of only a few meters, over which the electric-field magnitude decreases by about a factor of 2. The importance of the phenomenon lies in its ability to separate substantial amounts of charge near the Earth's surface. In conditions of low turbulence this leads to high space-charge densities in shallow layers, which can produce high electrical noise levels as intermittent eddies move this charge around. In strongly unstable conditions, on the other hand, it provides a source of charge to be carried deep into the interior of the PBL by convection currents.

The theory of the nonturbulent electrode effect is fully developed and has been verified over water and, at least qualitatively, over land. The simplest case, the so-called "classical" electrode effect, is illustrated schematically in the first frame of Figure 11.14. The thickness of the layer is determined by the lifetime of the small ions and their drift velocity in the ambient electric field. Since aerosol particles act as recombination sites for small ions, they reduce the ion lifetimes and, hence, the thickness of the nonturbulent electrode layer, as illustrated in the second frame of Figure 11.14. A shallow layer of enhanced ionization, which can arise from surface radioactivity or trapping of radioactive emanations from the soil, can cause a reversed electrode effect, as illustrated in the third frame of the figure. Here a stratum of negative space charge is developed owing to the sweeping of negative ions upward out of the highly ionized layer, causing the electric-field magnitude to in-

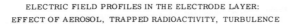

ELECTRIC FIELD PROFILES IN THE ELECTRODE LAYER:
EFFECT OF AEROSOL, TRAPPED RADIOACTIVITY, TURBULENCE

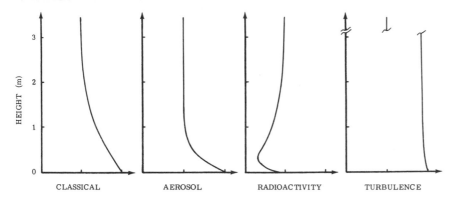

FIGURE 11.14 Heuristic profiles of electric field in the electrode layer illustrate the effects of various physical processes. From left to right are the "classical" electrode effect in clean, nonturbulent air with uniform ionization, as in Figure 11.3; the effect of aerosol attachment on the nonturbulent case; the effect of a shallow layer of high ionization rate on the nonturbulent case; and the effect of strong turbulence.

crease with height. Unfortunately, these relatively simple cases rarely occur in the real atmosphere because of pervasive turbulent mixing.

The most obvious effect of turbulence on the electrode layer is to increase its thickness by mixing the space charge upward. The impact of nonuniform ionization is reduced as the turbulence intensity increases, both because trapping of radioactive emanations is eliminated and because the thicker layer appears to be less sensitive to surface radioactivity. The presence of aerosol particles thickens the layer further, in contrast to their effect in the nonturbulent case, by increasing the electrical relaxation time. These processes may increase the height scale of the electrode layer so much as to make it virtually undetectable with surface-based measurements, as illustrated in the fourth frame of Figure 11.14. For this reason turbulence blurs the distinction between the electrode effect proper and convective currents in the interior of the PBL. Turbulence can also cause significant loss of ions and space charge by diffusion to the surface.

The theory of the turbulent electrode effect is not so fully developed as that of the nonturbulent case, owing primarily to the difficulty of parameterizing the lower boundary conditions at an aerodynamically rough surface (Willett, 1983). Hoppel and Gathman (1972) obtained reasonable agreement between experimental observations and a numerical model of the turbulent electrode layer in clean maritime air over the tropical ocean (see Figure 11.4). At present, however, there is no satisfactorily verified model that applies over land. In view of the importance of the electrode effect as a charge source for convection currents throughout the PBL, the development and testing of such a model should be a high priority for future research.

Convection Currents in the Planetary Boundary Layer

The downward conduction-current density is often observed to vary with altitude in the PBL. Based on the reasonable assumption of a steady, horizontally homogeneous, mean charge-density distribution, Kraakevik (1958) concluded that these deviations from vertical uniformity imply the existence of a height-dependent "convection-current density" such that the total current density is constant with altitude. He speculated that this convection current is produced by the upward turbulent transport of space charge produced near the surface.

Convection currents can be modeled relatively easily in many circumstances using only the mean charge-conservation equation

$$\frac{\partial \bar{\rho}}{\partial t} = -\frac{\bar{\lambda}}{\epsilon_0}\bar{\rho} - \bar{E}\frac{\partial \bar{\lambda}}{\partial z} - \frac{\partial}{\partial z}(\overline{\lambda' E'})$$
$$- \frac{\partial}{\partial z}(\overline{w'\rho'}) \quad (11.7)$$

and Poisson's equation. This assumes that the mean conductivity profile is not influenced by the electric field, which appears to be the case under conditions of strong turbulent mixing. The first term on the right-hand side of Eq. (11.7) represents local electrical relaxation due to the mean conductivity. The effect of mean conduction down the conductivity gradient causes "piling up" of space charge and is represented by the second term. The third term is usually negligible compared with one of the first two. The convergence of convection current is, of course, represented by the final term.

Recent modeling of convection currents has shown that they only become important in unstable mixed layers, where the turbulent transport time across the entire PBL can be comparable with the electrical relaxation time. The controlling meteorological variables are the surface fluxes of momentum and buoyancy and the mean conductivity and thickness of the mixed layer. It appears that the convection of electrode-effect space charge can have a major impact on the electrical structure of the boundary layer. It can even reduce the magnitude of the total downward current density locally on the order of 50 percent owing to a mechanically generated electromotive force (EMF) of more than 100 kV in extreme cases. Although these theoretical predictions are not inconsistent with existing data, they remain to be tested thoroughly in the field.

Effect of the Planetary Boundary Layer on the Fair-Weather Electrical Circuit

Two boundary-layer processes can have a substantial impact on the fields and currents appearing throughout the entire atmospheric column from the Earth to the ionosphere. These are variations in the columnar resistance and convection currents. To appreciate the importance of these effects, consider a steady convection-current density $J_c(z)$ below an inversion at height H when a steady ionospheric potential V_∞ is applied from above. Since the total current density J_t must be independent of height, it is easy to show that

$$V_\infty = -J_t R_\infty + \int_0^H \frac{J_c}{\lambda}\, dz, \qquad (11.8)$$

where R_∞ is the total columnar resistance. The second term on the right may be considered an EMF generated by boundary-layer convection. This steady-state analysis is valid as long as λ, J_c, and V_∞ change slowly compared to the electrical relaxation time near the ground.

If we assume V_∞ to be constant, Eq. (11.8) shows that the magnitude of J_t is inversely proportional to R_∞ and decreases linearly as the boundary-layer EMF increases. An aerosol-related increase in columnar resistance of 40 percent can therefore produce a similar decrease in the total current density. A simultaneous 100-kV increase in the PBL EMF can cause a further 30 percent decrease, for a total reduction in J_t of 52 percent. This makes J_t alone a relatively poor indicator of global processes.

NEEDED RESEARCH AND POTENTIAL APPLICATIONS

Measurement of Global-Scale Phenomena

As discussed in detail in other chapters of this volume, there are ample reasons for interest in global-scale atmo-spheric-electrical phenomena. For example, valuable information about the distribution and temporal variability of horizontal potential differences in the ionosphere could be provided by monitoring the ionospheric potential simultaneously in different locations. Furthermore, the widely accepted relationship between global thunderstorm activity and ionospheric potential has yet to be verified on any but the crudest statistical basis. From the present perspective, finally, a detailed knowledge of the forcing from the global circuit would be useful in evaluating the electrical response of the PBL.

Unfortunately, the measurement of global-circuit parameters is complicated by the action of boundary-layer processes. Although local PBL structure cannot appreciably affect the total current in the global circuit, or even the local ionospheric potential, it can cause a redistribution of that current and alter the vertical profile of electric field. Therefore, the proper interpretation of local measurements in terms of global parameters requires a thorough understanding of noise sources in the PBL.

The electrostatic potential of the upper atmosphere with respect to Earth is the single parameter most indicative of the electrical state of the global circuit. Yet the temporal variability of this ionospheric potential is largely unknown outside of its average diurnal variation. Methods of measuring the ionospheric potential (such as aircraft and balloon soundings) have for the most part systematically excluded the detection of any shorter-term variations. Fluctuations in electric-field and current-density measurements in the PBL with periods shorter than a few hours are usually attributed entirely to local sources, primarily turbulence and pollution.

One way to separate local and global sources is to correlate measurements at widely separated stations or to make instantaneous measurements averaged over large horizontal areas. A preliminary attempt (Ruhnke *et al.*, 1983) to detect global variations with periods of seconds to minutes in the total Maxwell current (conduction, convection, and displacement) measured simultaneously in the United States and the Soviet Union revealed an apparent correlation that is difficult to attribute solely to chance. This approach deserves further attention as a relatively simple prospective method of monitoring short-period variations in the global circuit.

If it can be demonstrated that short-period and day-to-day global variations do indeed exist, then not only is the source of these variations of importance but also the usual interpretation of local variations in terms of turbulence must be re-evaluated. In light of the importance of the ionospheric potential as an indicator of the electrical state of the global circuit and the need to separate its variation from local fluctuations in the PBL, the ionospheric potential should be measured continuously and simultaneously at two or more locations for a period of days and with a time resolution of seconds, preferably in conjunction with observations of global thunderstorm activity and of upper-atmospheric disturbances. The direct measurement of potential in the lower atmosphere using a tethered balloon has been attempted (Willett and Rust, 1981; Holzworth et al., 1981). Extension of these techniques to higher altitudes and faster time resolution should be encouraged.

Potential Tool for Study of Planetary-Boundary-Layer Turbulence

We have seen how strongly the electrical structure of the PBL is influenced by turbulent mixing. Space charge is unique among natural scalar contaminants in having a lifetime (the electrical relaxation time) comparable in magnitude with the time scales of the largest eddies in the PBL. Charge density, used in conjunction with a conservative tracer like water vapor, therefore offers the possibility of useful information about the structure of these energy-containing motions, which the natural radioactive tracers, thoron (54-sec half-life) and radon (3.8-day half-life), cannot provide. If the sources of space charge and moisture are understood, comparison of their relative distribution through the boundary layer might be useful in determining the Lagrangian time scale of the transport process.

Another important consequence of the finite lifetime of charge density is that the convergence of its turbulent flux can be deduced from its mean distribution and the convergence of conduction-current density under steady-state conditions. Since these functions depend only on the mean profiles of electric field and conductivity, they are readily measured. Thus, convection-current density is one of the few turbulent fluxes the profile of which can be observed without recourse to the complex and technology-intensive eddy-correlation method.

Because the lifetime of space charge is comparable with that of the largest eddies, it resists becoming well mixed in an unstable PBL, where conservative scalars tend to be uniformly distributed. This fact and the ease of measuring the turbulent transport have recently been exploited to obtain profiles of the eddy-diffusion coefficient for space charge through the boundary layer from individual aircraft soundings (Markson et al., 1981).

To realize this potential, it will be necessary to develop a more thorough understanding of the sources and turbulent transport of electrical charge within the PBL. The most urgent need is for a field program to gather data on the dependence of these phenomena on the me-

teorological structure of the boundary layer. Further theoretical modeling will then be required to integrate these data into a coherent understanding of the processes involved. Areas of particular ignorance at present are the ionization rate within the plant canopy and the disposition of space charge accumulating at an inversion because of the discontinuity of conductivity usually found there. Further research into these areas may eventually lead to the use of atmospheric-electrical measurements to observe, perhaps remotely, the meteorological structure of the PBL.

Ion Physics and Balance in the Planetary Boundary Layer

Small atmospheric ions, existing by virtue of a balance between ionization of the neutral gas and recombination and attachment to aerosol particles, cause the conductivity of the air. Yet, many facets of the nature and behavior of these particles are still poorly understood. More research is needed in the area of ion physics, especially (1) identification of the terminal positive and negative species in the PBL, (2) determination of the dependence of ion chemistry on trace gases, (3) measurement of the attachment coefficients of ions to charged and uncharged aerosol particles of various sizes, and (4) evaluation of the resulting charge distribution on the aerosols. (1) and (2) show promise of becoming sensitive methods for detecting certain trace gases. Further identification of exact ion chemistry is required by physiologists before they can evaluate claims of physiological effects of air ions (MEQB, 1982).

Values of ion-aerosol attachment coefficients as a function of particle radius and charge are necessary to determine accurately the loss of ions (or conductivity) as a function of aerosol load. Few measurements of absolute values of attachment coefficients have ever been attempted. Usually only ratios of attachment coefficients are measured and compared with theoretically predicted values of the ratios. Better measurement of the absolute values of the coefficients are necessary to predict ion loss and validate theory.

The use of conductivity or columnar resistance as a pollution monitor depends on the inverse relationship between conductivity and aerosol burden. The conductivity is sensitive to the aerosol concentration, and measurements spanning several decades have been used to evaluate changes in global particulate pollution (Cobb and Wells, 1970). However, the quantitative reliability of these indicators of particulate burden should be more fully investigated. The use of conductivity measurements for deducing aerosol burden is complicated by their sensitivity to the ionization rate. This latter sensitivity has even led to the suggestion that conductivity be used as a monitor of nuclear operations and accidents.

The electrical state of the atmosphere depends critically on the ionization profile. Over the ocean ionization is due only to cosmic rays, and oceanic measurement of the electrode effect are in satisfactory agreement with numerical solutions of the governing equations. Over land, the ionization profile is complicated owing to ionization from ground radioactivity and radioactive gases as discussed earlier. Simultaneous measurement of all contributions to the ionization profile has never been accomplished. It is imperative that future studies of atmospheric-electrical profiles in the PBL over land include such measurements.

A theoretical explanation is needed for the atmospheric-electric fog effect. This is most likely to be found in the dependence of the conductivity on changes in the aerosol size distribution with changes in relative humidity. Measurements of all pertinent parameters are needed during a fog event to formulate a physical theory that adequately accounts for the observations. The possible usefulness of the observed precursor phenomenon could then be evaluated.

Finally it should be mentioned that there are some experimental techniques that do not yet exist in satisfactory form for proper studies of PBL electrical processes. Briefly, these include a continuous measurement of ionospheric potential with fine time resolution, ion-sampling techniques with spatial resolutions suitable for profile determinations in the lowest few centimeters of the PBL (and within the plant canopy), and an adequately instrumented platform for making accurate atmospheric-electrical and micrometeorological profiles throughout the PBL.

REFERENCES

Anderson, R. V. (1967). Measurement of worldwide diurnal atmospheric electricity variations, *Mon. Weather Rev. 95*, 899.

Anderson, R. V., and E. M. Trent (1969). Atmospheric electricity measurements at five locations in eastern North America, *J. Appl. Meteorol. 8*, 707.

Arnold, F., and E. F. Ferguson (1983). Ions of the middle atmosphere: Their composition, chemistry and role in atmospheric processes, in *Proceedings in Atmospheric Electricity*, L. H. Ruhnke and J. Latham, eds., A. Deepak Publishing, Hampton, Va., p. 14.

Blanchard, D. C. (1963). Electrification of the atmosphere by particles from bubbles in the sea, *Prog. Oceanogr. 1*, 71-202.

Clark, J. F. (1958). The fair-weather atmospheric electric potential and its gradient, in *Recent Advances in Atmospheric Electricity*, L. G. Smith, ed., Pergamon Press, New York, p. 61.

Cobb, W. E., and H. J. Wells (1970). The electrical conductivity of ocean air and its correlation to global atmospheric pollution, *J. Atmos. Sci. 27*, 814.

Crozier, W. D., and N. Biles (1966). Measurements of Radon 220

(thoron) in the atmosphere below 50 centimeters, *J. Geophys. Res. 71*, 4735.

Dolezalak, H. (1963). The atmospheric electric fog effect, *Rev. Geophys. 1*, 231-282.

Dolezalek, H. (1978). On the application of atmospheric electricity concepts and methods to other parts of meteorology, *Technical Note No. 162*, World Meteorological Organization, Geneva, Switzerland, 130 pp.

Gathman, S. G., and W. A. Hoppel (1970). Electrification processes over Lake Superior, *J. Geophys. Res. 75*, 1041.

Haugen, D. A. (1973). *Workshop on Micrometeorology*, American Meteorological Society, Boston, Mass., 392 pp.

Holzworth, R. H., M. H. Daley, E. R. Schnaus, and O. Youngblath (1981). Direct measurement of lower atmospheric vertical potential differences, *Geophys. Res. Lett. 8*, 783.

Hoppel, W. A. (1967). Theory of the electrode effect, *J. Atmos. Terrest. Phys. 29*, 709.

Hoppel, W. A., and S. G. Gathman (1972). Experimental determination of the eddy diffusion coefficient over the open ocean from atmospheric electric measurements, *J. Phys. Oceanogr. 2*, 248.

Huertas, M. L., J. Fontan, and J. Gonzalez (1978). Evolution times of tropospheric ions, *Atmos. Environ. 12*, 2351.

Huzita, A. (1969). Effect of radioactive fallout upon the electrical conductivity of the lower atmosphere, in *Planetary Electrodynamics*, S. C. Coroniti and J. Hughes, eds., Gordon & Breach Science Publishers, New York, Vol. 1, p. 49.

Ikebe, Y. (1970). Evaluation of the total ionization in the lower atmosphere, *J. Earth Sci., Nagoya Univ. 18*, 85.

Ikebe, Y., and M. Shimo (1972). Estimation of the vertical turbulent diffusivity from thoron profiles, *Tellus 24*, 29.

Israël, H. (1958). The atmospheric electric agitation, in *Recent Advances in Atmospheric Electricity*, L. G. Smith, ed., Pergamon Press, New York, p. 149.

Israël, H. (1961). *Atmosphärische Elektrizität*, Teil II, Akademische Verlagsgesellschaft, Leipzig, 503 pp.

Israël, H. (1973). *Atmospheric Electricity*, Volume II, NTIS, U.S. Department of Commerce, Springfield, Va., 570 pp.

Kraakevik, J. H. (1958). Electrical conduction and convection currents in the troposphere, in *Recent Advances in Atmospheric Electricity*, L. G. Smith, ed., Pergamon Press, New York, p. 75.

Markson, R. (1975). Atmospheric electrical detection of organized convection, *Science 188*, 1171-1177.

Markson, R. (1977). Airborne atmospheric electrical measurements of the variation of ionospheric potential and electrical structure in the exchange layer over the ocean, in *Electrical Processes in Atmospheres*, H. Dolezalek and R. Reiter, eds., Steinkopff, Darmstadt, p. 450.

Markson, R., J. Sedlacek, and C. W. Fairall (1981). Turbulent transport of electric charge in the marine atmospheric boundary layer, *J. Geophys. Res. 86*, 12115.

MEQB (1982). A health and safety evaluation of the + 400 kV powerline: Report of science advisors to the Minnesota Environmental Quality Board, Minnesota Environmental Quality Board, Saint Paul, Minn.

Mohnen, V. A. (1977). Formation, Nature and Mobility of ions of atmospheric importance, in *Electrical Processes in Atmospheres*, H. Dolezalek and R. Reiter, eds., Steinkopff, Darmstadt, p. 1.

Moses, H., A. F. Stehney, and H. F. Lucas, Jr. (1960). The effect of meteorological variables upon the vertical and temporal distributions of atmospheric radon, *J. Geophys. Res. 65*, 1223.

Mühleisen, R. (1977). The global circuit and its parameters, in *Electrical Processes in Atmospheres*, H. Dolezalek and R. Reiter, eds., Steinkopff, Darmstadt, p. 467.

Nolan, P. J. (1943). The recombination law for weak ionization, *Proc. R. Irish Acad.* 49, 67.

Ruhnke, L. H., H. F. Tammet, and M. Arold (1983). Atmospheric electric currents at widely spaced stations, in *Proceedings in Atmospheric Electricity*, L. H. Ruhnke and J. Latham, eds., A. Deepak Publishing, Hampton, Va., p. 76.

Sagalyn, R. C., and G. A. Faucher (1956). Space and time variations of charged nuclei and electrical conductivity of the atmosphere, *Q. J. R. Meteorol Soc. 82,* 128.

Serbu, G. P., and E. M. Trent (1958). A study of the use of atmospheric electric measurements in fog forecasting, *Trans. Am. Geophys. Union 39,* 1034.

Takagi, M., and N. Toriyama (1978). Short-period fluctuations in the atmospheric electric field over the ocean, *Pure Appl. Phys. 116,* 1090.

Torreson, O. W., *et al.* (1946). *Scientific Results of Cruise VII of the Carnegie, Oceanography III, Ocean Atmospheric Electricity Results,* Carnegie Institution Publication 568, Washington, D.C.

Willett, J. C. (1983). The turbulent electrode effect as influenced by interfacial ion transfer, *J. Geophys. Res. 88,* 8453.

Willett, J. C., and W. D. Rust (1981). Direct measurements of atmospheric electric potential using tethered balloons, *J. Geophys. Res. 86,* 12139.

Wyngaard, J. C. (1980). *Workshop on the Planetary Boundary Layer,* American Meteorological Society, Boston, Mass., 322 pp.

Electrical Structure from 0 to 30 Kilometers

12

WOLFGANG GRINGEL
Universität Tübingen

JAMES M. ROSEN *and* DAVID J. HOFMANN
University of Wyoming

INTRODUCTION

This chapter deals with the electrical structure of the lower atmosphere, i.e., the troposphere and the portion of the stratosphere below about 30 km. Here the principal observing platforms (not including surface measurements) are balloons. Their limited height range, rather than other physical considerations, is the main reason that the electrical structure above 30 km will be discussed separately in the following chapter.

For better understanding of the electrical phenomena taking place in the lower atmosphere and the coupling between them, the concept of a "global circuit" will be briefly touched on—a complete discussion is presented by Roble and Tzur (Chapter 15, this volume).

The discovery of the atmospheric conductivity raised a question concerning the origin of the electric fields and the electric currents that were known to exist and flow continuously in the atmosphere. According to the classical picture of the global circuit (Dolezalek, 1972), the total effect of all thunderstorms acting at the same time can be regarded as the global generator, which charges the ionosphere to several hundred kilovolts with respect to the Earth's surface. This potential difference drives the air-earth current downward from the ionosphere to the ground in the nonthunderstorm areas through the conductive atmosphere. The value of this air-earth cur-

rent density varies according to the ionospheric potential and the total columnar resistance between ionosphere and ground. Finally the local atmospheric electric field must be consistent with this current flowing through a resistive medium, i.e., the atmosphere.

In addition to the global generator there also exist effective local generators such as precipitation, convection currents (charges moved by other than electrical forces), and blowing snow or dust. The latter create their own local current circuits and electric fields superimposed on parts of the global circuit. Generators can be regarded as local generators (Dolezalek, 1972) if the resistance from the upper terminal to the ionosphere is much greater than the resistance from that point to the Earth's surface along the shortest possible path and with the consequence that almost no current flows to the ionosphere from this generator.

In the following sections we discuss initially the sources of ionization in the lower atmosphere together with solar-induced and latitudinal variations. In the next section a brief review of aerosol distributions in the troposphere and lower stratosphere is presented. Variations following major volcanic eruptions are emphasized. Atmospheric conductivity, small ion concentrations, and ion-mobility measurements are the subject of the third section. Here the influence that solar activity or aerosols have on the conductivity, and therefore on parts of the columnar resistance, are discussed. In the final section the air-earth current and electric fields in the lower atmosphere are considered. Here the results are interpreted from a global viewpoint with perturbations from local generators. Examples of anthropogenic influences on the electric field near the ground are also presented and discussed briefly.

ION PRODUCTION IN THE LOWER ATMOSPHERE

The electric structure of the troposphere and lower stratosphere depends strongly on the ion-pair production rate and the physical properties of the ions produced. Cosmic rays are the primary source of ionization in the atmosphere range under consideration. Near the Earth's surface over the continents there is an additional component due to ionization by radioactive materials exhaling from the soil. This radioactive ionization component depends on different meteorological parameters and can exceed the cosmic-ray component by an order of magnitude as discussed in Chapter 11. It decreases rapidly with increasing height, and at 1 km it is already significantly less than the contribution due to cosmic rays (Pierce and Whitson, 1964). The ion-production rate by cosmic rays is shown in Figure 12.1 for different geomagnetic latitudes during the years of solar mini-

mum (1965) and solar maximum (1958) based on balloon measurements by Neher (1961, 1967). The existence of the geomagnetic field gives rise to a pronounced latitude effect. Only at latitudes higher than about 60° can the full energy spectrum of the cosmic rays reach the Earth and the depth of penetration be limited only by the increasing atmospheric density for low-energy particles. At high latitudes 100-MeV protons can penetrate to about 30 km height, for example. Moving downward to lower latitudes more and more particles with lower energies are deflected by the geomagnetic field and, therefore, are excluded. The geomagnetic equator itself can only be reached by particles with energies greater than about 15 GeV. The hardening of the cosmic-ray spectrum with decreasing latitude is indicated in Figure 12.1 by the lowering of the height at which the maximum ionization rate occurs. Near the equator this maximum ionization rate is observed around 10 km.

Furthermore, the ionization rate depends strongly on solar activity in a sense that at a particular height the ion-production rate is lower during the sunspot maximum and higher during the sunspot minimum, as illustrated in Figure 12.1. The mechanisms are not fully understood, but it appears that irregularities and enhancements of the interplanetary magnetic field tend to exclude part of the lower-energy cosmic rays from the inner solar system (Barouch and Burlaga, 1975). The effect becomes more pronounced with increasing height and/or increasing geomagnetic latitude. At geomagnetic latitudes around 50° the reduction of the ion-production rate during the periods of sunspot maximum is about 30 percent at 20 km and about 50 percent at 30 km. More recently this solar-cycle dependence was confirmed by measurements with open balloonborne ionization chambers by Hofmann and Rosen (1979). Analytical expressions for computing the ionization rates dependent on latitude and solar-cycle period are given by Heaps (1978). Superimposed on the 11-yr solar-cycle variation are so-called Forbush decreases (Forbush, 1954), which are somehow related to solar flares and exhibit a temporary reduction of the incoming cosmic-ray flux for periods of a few hours to a few days or weeks (Duggal and Pomerantz, 1977).

On the other hand, solar proton events (SPE) can drastically increase the ion-production rate within the stratosphere and, for high-energy solar protons, sometimes even near the ground. The duration of such SPEs is of the order of hours, and they are normally restricted to high-latitude regions, as discussed in more detail in the following chapters.

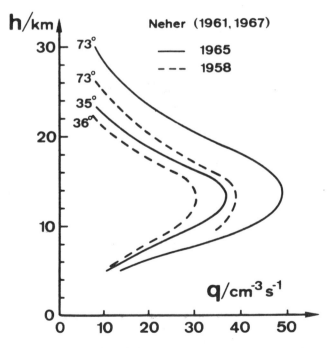

FIGURE 12.1 Profiles of the ionization rate at different latitudes in years of the minimum (1965) and maximum (1958) of the 11-yr solar sunspot cycle (Neher, 1961, 1967).

AEROSOLS IN THE LOWER ATMOSPHERE

Within the portion of the atmosphere under consideration in this chapter there are three primary regions of interest to atmospheric electricity: the boundary layer (approximately the first 3 to 5 km), the remaining portion of the troposphere, and the stratosphere. The characterizing aerosol parameters of central importance here are concentration, size distribution, and vertical structure. In addition, variability of the aerosol in each of these regions must be recognized. Such variations may influence electrical parameters and significantly detract from the apparent repeatability of various atmospheric electrical measurements.

The boundary layer is generally thought of as a relatively well-mixed region capped on the upper side by a temperature inversion. Since mixing across the inversion tends to be inhibited, a potential exists for the accumulation of aerosols within the boundary layer if a significant source is present. The buildup of aerosols is frequently of sufficient magnitude to cause a notable reduction in visibility. Even when there is no apparent loss in visibility as observed from the surface, the upper boundary may still be apparent even to an airline passenger at the moment when the aircraft passes through inversion. The buildup of aerosols in the boundary layer and the capping effect of the inversion have been observed in a more quantitative sense by airborne lidar. The ion concentration and conductivity can be greatly affected by the aerosol buildup in the boundary layer and a dramatic change in these quantities is often observed at or near the altitude of the defining inversion.

The size distribution of aerosols near the surface of the Earth has frequently been approximated with a power-law function (Junge, 1963). However, it has more recently been suggested that the near-surface aerosol is actually made up of two or three size modes that when added together approximate a power-law function over a limited range of sizes (Willeke and Whitby, 1975). This interpretation of the size distribution seems to better reflect the physical processes affecting the aerosol concentration in the atmosphere. Some examples of typical size distributions for a variety of conditions and locations can be found in the work of Willeke and Whitby (1975) and Patterson and Gillette (1977).

The aerosol mixing ratio profile (and usually the concentration profile) typically show a relative minimum in the upper troposphere for particle radii greater than about 0.1 μm. In contrast the condensation nuclei (cn) profile (corresponding to particle radii of about 0.01 μm) is more or less constant throughout the entire troposphere above the boundary layer. Near the surface of the Earth the cn concentration may be relatively high

owing to local contamination (10^4 per cm^3 or more), but above the boundary layer the concentration ranges from about 100 to 10^3 per cm^3 with a global average of approximately 300 to 500 per cm^3 (Rosen *et al.*, 1978a, 1978b).

The size distribution of aerosols in this region of the atmosphere can usually be approximated by a power-law function between about 0.01- and 10-μm particle diameter (Junge, 1963). Below the minimum size of 0.01 μm there are relatively few particles owing to coagulation, and above 10 μm the particle concentration drops quickly from sedimentation effects. Thus limits of the power-law distribution must always be specified.

The character of upper tropospheric aerosols can be temporarily disturbed by volcanic eruptions, forest fires, biomass burning, and large dust storms. In addition periodic annual variations of concentration have also been observed (Hofmann *et al.*, 1975). Another important temporal variation of tropospherical aerosols is associated with the so-called arctic haze events. New evidence suggests that these events are characterized by high particle loading throughout a large portion of the troposphere. The impact of this extensive aerosol loading on atmospheric electrical parameters is yet to be determined.

Hogan and Mohnen (1979) reported the results of a global survey of aerosols in the troposphere and lower stratosphere. They found that the concentrations were more or less symmetrically distributed about the Earth. Measurements of this type could provide the basis for extrapolating local or isolated observations to characteristic worldwide values.

The morphology of stratospheric aerosols is dominated by a persistent structure frequently referred to as the 20-km sulfate layer, or Junge layer. It is now known that the character of this layer is highly affected by large volcanic eruptions. For several years prior to 1980, stratospheric aerosols were in a quasi-steady-state condition, not being under the influence of any significant recent volcanic eruptions. During that period the size distribution appeared to be consistent with a single mode log-normal distribution (Pinnick *et al.*, 1976), although other types of single-mode distribution were also employed (Russell *et al.*, 1981) with similar results. The composition was thought to be primarily sulfuric acid droplets.

The appearance of the 20-km layer is not evident in the vertical profile of all particle size ranges. The cn profile, for example, which is representative of particles with sizes in the neighborhood of 0.01-μm radius usually shows a dramatic drop in concentration above the tropopause and no relative maximum at the altitude of the stratospheric aerosol layer. This would be consistent with a tropospheric cn source and a vertical profile dictated by diffusion and coagulation (Rosen *et al.*, 1978a).

After a large volcanic eruption the size distribution may be greatly disturbed and highly altitude dependent. Following the eruption of El Chichon in April 1982, Hofmann and Rosen (1983a, 1983b) reported a size distribution that could be approximated by a sum of multiple log-normal distributions, each with a different mode. The source of the smallest particle mode (~ 0.01 μm) appeared to be associated with the homogeneous condensation of highly saturated sulfuric acid vapors. This production ceased a few months after the eruption, and consequently the mode disappeared. A midsize mode (~ 0.1 μm) was also observed, which may have resulted from coagulation and growth by condensation of the initial particles formed by the homogeneous condensation. Another mode near 1-2 μm diameter was also observed that may have formed from the condensation of sulfuric acid vapors onto the solid silicate ash particles as well as on particles already present in the stratosphere. Further assessments of the aerosol injected by the El Chichon eruption have been described in a collection of papers introduced by Pollack *et al.* (1983).

An unexpected influence of the El Chichon eruption was the enhancement of an annually appearing cn layer near 30 km altitude. Although the phenomenon was observed in previous years (Rosen and Hofmann, 1983) the concentration of cn associated with these event layers was enhanced by at least 2 orders of magnitude. Even though the sizes of the particles were quite small (~ 0.01-μm diameter), the concentration was large enough to measurably affect the ambient ion concentration and conductivity (Gringel *et al.*, 1984). The event particles are thought to have formed in polar regions from highly supersaturated sulfuric acid vapors. The affect of this phenomenon on the various aspects of atmospheric electricity at high latitude has not been assessed at the time of this writing. Of particular interest would be the influence of the highly supersaturated vapors on the ambient ion mass (and therefore mobility). Careful measurements at appropriate locations may provide an unexpected means of finding supporting experimental evidence for some of the various models of ion composition (and mass) that have been proposed (Arnold, 1983; Arnold and Bührke, 1984).

Figures 12.2 to 12.4 illustrate several of the characteristics of atmospheric aerosols that have been discussed above. The long-term influence of volcanic aerosols on the stratosphere are illustrated in Figure 12.2, which compares the quasi-steady-state period with the disturbed conditions still observed some 18 months after the eruption of El Chichon. Note also the presence of the boundary layer near the surface (as evidenced by a

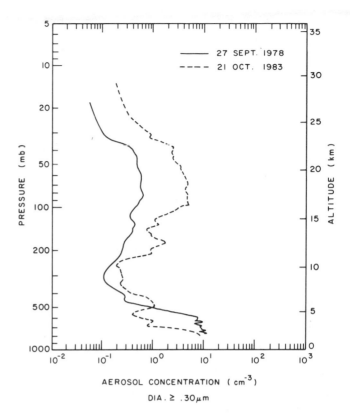

FIGURE 12.2 A comparison of aerosol profiles (particle diameter greater than 0.30 μm) for the quasi-steady-state period (September 27, 1978) with the decay period following the April 1982 eruption of El Chichon.

sharp drop in concentration) and a relative minimum in the aerosol concentration occurring in the upper troposphere.

An example of a normal and disturbed cn profile is illustrated in Figure 12.3. A significant cn event layer is evident at about 30 km altitude. Both profiles show a relatively large drop in the cn concentration just above the surface, a relatively constant mixing ratio throughout most of the troposphere, and a noticeable drop in concentration near the tropopause.

An example of the influence of volcanic eruptions on the size distributions of stratospheric aerosols is illustrated in Figure 12.4. As previously discussed, the size distribution during the quasi-steady-state period could be described quite well by a single-mode log-normal distribution. At the time of this writing some 18 months after the eruption of El Chichon, the size distribution appears still to be quite disturbed.

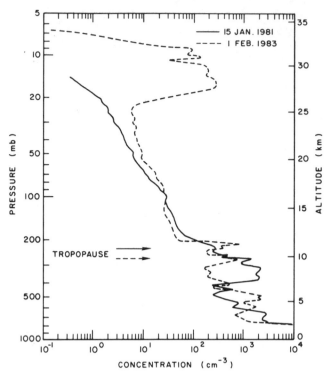

FIGURE 12.3 A comparison of a normal cn profile (January 15, 1981) with one obtained during a period of the 30-km cn event.

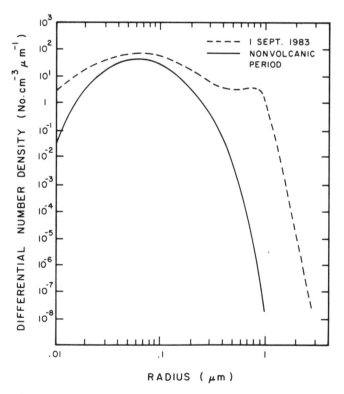

FIGURE 12.4 A comparison of the single-mode stratospheric aerosol size distribution (solid line) typical of the quasi-steady-state period with a recently obtained distribution (dashed line) that is believed to be still under the influence of the April 1982 eruption of El Chichon.

ATMOSPHERIC CONDUCTIVITY, SMALL ION CONCENTRATION, AND MOBILITY

The atmospheric conductivity depends on the existence of positive and negative ions, whereas the contribution of free electrons can be neglected below about 45 km. There is normally a mixture of ions and the resulting conductivity can be expressed in terms of the number densities and mobilities of the individual species as

$$\sigma = \sigma_+ + \sigma_- = e\sum_i n_{i+}k_{i+} + e\sum_i n_{i-}k_{i-}$$

$$(12.1)$$

with e the electronic charge and n and k the number densities and mobilities of the particular positive and negative ions. The ion mobility is defined by the drift velocity v of the ions in an electric field E as

$$v = kE \qquad (12.2)$$

and depends on the mass, the collisional cross section, and charge of the individual ions (ions are believed to be singly charged in the lower atmosphere) as well as on the density and polarizability of the surrounding gas. It is inversely proportional to the air density and can be expressed by the reduced mobility k_0 as

$$k(h) = k_0\,\frac{p_0\,T(h)}{p(h)\,T_0}, \qquad (12.3)$$

where p_0 and T_0 are STP pressure and temperature (1013 mbar, 273 K) and p and T are pressure and temperature at height h. An empirical relationship between reduced mobility and ion mass in nitrogen is given by Meyerott *et al.*, (1980) and shown later in Figure 12.13.

In the troposphere and lower stratosphere the atmospheric conductivity is maintained by the so-called small ions with reduced mobilities around 1.5 cm²/V sec, whereas the mobilities of large ions (charged aerosol particles or large molecular clusters) are too small to contribute directly to the conductivity (e.g., Israël, 1973b). As will be discussed later, large ions as well as attachment by aerosol particles can reduce the conductivity significantly. The production and annihilation of small ions is shown schematically in Figure 12.5. The molecular ion and remaining electron created by the ionizing process form charged molecular clusters (the small ions) after several reactions. These small ions are annihilated by mutual recombination and neutralization, or they become almost immobile by attachment to aerosol particles or large ions. Under steady-state conditions and assuming equal densities of positive and negative small ions, the fundamental balance equation becomes in its most simplified form

$$dn/dt = 0 - q - \alpha n^2 - \beta nZ, \qquad (12.4)$$

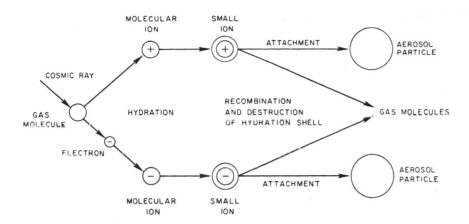

FIGURE 12.5 Schematic representation of the production and annihilation of atmospheric small ions.

where q is the ionization rate, α the recombination coefficient, β the attachment coefficient, and Z the aerosol concentration. An expression for (r) that is relevant to stratospheric conditions has been developed by Zikmunda and Mohnen (1972) and is approximately $\beta(r) = 2.35 \times 10^{-4} r^{1.457}$, where r is the radius of the aerosol particle in micrometers. Recombination coefficients were reported by Smith and Church (1977) and experimentally derived from nearly simultaneous balloon measurements of the ionization rate, the small ion density, and the conductivity by Rosen and Hofmann (1981a, 1981b) and Gringel et al. (1983).

The atmospheric conductivity within balloon altitudes is measured mostly with the Gerdien condenser technique (e.g., Israël, 1973a). In contrast to ion-density measurements, conductivity measurements are largely independent of the airflow rate through the Gerdien tube. This seems to be the main reason for the relatively good agreement among different conductivity profiles as reported by Meyerott et al. (1980). A direct comparison of three different Gerdien-type sondes has shown in fact an agreement of the resulting conductivity profiles within 10 percent to 32 km height (Rosen et al., 1982).

Figure 12.6 shows a mean profile (21 soundings) of the positive polar conductivity for quiet solar conditions and a geomagnetic latitude around 50° (Gringel, 1978). This profile can be analytically expressed by the relation

$$\sigma_+ (10^{-14}\, \text{mho/m}) = \exp(\sum_{i=0}^{3} a_i z_i), \qquad (12.5)$$

with z the altitude (in kilometers) and $a_0 = 6.363 \times 10^{-1}$, $a_1 = 3.6008 \times 10^{-1}$, $a_2 = -8.605 \times 10^{-4}$, and $a_3 = 1.0331 \times 10^{-1}$.

The scattering of the mean values as shown for kilometer intervals is considerably below 15 km, indicating a highly variable aerosol density throughout the troposphere. The noticeable increase of the conductivity

around 13 km can also be attributed to a sharp decrease of aerosol particle concentrations, especially condensation nuclei, above the tropopause.

For comparison, Figure 12.6 shows the mean of three conductivity measurements conducted within 1 week following solar flares. One profile was obtained on August 8, 1972, during a period of intense solar events. The other two flights were conducted 3 and 7 days, respectively, after a solar flare associated with an intense type-IV radio burst on April 11, 1978. For all three flights the

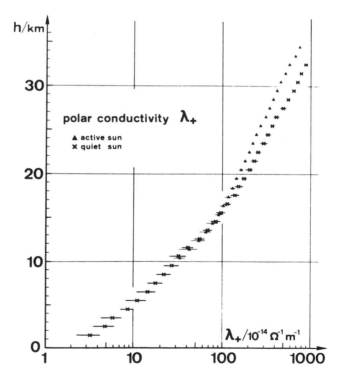

FIGURE 12.6 Profiles of the positive polar conductivity versus altitude ($\sigma_+ = \sigma_-$) during quiet and active solar conditions. Mean values and standard deviations are shown for altitude intervals in kilometers (Gringel, 1978).

ionization rate was appreciably reduced by the flare-related Forbush decrease, resulting in a remarkable conductivity reduction throughout the lower stratosphere. For the August 8, 1972, flight the conductivity reduction reached deep into the troposphere (see also Figure 12.8 below).

The ratio between the values of positive and negative polar conductivity exhibit a noticeable altitude dependence as shown in Figure 12.7 (Gringel, 1978). In the troposphere the negative conductivity values exceed the positive ones by about 12 percent. Between 15 and 20 km both polarities are about equal, whereas above 20 to 25 km the negative conductivity becomes 5 to 10 percent smaller than the positive on the average. Assuming equal densities of positive and negative small ions, the same can be concluded for the corresponding average small-ion mobilities.

As shown by Roble and Tzur (Chapter 15, this volume) the vertical columnar resistance is an important parameter within the scope of the global atmospheric electrical circuit. This columnar resistance R_c is related to the atmospheric conductivity by

$$R_c = \int_{h_1}^{h_2} (\sigma_+ + \sigma_-)dh \qquad (12.6)$$

and denotes the resistance of a vertical air column with a 1-m^2 base between the ground and the equalization

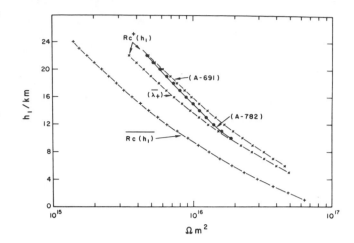

FIGURE 12.8 Mean columnar resistance $[R_c(h_1)]$ between the lower-altitude h_1 and 60 km. The profiles $R_c^+(h_1)$ are calculated from positive polar conductivity profiles only for quiet solar conditions ($\bar{\lambda}_+$) and after solar flares measured on August 8, 1972 (A-691) and on April 14, 1978 (A-782).

layer. From Eq. (12.6) R_c is determined mainly by the low conductivity values near the ground. Figure 12.8 shows the columnar resistance $R_c(h_1)$ between the lower height h_1 and 60 km as function of h_1 under quiet solar conditions as calculated from the conductivity profile in Figure 12.6 and using the ratio between positive and negative conductivity shown in Figure 12.7. The mean value from the ground to 60 km was found to be 1.3 × 10^{17} Ω m^2 at Weissenau (South Germany, 450 m above sea level) during fair-weather conditions. The variations are on the average about 30 percent and are caused by a changing ionization from radioactive materials near the ground and by varying aerosol concentrations in the lower troposphere. The first 2 km of the atmosphere contribute about 50 percent and the first 13 km about 95 percent to the total columnar resistance.

The right side of Figure 12.8 shows three profiles of the positive polar columnar resistance R_c^+ as a function of the lower height h_1 for quiet solar conditions and two profiles following solar flares. These three profiles were calculated using only positive conductivity values and therefore show about twice the value of the actual R_c obtained from both polarities. Whereas the Forbush decrease observed on April 14, 1978, influenced mainly the part of the columnar resistance above the troposphere, the enhancement reached deep into the troposphere on August 8, 1972, at a geomagnetic latitude of 48°. The part of R_c above 13 km was found to be higher by 13 and 28 percent, respectively, when compared with R_c obtained under quiet solar conditions. As proposed by Markson (1978) this could cause a reduction of the current flowing from the top of the thunderclouds to

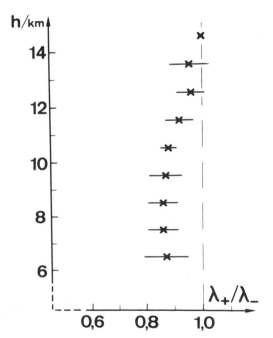

FIGURE 12.7 The ratio of positive to negative polar conductivity as deduced from 10 balloon flights (Gringel, 1978).

the ionosphere, resulting in a lower value for the ionospheric potential reported for periods of high solar activity by Fischer and Mühleisen (1972).

Layers of an increased aerosol density can significantly reduce the atmospheric conductivity owing to attachment of small ions on these aerosol particles. Figure 12.9, as an extreme, shows a conductivity profile through a Sahara dust layer between 1.7 and 3.7 km height and 2200 km west of the West African coast (Gringel and Mühleisen, 1978). The dust concentration Z^* responsible for the conductivity decrease is also shown. The authors report a mean mass concentration of 1200 g^{-3} throughout the layer. Owing to the low altitude of the layer the total columnar resistance was increased about 30 to 50 percent within these large-scale areas of Sahara dust transport across the North Atlantic. It is estimated that about the same increase of R_c occurs

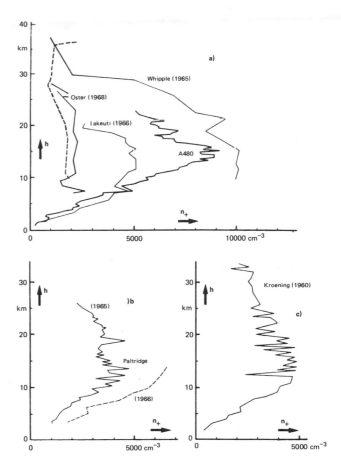

FIGURE 12.10 Small-ion-density profiles by different authors taken from Riekert (1971), who measured the profile A-480.

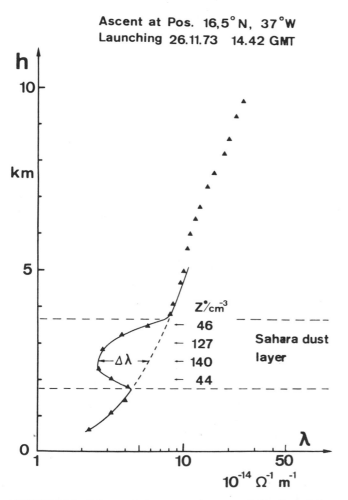

FIGURE 12.9 Polar conductivity as a function of altitude and the concentration of mineral dust particles Z^* derived from the conductivity decrease $\Delta\lambda$ in the main Sahara dust transport layer at 2200 km distance from West Africa (Gringel and Mühleisen, 1978).

under stratus clouds extending from 2 to 4 km height. The influence of altostratus and cirrus clouds on R_c is negligible compared with the above. Up to several hundred percent increase of the columnar resistance can occur in polluted areas where high aerosol concentrations (in excess of 10^4 cm^{-3}) or even smog reduce the conductivity drastically within the first few kilometers of the troposphere.

The concentration of small ions is one of the more fundamental electrical properties of the atmosphere. The closely related parameter conductivity, although highly important for the electrical structure of the atmosphere, depends on the product of the ion concentration and ion mobility and is therefore of a somewhat less basic nature. Small-ion concentrations are also measured with the Gerdien condenser technique. In contrast to conductivity measurements, the measured ion current is proportional to the airflow rate through the Gerdien chamber, and the latter must be well known in evaluating the appropriate ion density values. Figure 12.10 shows some ion-density profiles measured by different authors as reported by Riekert (1971). The profiles show large variations among each other and differ by almost a factor of 5 at the ion density maximum near 15 km. The small-scale fluctuations shown in the profiles of Kroening (1960) and Paltridge (1965) have been attributed to the attachment of small ions on particles. However, it is not clear that the known stratospheric aerosol size and concentration can quantitatively account for the required amount of ion depletion, except under rare circumstances.

If, on the other hand, the airflow through the Gerdien chamber is maintained by the rising balloon, it becomes difficult to estimate the actual airflow rate from the balloon rise rate as shown by Riekert (1971) and Morita *et al.* (1971). The potential seriousness of these uncertainties has led Rosen and Hofmann (1981a, 1981b) to develop a new Gerdien-type ion-density sonde, which is force ventilated by a lobe pump producing a steady and well-defined airflow. A typical ion-density profile measured by these authors is shown in Figure 12.11. Over a 2-yr period their measurements show only small variations from flight to flight, and the fluctuations in the individual profiles are considerably less than those obtained with Gerdien tubes that are not force ventilated.

The ion-mobility values among different researchers show also, similarly to the ion density, large variations

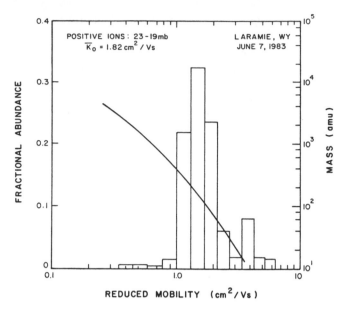

FIGURE 12.12   Reduced mobility spectrum (histogram) for positive ions around 26 km (Gringel, in preparation). The relationship between the mass of molecular ions and their reduced mobility in nitrogen as reported by Meyerott *et al.* (1980) is also shown.

as discussed in detail by Meyerott *et al.* (1980). The reduced positive mobility values up to 30 km range from 1 cm²/V sec (Riekert, 1971) to 2.7 cm²/V sec (Widdel *et al.*, 1976), corresponding to average ion masses of approximately 400 and 30 amu, respectively (see the ion mass/mobility relation shown in Figure 12.12). Ion mobilities obtained by Mitchell *et al.* (1977) are even larger, around 5 cm²/V sec at 30 km and above, which would indicate ion masses less than 10 amu if the theoretical curve in Figure 12.12 could be extrapolated. Meyerott *et al.* (1980) point out that these small ion masses might be attributed to rather large field strengths of $E/p \simeq$ 20-30 V/cm Torr used in some of the instruments, which can cause an ion breakup during the sampling procedure. If complementary measurements of conductivity and ion concentration are used to determine the ion mobility, a poorly known and variable airflow rate through the ion-density chamber might explain some of the above discrepancies.

Three simultaneous balloonborne measurements of ion density [force-ventilated Gerdien tube by Rosen and Hofmann (1981a)] and conductivity [Gerdien tube ventilated by the rising balloon (Gringel, 1978)] yielded a reduced positive ion mobility of 1.3 ± 0.15 cm²/V sec between 4 and 34 km (Gringel *et al.*, 1983). Figure 12.13 shows the result from flight W-204 conducted on May 15, 1979, at Laramie, Wyoming. The reduced positive mobility from this particular flight indicates an "average" ions mass of about 180 amu throughout the

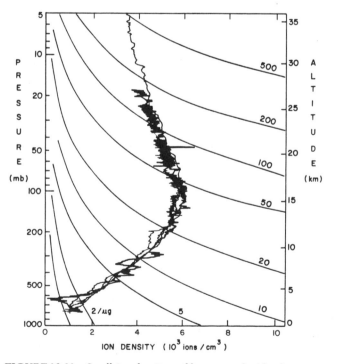

FIGURE 12.11   Small-ion-density profile measured with a force ventilated Gerdien condenser (from Rosen and Hofmann, 1981a).

entire altitude range. Following these measurements a balloonborne ion-mobility spectrometer was developed and for the first time was flown successfully at Laramie, Wyoming, on June 6, 1983 (Gringel, in preparation). The spectrometer consists of a voltage-stepped Gerdien condenser with divided collector electrode, force ventilated by a lobe pump. The maximum electric-field strength is kept well below 1 V/cm Torr in order to avoid a breakup of the weakly bonded small ions. As preliminary results, the fractional abundances of positive and negative small ions with respect to the reduced mobility are shown in Figures 12.13 and 12.14 for altitudes of 26 and 24.5 km, respectively. These results indicate that most of the positive ($\sim$78 percent) and negative ($\sim$90 percent) ions have reduced mobilities between $\sim$1 and $\sim$2.2 cm^2/V sec. The mass-mobility relation by Meyerott *et al.* (1980) indicates that most of the ions of both polarities have masses between 55 and 400 amu. Whereas only 20 percent of the negative ions show reduced mobilities larger than $\sim$1.7 cm^2/V sec, approximately 43 percent of the positive ions exceed this value and have masses smaller than about 100 amu. The re-

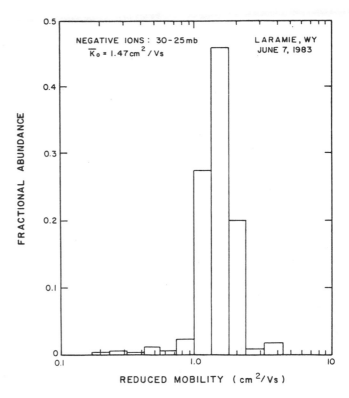

FIGURE 12.14 Reduced mobility spectrum of negative ions around 24.5 km (Gringel, in preparation).

duced average mobility values are 1.5 cm^2/V sec for the negative ions and 1.8 cm^2/V sec for the positive ions with an accuracy range of $+10$ to -20 percent. It cannot be decided at this time whether the reduced mobilities measured in 1979 (see Figure 12.13) and 1983 are really different from present values.

The negative ion-mobility spectrum shown in Figure 12.14 is quite consistent with ion mass spectrometer measurements reported recently by Viggiano *et al.* (1983). These authors found that the main mass peaks of negative ions between 125 and 489 amu belong to the main ion families $NO_3^-(HNO_3)_n$ and $HSO_4^- H_2SO_4)_m(HNO_3)_n$. The heavy HSO_4^- core ion family was found during their flights (September/October 1981) mainly at altitudes above 30 km. However the major volcanic eruption of El Chichon in early April 1982 has changed the H_2SO_4 content of the stratosphere dramatically, as reported among others by Hofmann and Rosen (1983a, 1983b). Their aerosol measurements as well as the negative-ion mobility spectra indicate the presence of the heavy HSO_4^- core ion family well below 30 km.

In contrast to negative-ion mass spectrometer measurements, as of this writing the presence of positive ions having masses greater than 140 amu in the lower strato-

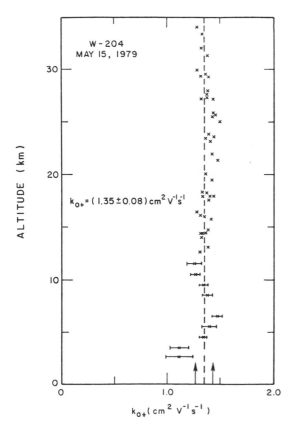

FIGURE 12.13 Reduced mobilities of positive small ions versus altitude calculated from simultaneous measurements of small-ion density and conductivity (Gringel *et al.*, 1983).

sphere (e.g., Arnold *et al.*, 1981) have not been detected. This is not consistent with ion-mobility measurements and clearly needs further investigation.

AIR-EARTH CURRENT AND ELECTRIC FIELDS IN THE LOWER ATMOSPHERE

The atmospheric electric circuit is characterized by a difference in voltage (on the order of 300 kV) between the highly conductive ionosphere (commonly referred to as the equalization layer) and the Earth's surface, which is also a relatively good conductor. This voltage V_I is thought to be maintained principally by thunderstorms acting as the generators and the atmospheric conductivity acting to discharge the ionosphere through continuous flow of current. The value of this air-earth current density in fair-weather areas depends on the voltage V_I and the columnar resistance R_c and is, according to Ohm's law,

$$j_v = V_I/R_c. \qquad (12.7)$$

The value of the atmospheric electric field $E(h)$ depends on the air-earth current density and the electrical conductivity of the air according to the following Ohm's law relationship:

$$j_v(h) = E(h)\,[\sigma_+(h) + \sigma_-(h)]. \qquad (12.8)$$

Under steady-state conditions it is expected for reasons of continuity that the air-earth current density is constant with altitude if large-scale horizontally homogeneous conditions exist and if no charged clouds or other disturbances alter the so-called fair-weather conditions.

Figure 12.15 shows the atmospheric electric field and both polar conductivities as measured simultaneously over the North Atlantic by Gringel *et al.* (1978). The field and conductivity profiles show clearly the inverse pattern to each other, which is expected for a constant air-earth current density through the atmosphere. Even a thin cloud layer at 6 km did not disturb this inverse pattern. The mean vertical air-earth current density for this particular balloon flight was calculated to be $j_v = (2.35 \pm 0.15)$ pA/m^2, a typical value for our oceanic measurements. Another profile of j_v obtained at Laramie, Wyoming (Rosen *et al.*, 1982) to 31 km height is shown in Figure 12.16. Again the profiles of both polar current densities show the expected constancy with altitude. At the same time of this flight at Laramie, the ionospheric potential V_I was determined over Weissenau, Germany, by integration of the measured electric-field strength (Fischer and Mühleisen, 1975) and found to be 330 kV, a value that should be the same as over Laramie according to the classical picture of the global

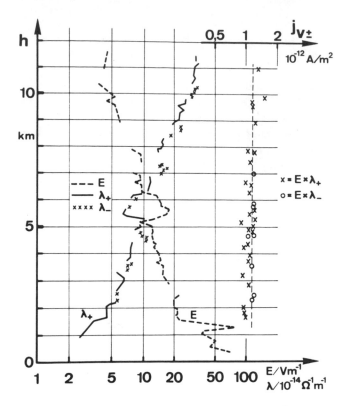

FIGURE 12.15   Polar air-earth current densities j_{v+} and j_{v-} versus altitude calculated from simultaneously measured electric-field strength E and both polar conductivities ($\lambda = \sigma$) over the North Atlantic (Gringel *et al.*, 1978).

electrical circuit. In addition, the columnar resistance was obtained from an atmospheric conductivity profile and found to be $R_c = 0.65 \times 10^{17}\ \Omega\ \text{m}^2$. This implies that the total conduction current density, calculated from V_I and R_c, is 5.1 pA/m^2. The good agreement with the mean value calculated from the conductivity and electric-field profiles proves that the Earth surface and the ionosphere can be regarded as good conductors where charges are distributed worldwide within short times. The fact that j_v over Laramie shows about twice the value as over the Atlantic is explained by the high altitude of Laramie (2150 m above sea level) resulting in the low R_c value given above. These 2 km normally contribute about 50 percent to the total columnar resistance of around $1.3 \times 10^{17}\ \Omega\ \text{m}^2$ between sea level and the ionosphere.

Direct measurements for the air-earth current density in the free atmosphere have been carried out also with long-wire antenna sondes described by Kasemir (1960). Whereas Ogawa *et al.* (1977) reported a constant air-earth current throughout the troposphere and lower stratosphere, the measurements by Cobb (1977) at the South Pole indicated a slight decrease of the j_v values above the tropopause. The reasons for this current decrease are unknown.

At globally representative stations the air-earth current density shows a diurnal variation versus universal time with a minimum at around 0300 GMT and a maximum near 1800 GMT, reflecting the diurnal variations of the ionospheric potential. Figure 12.17 shows this diurnal variation measured near the surface in the North Atlantic (Gringel *et al.*, 1978) and near the ground at the South Pole (Cobb, 1977) during fair weather. The larger variation over the North Atlantic is probably due to the relatively short observation time of only 15 days. The agreement of the two curves also strongly supports the concept of an universally controlled global circuit. Direct-current density measurements near the ground at a continental station have been reported by Burke and Few (1978). They observed a typical sunrise effect that is characterized by a gradual increase of the atmospheric conduction current within an hour after sunrise, reaching a peak about 2 hours after sunrise. The current density then gradually decreased but usually remained at a higher value than was observed before sunrise. This sunrise effect is thought to be caused by mechanical transport of positive charges following the onset of con-

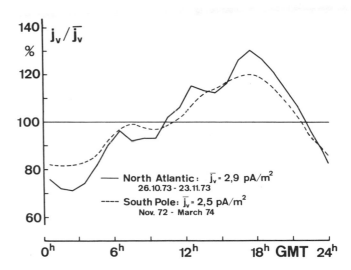

FIGURE 12.17 Mean diurnal variations of the air-earth current density in relative units over the North Atlantic (Gringel *et al.*, 1978) and at the South Pole (Cobb, 1977).

vection. During fog they observed low values of j_v that can be attributed to an enhancement of the columnar resistance, caused by the attachment of small ions to the fog droplets. Beneath low clouds without precipitation they usually found negative current readings, which are interpreted by Burke and Few (1978) as charge-separation processes occurring in most of the low clouds, whether or not the clouds finally produced local precipitation or developed into thunderclouds.

The electric field in the lower atmosphere is vertical and directed downward during fair-weather conditions and large-scale atmospheric homogeneity. In the literature of atmospheric electricity, that direction is defined as the direction a positive charge moves in the electric field (e.g., Chalmers, 1967). At globally representative stations, such as ocean or polar stations, the vertical columnar resistance remains nearly constant during fair weather (Dolezalek, 1972). Here the electric-field strength near the ground shows a diurnal variation with universal time similar to that shown for the air-earth current density in Figure 12.17, both reflecting variations of the ionospheric potential. Over the continents consideration must be given to a varying ionization rate in the first few hundred meters caused by the exhalation of radioactive materials from the Earth. This ionization rate depends strongly on different meteorological parameters, such as convection, and therefore the columnar resistance can no longer be regarded as constant. As shown by Israël (1973b) the global diurnal variation of the electric field is normally masked by local variations at these stations. If local generators, such as precipitation, convection currents, and blowing snow or dust,

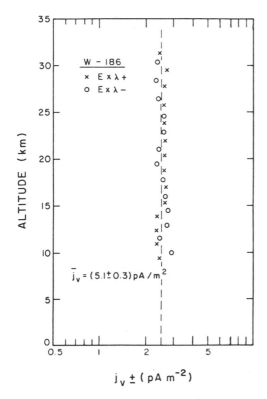

FIGURE 12.16 Polar air-earth current densities j_{v+} and j_{v-} measured on August 4, 1978, at Laramie, Wyoming. The mean total air-earth current density is $j_v = (5.1 \pm 0.3)$ pA/m^2.

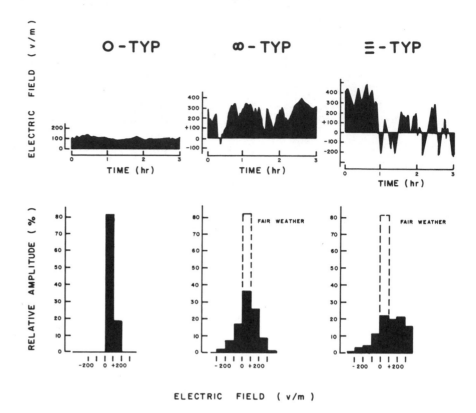

FIGURE 12.18 Typical variations of the vertical electric field near the ground and their relative amplitude distribution during fair weather (0), haze (∞), and fog (≡) (from Fischer, 1977).

also become active, the description becomes increasingly complicated and the vertical electric-field strength E_z can vary considerably.

Figure 12.18 shows typical variations of E_z at a continental mid-latitude station during fair weather and also during haze and fog (Fischer, 1977). The relative occurrence of the field amplitudes is also shown. During fair weather the variation is small with a mean value of E_z about 120 V/m with no negative fields. During haze and fog the variations become much larger and even negative values of E_z occur indicating the presence of space charges around the station. The higher positive E_z values are mainly caused by a drastic reduction of the atmospheric conductivity due to the attachment of small ions to haze or fog droplets.

The highest values of E_z are measured during rain or snow showers and thunderstorms as shown in Figure 12.19 (Fischer, 1977). For both cases the amplitude distribution shows a typical U pattern with mainly high positive or negative field values. The highest field values can reach 5000 V/m at the ground. This seems to be an upper limit because corona discharges build up a space-charge layer with the appropriate sign so as to reduce the original field values. Examples of anthropogenic influences are shown in Figures 12.20 and 12.21. Figure

12.20 shows the undisturbed and disturbed electric-field values near the ground on the upwind and downwind side of a high-voltage power line. Figure 12.21 shows the undisturbed and disturbed fields values near a large city in Germany (Fischer, 1977). Whereas the station at the south shows almost a fair-weather field pattern, the values at the northern station exhibit large variations, and even negative values of E_z occur. The reasons for the large variations at the disturbed stations are in both cases drifting pollution and/or space charges. Above the ground the vertical electric field E_z drops rapidly with increasing altitude owing to the increasing atmospheric conductivity. Figure 12.22 shows the decrease of the vertical electric field with altitude during fair weather, during cloudiness without precipitation, and during haze and fog as measured again over Weissenau, Germany (Fischer, 1977). The positive sign of E_z means that the field vector is pointed downward again. The variations of different profiles, shown by the hachured areas, are mainly caused by conductivity variations, especially in the lower troposphere, rather than by variations of the ionospheric potential itself. The scatter is greatly increased during periods of cloudiness and haze or fog, whereas the mean profile of the same 20 balloon flights (indicated by the thick curve in Figure

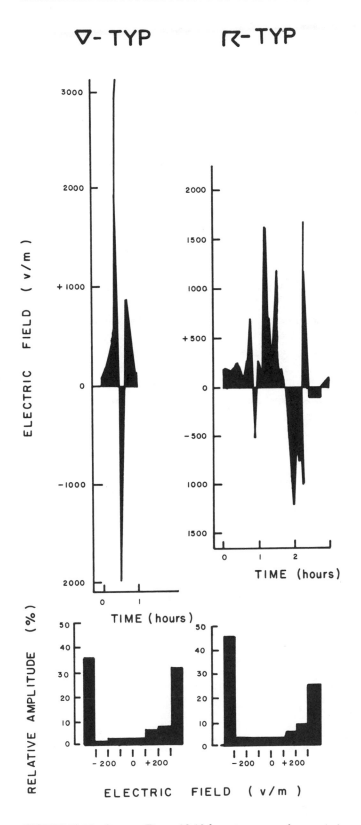

FIGURE 12.19   Same as Figure 12.18 for rain or snow showers (▽) and for thunderstorms (↖) (from Fischer, 1977).

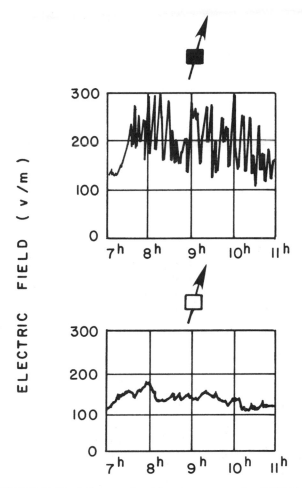

FIGURE 12.20   Influence of a high-voltage power line (220 kV) on the electric field near the ground (black station) compared with the electric field at the undisturbed station (Fischer, 1977).

12.22) still shows a pattern typical for fair-weather conditions. During rain or snow and especially in thunderclouds, the scatter in the field values becomes much larger, including regions with large negative field values. Temporal variations are fast, and the horizontal components of the electric-field strength can reach the same order of magnitude as the vertical components as measured by Winn *et al.* (1978). These large variations of the electric-field strength in shower and thunderclouds are caused by regions of high-space-charge density of both signs in these clouds.

Above the tropopause the vertical electric-field strength continues to decrease nearly exponentially as the atmospheric conductivity increases and normally drops to around 300 mV/m at 30 km at mid-latitudes. Holzworth and Mozer (1979) showed that solar proton events can cause large reductions of the stratospheric electric-field values at high latitudes by more than an

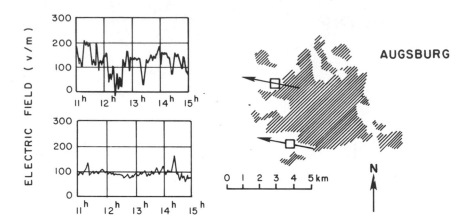

FIGURE 12.21 Disturbed electric field near the ground in the neighborhood of a big city (black station). The electric-field pattern at the south is not disturbed and shows typical fair-weather values (Fischer, 1977).

order of magnitude. The authors could explain their behavior with the greatly enhanced ionization by solar protons, which in turn enhances the stratospheric conductivity and thereby reduces the local electric fields.

CONCLUSION

Galactic cosmic rays are the primary source of ionization in the lower atmosphere. They control the bulk atmospheric conductivity parameter, which in turn is linearly related to the small-ion concentration and small-ion mobility. Although the existence of a solar-induced modulation of the ionization rate and of the atmospheric conductivity is evident, the basic physical mechanisms are not fully understood. More measurements of both parameters (if possible simultaneously) are desirable in order to establish cyclic and transient solar modulation effects. The accuracy of direct measurements of small-ion concentrations and mobilities have shown improvements, but the results among different researchers are still contradictory. Almost nothing is known about the mobility distribution of atmospheric ions throughout the troposphere and lower stratosphere. The existence of heavy ions with masses of several hundred amu, as inferred from mobility measurements, has only just recently been established by mass spectrometer measurements. Problems with the sampling procedure employed can not be overlooked. Simultaneous measurements of ion masses, ion concentrations, and ion mobilities, together with conductivity

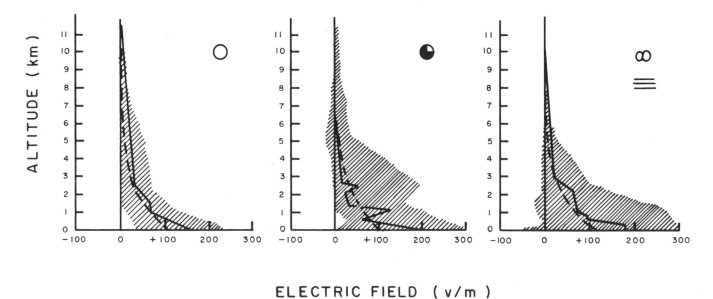

FIGURE 12.22   The atmospheric electric field versus altitude during fair weather (○), cloudiness without precipitation (◕), and during haze and fog (∞, ≡). The black curves show typical measurements, the white curves show mean profiles and the hachured areas show the scattering of the values (Fischer, 1977).

and aerosol measurements, are clearly needed to gain a deeper insight into the physics and chemistry of atmospheric ions. Volcanic eruptions can result in a rather dramatic increase of the aerosol content of the lower atmosphere. The extent to which these aerosols directly affect the atmospheric conductivity, small-ion concentration, and mobility should be investigated to a greater extent. On the other hand, the sulfuric acid content of the lower atmosphere is also drastically enhanced following volcanic eruptions and might considerably influence the ion composition. Air-earth current-density measurements seem to be consistent with the classical picture of the global circuit with some exceptions. The question of whether there are other global generators in the lower atmosphere in addition to thunderstorms could probably be unraveled by ground-based and balloonborne current measurements at different locations. The electric-field strength in the lower atmosphere, which is closely related to the air-earth current and the atmospheric conductivity, can undergo considerable fluctuations near the ground owing to conductivity variations and the influence of a local generator.

REFERENCES

Arnold, F. (1983). Ion nucleation—A potential source for stratospheric aerosols, *Nature 299*, 134-137.

Arnold, F., and Th. Bührke (1984). New H_2SO_4 and HSO_3 vapor measurements in the stratosphere—Evidence for a volcanic influence, *Nature 301*, 293-295.

Arnold, F., G. Henschen, and E. E. Ferguson (1981). Mass spectrometer measurements of fractional ion abundances in the stratosphere—Positive ions, *Planet. Space Sci. 29*, 185-193.

Barouch, E., and L. F. Burlaga (1975). Causes of Forbush decreases and other cosmic ray variations, *J. Geophys. Res. 80*, 449-456.

Burke, H. K., and A. A. Few (1978). Direct measurements of the atmospheric conduction current, *J. Geophys. Res. 83*, 3093-3098.

Chalmers, J. A. (1967). *Atmospheric Electricity*, Pergamon Press, London.

Cobb, W. E. (1977). Atmospheric electric measurements at the South Pole, in *Electrical Processes in Atmospheres*, H. Dolezalek and R. Reiter, eds., Steinkopff, Darmstadt, pp. 161-167.

Dolezalek, H. (1972). Discussion of the fundamental problem of atmospheric electricity, *Pure Appl. Geophys. 100*, 8-43.

Duggal, S. P., and M. A. Pomerantz (1977). The origin of transient cosmic ray intensity variations, *J. Geophys. Res. 82*, 2170-2174.

Fischer, H. J. (1977). Das luftelektrische Feld in Abhangigkeit von Luftverunreinigung und Wetterlage, *Prometheus 7/2*, 4-12.

Fischer, H. J., and R. Mühleisen (1972). Variationen des Ionospharenpotentials und der Weltgewittertatigkeit im 11-jahrigen solaren Zyklus, *Meteorol. Rundsch. 25*, 6-10.

Fischer, H. J., and R. Mühleisen (1975). A method for precise determination of the voltage between ionosphere and ground, Report, Astron. Inst., Universität Tübingen, FRG.

Forbush, S. E. (1954). World-wide cosmic-ray variations, 1937-1952, *J. Geophys. Res. 59*, 525-542.

Gringel, W. (1978). Untersuchungen zur elekrischen Luftleitfahigkeit unter Berucksichtigung der Sonnenaktivitat und der Aerosolteilchenkonzentration bis 35 km Hohe, Dissertation, Universität Tübingen, FRG.

Gringel, W., and R. Mühleisen (1978). Sahara dust concentration in the troposphere over the North Atlantic derived from measurements of air conductivity, *Beitr. Phys. Atmos. 51*, 121-128.

Gringel, W., J. Leidel, and R. Mühleisen (1978). The air-earth current density at the water surface and in the free atmosphere above the ocean, *Meteor Forschungsergebn. Reihe B 13*, 41-52.

Gringel, W., D. J. Hofmann, and J. M. Rosen (1983). Measurements of mobility, small ion recombination and aerosol attachment coefficients to 33 km, unpublished manuscript.

Gringel, W., D. J. Hofmann, and J. M. Rosen (1984). Stratospheric conductivity reductions related to El Chichon aerosol layers, presented at the VII International Conference on Atmospheric Electricity.

Heaps, M. G. (1978). Parameterization of the cosmic ray ion-pair production rate above 18 km, *Planet. Space Sci. 26*, 513-517.

Hofmann, D. J., and J. M. Rosen (1979). Balloon-borne measurements of atmospheric electrical parameters. I: The ionization rate, Atmos. Phys. Rep. AP-54, Univ. of Wyoming, Laramie.

Hofmann, D. J., and J. M. Rosen (1983a). Stratospheric sulfuric acid fraction and mass estimate for the 1982 volcanic eruption of El Chichon, *Geophys. Res. Lett. 10*, 313-316.

Hofmann, D. J., and J. M. Rosen (1983b). Sulfuric acid droplet formation and growth in the stratosphere after the 1982 eruption of El Chichon, *Science 222*, 325-327.

Hofmann, D. J., J. M. Rosen, T. J. Pepin, and R. G. Pinnick (1975). Stratospheric aerosol measurements I: Time variations at northern mid-latitudes, *J. Atmos. Sci. 32*, 1446-1456.

Hogan, A. W., and V. A. Mohnen (1979). On the global distributions of aerosols, *Science 205*, 1373-1375.

Holzworth, R. H., and F. S. Mozer (1979). Direct evidence of solar flare modification of stratospheric electric fields, *J. Geophys. Res. 84*, 363-367.

Israël, H. (1973a). *Atmospheric Electricity, Vol. I*, Israel Program for Scientific Translations, Jerusalem.

Israël, H. (1973b). *Atmospheric Electricity, Vol. II*, Israel Program for Scientific Translations, Jerusalem.

Junge, C. E. (1963). *Air Chemistry and Radioactivity*, Academic Press, New York.

Kasemir, H. W. (1960). A radiosonde for measuring the air-earth current density, USASRDL Tech. Rep. 2125.

Kroening, J. L. (1960). Ion density measurements in the stratosphere, *J. Geophys. Res. 65*, 145-151.

Markson, R. (1978). Solar modulation of atmospheric electrification and possible implications for the sun-weather relationship, *Nature 273*, 103-109.

Meyerott, R. E., J. B. Reagen, and R. G. Joiner (1980). The mobility and concentration of ions and the ionic conductivity in the lower stratosphere, *J. Geophys. Res. 85*, 1273-1278.

Mitchell, J. D., R. S. Sagar, and R. S. Olsen (1977). Positive ions in the middle atmosphere during sunrise conditions, Rep. ECOM-5819, U.S. Army Electron. Command, Fort Monmouth, N.J.

Morita, Y., H. Ishikawa, and M. Kanada (1971). The vertical profiles of the small ion density and the electric conductivity in the atmosphere up to 19 km, *J. Geophys. Res. 76*, 3431-3436.

Neher, H. V. (1961). Cosmic-ray knee in 1958, *J. Geophys. Res. 66*, 4007-4012.

Neher, H. V. (1967). Cosmic ray particles that changed from 1954 to 1958 to 1965, *J. Geophys. Res. 72*, 1527-1539.

Ogawa, T., Y. Tanaka, A. Huzita, and M. Yasuhara (1977). Three dimensional electric fields and currents in the stratosphere, in *Elec-*

trical Processes of Atmospheres, H. Dolezalek and R. Reiter, eds., Steinkopff, Darmstadt, pp. 552-556.

Paltridge, G. W. (1965). Experimental measurements of the small ion density and electrical conductivity of the stratosphere, *J. Geophys. Res. 70*, 2751-2761.

Patterson, E. M., and D. A. Gillette (1977). Commonalities in measured size distribution for aerosols having a soil-derived component, *J. Geophys. Res. 82*, 2074-2081.

Pierce, E. T., and A. L. Whitson (1964). The variation of potential gradient with altitude above ground of high radioactivity, *J. Geophys. Res. 69*, 2895-2898.

Pinnick, R. G., J. M. Rosen, and D. J. Hofmann (1976). Stratospheric aerosol measurements III: Optical model calculations, *J. Atmos. Sci. 33*, 304-314.

Pollack, J. B., O. B. Toon, E. F. Danielson, D. J. Hofmann, and J. M. Rosen (1983). The El Chichon volcanic cloud: An introduction, *Geophys. Res. Lett. 10*, 989-992.

Riekert, H. (1971). Untersuchungen zur Beweglichkeit der Kleinionen in der freien Atmosphere, Dissertation, Universität Tübingen, FRG.

Rosen, J. M., and D. J. Hofmann (1981a). Balloon borne measurements of the small ion concentration, *J. Geophys. Res. 86*, 7399-7405.

Rosen, J. M., and D. J. Hofmann (1981b). Balloon borne measurements of electrical conductivity, mobility, and the recombination coefficient, *J. Geophys. Res. 86*, 7406-7410.

Rosen, J. M., and D. J. Hofmann (1983). Unusual behavior in the condensation nuclei concentration at 30 km, *J. Geophys. Res. 88*, 3725-3731.

Rosen, J. M., D. J. Hofmann, and K. H. Kaselau (1978a). Vertical profiles of condensation nuclei, *J. Appl. Meteorol. 17*, 1737-1740.

Rosen, J. M., D. J. Hofmann, and S. P. Singh (1978b). A steady-state aerosol model, *J. Atmos. Sci. 35*, 1304-1313.

Rosen, J. M., D. J. Hofmann, W. Gringel, J. Berlinski, S. Michnowski, Y. Morita, T. Ogawa, and D. Olson (1982). Results of an international workshop on atmospheric electrical measurements, *J. Geophys. Res. 87*, 1219-1227.

Russell, P. B., T. J. Swissler, M. P. McCormick, W. P. Chu, J. M. Livingston, and T. J. Pepin (1981). Satellite and correlative measurements of the stratospheric aerosol I: An optical model for data conversions, *J. Atmos. Sci. 38*, 1279-1294.

Smith, D., and M. J. Church (1977). Ion-ion recombination rates in the Earth's atmosphere, *Planet. Space Sci. 25*, 433-439.

Viggiano, A. A., H. Schlager, and F. Arnold (1983). Stratospheric negative ions—Detailed height profiles, *Planet. Space Sci. 31*, 813-820.

Widdel, H. U., G. Rose, and R. Borchers (1976). Experimental results on the variation of electrical conductivity and ion mobility in the mesosphere, *J. Geophys. Res. 81*, 6217-6220.

Willeke, K, and K. T. Whitby (1975). Atmospheric aerosols: Size distribution interpretation, *J. Air Pollut. Control Assoc. 25*, 529-534.

Winn, W. P., C. B. Moore, C. R. Holmes, and L. G. Byerley III (1978). Thunderstorm on July 16, 1975, over Langmuir Laboratory: A case study, *J. Geophys. Res. 83*, 3079-3092.

Zikmunda, J., and V. A. Mohnen (1972). Ion annihilation by aerosol particles from ground level to 60 km height, *Meteorol. Rundsch. 25*, 10-14.

Electrical Structure of the Middle Atmosphere

13

GEORGE C. REID
NOAA Aeronomy Laboratory

INTRODUCTION

Conventional usage divides the atmosphere into layers on the basis of the average temperature profile. The stratosphere is the region of positive vertical temperature gradient extending from the tropopause to a height of about 50 km, and the overlying region of negative temperature gradient is the mesosphere, extending to about 85 km altitude, where the lowest temperatures in the atmosphere are reached. The main heat source in both of these regions is provided by absorption of solar-ultraviolet radiation by ozone. At still greater heights lies the thermosphere, in which absorption of extreme-ultraviolet radiation causes the temperature to increase again with height. This chapter is concerned mainly with the electrical properties of the upper stratosphere and the mesosphere. The troposphere and lower stratosphere were considered in Chapters 11 and 12 (this volume) and the thermosphere is discussed in Chapter 14.

Both the composition and the temperature play important roles in determining the electrical structure. As noted above, the temperature in the stratosphere rises from typical tropopause values near 200 K to values of about 270 K at the stratopause (~50 km), above which the temperature decreases to mesopause values that are seasonally and latitudinally variable, occasionally dropping below 140 K in the high-latitude summer.

The principal atmospheric constituents are molecular oxygen and nitrogen, just as in the lower atmosphere; but there are a number of minor constituents that are important from the point of view of the electrical properties. Among these are nitric oxide (NO), which diffuses into the region from sources below and above; atomic oxygen (O) and ozone (O_3), which are formed locally by photodissociation of O_2; and water vapor, which can be transported from the troposphere as well as being locally produced. The role of aerosols in the atmosphere at heights above 30 km is uncertain and controversial and is an area of active study. The occasional presence of noctilucent clouds at the high-latitude summer mesopause and the more regular existence of a summertime polar scattering layer seen by satellites have certainly shown that aerosols (probably ice crystals) can exist near the top of the region, but the gap between the mesopause and the well-known aerosol layers of the lower stratosphere remains relatively unexplored.

In what follows, sources of ionization, the ion chemistry that determines the steady-state ion composition, and the present status of our knowledge of aerosol distribution are discussed. The final two sections discuss the theory and measurement of conductivity and electric fields in the middle atmosphere.

183

SOURCES OF IONIZATION

Figure 13.1 shows typical ion-pair production rates (q) at middle latitudes during daytime. Throughout the stratosphere, galactic cosmic rays provide the principal ionization source, as in most of the lower atmosphere. The cosmic-ray ionization rate does not vary diurnally but does vary with geomagnetic latitude and with the phase of the 11-year solar cycle. Heaps (1978) provided useful relations for computing the rate of ion production at any latitude and time. Roughly, the ion-production rate above 30 km increases by a factor of 10 in going from the geomagnetic equator to the polar caps at sunspot minimum (cosmic-ray maximum) and by a factor of 5 at sunspot maximum. The solar-cycle modulation is near zero at the equator, increasing to a factor of about 2 in the polar caps. The ionization rate above 30 km is approximately proportional to the atmospheric density. These properties are a result of (a) the shielding effect ot the geomagnetic field, which allows cosmic-ray particles to enter the atmosphere at successively higher latitudes for successively lower energies, and (b) the reduction in cosmic-ray flux in the inner solar system as solar activity intensifies.

Superimposed on these long-term global variations are brief reductions in cosmic-ray flux known as Forbush decreases, after their discoverer (Forbush, 1938). Forbush decreases occur in coincidence with geomagnetic storms and are of brief (hours) duration. However, as their magnitude can be as large as some tens of percent, they can change global electrical parameters significantly.

In the mesosphere the major daytime source of ionization in undisturbed conditions is provided by the NO molecule, whose low ionization potential of 9.25 electron volts (eV) allows it to be ionized by the intense solar Lyman-alpha radiation. The concentration of NO in the mesosphere is not well known and is almost certainly variable (Solomon et al., 1982a) in response to meteorological factors. The production-rate profile in Figure 13.1 is an estimate based on reasonable values for the NO concentration and the solar Lyman-alpha flux, which is itself a function of solar activity (Cook et al., 1980).

At the upper limit of the middle atmosphere, significant amounts of ionization are produced by solar x rays, forming the base of the E region of the ionosphere, and by ionization of O_2 in its metastable $^1\Delta$ state, which is a by-product of ozone photodissociation. While these sources are never competitive with the NO source in terms of ionization rates, they give rise to different primary positive-ion species (N_2^+ and O_2^+ as opposed to NO^+), and hence to different chemical reaction chains.

A sporadic and intense source of ionization at high latitudes is provided by solar-proton events (SPE) (Reid, 1974), and Figure 13.1 shows an ionization-rate profile calculated for the peak of a major SPE in May 1959. These events are caused by the entry into the atmosphere of particles accelerated during solar flares and traveling fairly directly from the Sun to the Earth. The particles are mostly protons, with much smaller fluxes of heavier nuclei and of electrons, having typical energies of 1 to 100 MeV and considerably less atmospheric penetration power than galactic cosmic rays. As a consequence, their effects are largely confined to high magnetic latitudes ($\gtrsim 60°$) and to altitudes well above the lower stratosphere. Solar-proton events typically reach their peak intensity within a few hours of a major solar flare and then decay exponentially over the following day or two. Their occurrence is a strong function of the phase of the solar cycle, as illustrated in Figure 13.2, which shows the distribution in the 1956-1973 period of polar-cap absorption (PCA) events and of ground-level events (Pomerantz and Duggal, 1974). Polar-cap absorption is the name given to the intense radio-wave absorption caused by the enhanced mesospheric ionization during an SPE, while ground-level events are the rare events with a large enough high-energy flux to cause an increase in cosmic-ray neutron monitors at the surface. The frequency of the events is related to the solar-activity cycle, which peaked about 1958 and 1969, but intense events can occur at any time, as evidenced by those of February 1956 and August 1972. Figure 13.1 shows clearly that SPEs cause major alterations in middle-atmospheric ionization rates, and hence in the electrical parameters.

Energetic electron precipitation from the radiation

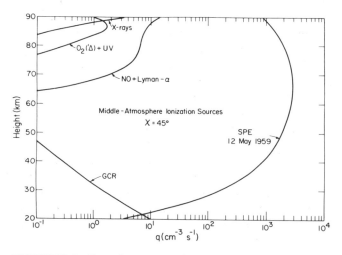

FIGURE 13.1   Typical ion-pair production rates in the middle atmosphere.

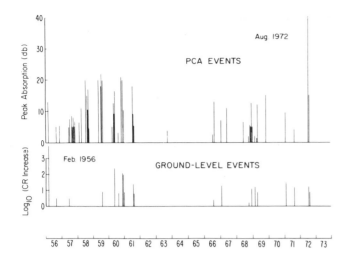

FIGURE 13.2 Distribution and intensity of solar energetic-particle events, 1956-1973. The peak absorption in the upper part is a measure of the intensity of the polar-cap absorption (PCA) events resulting from high-latitude ionization in the mesosphere; the lower part shows the intensity of the cosmic-ray (CR) increases recorded by neutron monitors and caused by solar-particle-induced nuclear reactions in the lower stratosphere.

belts also contributes to middle-atmosphere ionization but in a manner that is highly variable in both latitude and time. During major electron precipitation events this can become the dominant source of ionization above 70 km for brief periods, and ionization rates can be as high as 10^5 cm^{-3} sec^{-1} above 80 km (Reagan, 1977). Vampola and Gorney (1983) deduced zonally averaged ionization rates due to energetic electron precipitation at several magnetic latitudes. Maximum ionization rates occur between 80 and 90 km and vary between about 0.7 cm^{-3} sec^{-1} at 45° and 6 cm^{-3} sec^{-1} at 65° latitude. At the higher latitudes energetic electrons are competitive with solar Lyman-alpha as an ionization source even in daytime. They are probably the dominant source above 70 km at night, when the main competitor is photoionization of NO by the weak Lyman-alpha radiation scattered from the Earth's hydrogen geocorona (Strobel et al., 1974).

Bremsstrahlung x rays, generated by the energetic electrons, ionize weakly at heights below 60 km (Luhmann, 1977; Vampola and Gorney, 1983) but are probably rarely competitive with cosmic rays as a global ionization source.

ION CHEMISTRY IN THE MIDDLE ATMOSPHERE

The principal primary positive ions produced in the middle atmosphere are N_2^+, O_2^+, and NO^+, all of which participate in a wide range of ion-molecule reactions that lead to a rich spectrum of ambient ions. An equally rich spectrum of negative ions is generated by reactions that are initiated by the attachment of electrons to form the main primary species O_2^- and O^-. In this section the current state of our knowledge of this ion chemistry and of the steady-state ion composition that it produces are discussed. More detailed treatments can be found in review articles by Ferguson et al. (1979) and Ferguson and Arnold (1981).

Positive Ions

The first measurements of positive-ion composition in the mesosphere were made by a rocketborne mass spectrometer in 1963 (Narcisi and Bailey, 1965). The dominant species below the mesopause were found to be proton hydrates, i.e., members of the family $H^+(H_2O)_n$, with a sharp transition at about the mesopause to such simple species as O_2^+, NO^+, and several metallic species, probably of meteoric origin. Many subsequent measurements have verified these results and have shown that the size spectrum of the proton hydrates is very temperature sensitive. At the cold high-latitude summer mesopause, as many as 20 water molecules have been seen clustered in individual ions (Björn and Arnold, 1981).

The currently proposed positive-ion reaction scheme leading from the primary ions to the proton hydrates is illustrated in Figure 13.3. Since N_2^+ is rapidly converted into O_2^+ by charge exchange with O_2, the two primary ions of concern are O_2^+ and NO^+. The chain that converts O_2^+ into the proton hydrates was identified by Fehsenfeld and Ferguson (1969) and Good et al. (1970) and is fairly straightforward. Clustering of O_2^+ to O_2 forms O_4^+, which rapidly undergoes a switching reaction in which the O_2 molecule forming the cluster switches with an H_2O molecule to form $O_2^+(H_2O)$. When they are energetically allowed, such switching reactions are usually fast, occurring at virtually every collision between the two species. Subsequent collisions with water molecules lead rapidly to the proton hydrates.

The failure of this mechanism above the mesopause is probably due to a combination of factors: the decreasing water-vapor concentration, the increasing electron concentration leading to shorter ion lifetimes against recombination, and the increasing concentration of atomic oxygen. The latter attacks the O_4^+ clusters through the reaction

$$O_4^+ + O \longrightarrow O_2^+ + O_3. \qquad (13.1)$$

The chain of reactions leading from NO^+ to the proton hydrates is less certain but probably involves several

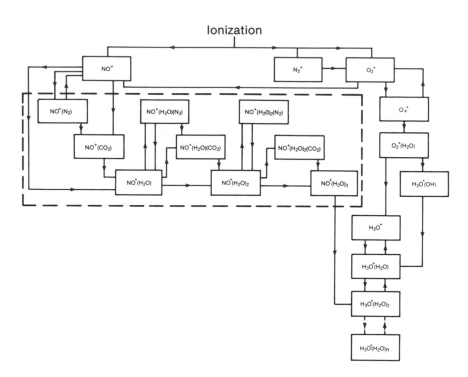

FIGURE 13.3 Schematic diagram of the principal positive-ion reactions in the mesosphere. The details of the NO^+ hydration scheme, enclosed by the broken lines, are not yet established.

steps of clustering and switching as shown in Figure 13.3. This mechanism, first proposed by Ferguson (1974), yields steady-state ion distributions that are reasonably close to those observed when appropriate reaction rates are used in model calculations (Reid, 1977). Most of the critical reaction rates are still unmeasured at mesospheric temperatures, however.

In the stratosphere, the picture is rather more complicated. Mass spectrometry has recently been developed for use at the high ambient gas pressures of the stratosphere, and measurements of positive-ion composition have been made from rockets and balloons (Arnold *et al.*, 1978). These experiments showed the existence of the proton hydrates, as in the mesosphere, and also that below about 40 km the proton hydrates are replaced as the dominant species by ions with a core of mass 42 amu. This has been tentatively identified as protonated acetonitrile, $H^+(CH_3CN)$ (Arnold *et al.*, 1978)—an identification that is reasonable in view of the high proton affinity of CH_3CN and its recent discovery in the troposphere (Becker and Ionescu, 1982).

It should be emphasized that our knowledge of stratospheric ion composition is very sketchy. Almost nothing is known of the composition at heights below 30 km or at locations other than continental middle latitudes.

Negative Ions

Our knowledge of negative-ion composition in the middle atmosphere is in an unsatisfactory state. Labora-

tory measurements of the negative-ion reactions thought to be the most important ones in the atmosphere have led to the reaction scheme shown in Figure 13.4 (Ferguson *et al.*, 1979). In this scheme, direct attachment of electrons takes place only to O_2 and O_3; associative detachment reactions occurring chiefly with atomic oxygen quickly destroy most of the resulting O_2^- and O^- ions in regions where O is present. The ions that escape destruction in this way, however, go on to form a wide variety of species whose electron affinity increases as we progress down the chain. In the absence of annihilation by positive ions, the dominant terminal species in the chain would be the nitrate ion, NO_3^-, with the high electron affinity of 3.9 eV (Ferguson *et al.*, 1972).

Mass-spectrometer measurements of negative-ion composition are much more difficult to make than the corresponding positive-ion measurements, largely owing to the problem of contamination by electrons. As a result, few measurements have been made in the mesosphere, and these have given somewhat conflicting results (Narcisi *et al.*, 1971; Arnold *et al.*, 1971, 1982). The predicted dominance of such species as NO_3^- and CO_3^- at heights below 80 km appears to be borne out, but many unidentified light ions have been seen in the mesosphere. Above 80 km, there appears to be a layer of heavy (>100 amu) ions (Arnold *et al.*, 1982) that may be a result of attachment to neutral species of meteoric origin, perhaps forming the very stable silicon species SiO_3^- (Viggiano *et al.*, 1982).

In the stratosphere, the first mass-spectrometer mea-

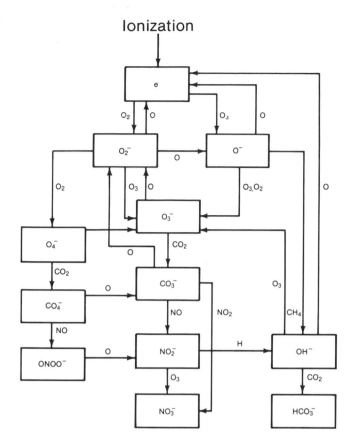

FIGURE 13.4 Schematic diagram of the principal negative-ion reactions in the mesosphere. The chain leading to the terminal species HCO_3^- is probably of minor importance.

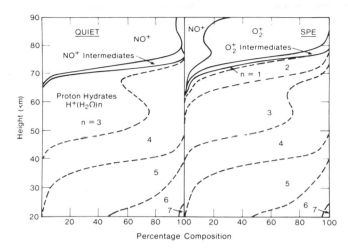

FIGURE 13.5 Model calculations of the steady-state positive-ion composition of the middle atmosphere, omitting the reactions leading to nonproton hydrates in the stratosphere. The left-hand panel represents quiet conditions, and the right-hand panel is for the case of an intense solar proton event.

Model Calculations

If the rates of production of the various ion species and the rates of the important chemical reactions are known, it is possible to calculate the steady-state ion composition. Many such calculations have been made, and examples are shown in Figures 13.5 and 13.6.

Figure 13.5 illustrates the positive-ion composition calculated for an intense solar-particle event (right-hand panel) and for undisturbed daytime conditions

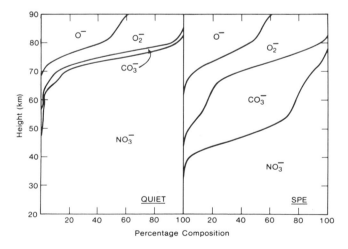

FIGURE 13.6 Model calculations of the steady-state negative-ion composition of the middle atmosphere, omitting reactions involving meteoric species at the higher altitudes and reactions involving sulfur species in the stratosphere. The left-hand panel represents quiet conditions, and the right-hand panel is for the case of an intense solar proton event.

surements (Arnold and Henschen, 1978) showed the dominance of heavy ions. Laboratory measurements by Viggiano *et al.* (1980) showed that in the presence of sulfuric acid the species HSO_4^- would become an important core ion. Sulfuric acid is known to be the major component of the stratospheric aerosol layer and is a by-product of volcanic activity, as discussed in Chapter 12 (this volume).

There thus appear to be three fairly distinct negative-ion strata in the middle atmosphere. The central region between about 55 and 80 km is formed mainly by ion-molecule reactions involving the commoner minor constituents, following initial attachment to O_2. This central layer has layers of heavier ions both above and below—the upper one probably a result of reactions involving meteoric species and the lower one built by clustering around HSO_4^-. At present this is a very sketchy and incomplete picture, and many more observations are needed to clarify it.

(left-hand panel). The stratospheric ion chemistry leading to the formation of nonproton hydrates has been omitted. In both cases the proton hydrates are dominant in the lower mesosphere with a fairly abrupt transition to NO^+ at greater heights in the quiet case and to mostly O_2^+ in the SPE case. A thin layer containing mostly the intermediate clusters of O_2^+ or NO^+ separates the two regions. The tendency toward horizontal layering of the principal proton-hydrate species is caused by the shift in equilibrium toward lighter species as the temperature increases.

Figure 13.6 shows the result of similar calculations of negative-ion composition, again omitting both the upper meteoric layer above 80 km and the stratospheric region of heavy HSO_4^- derived ions. In the undisturbed case NO_3^- is dominant in most of the middle atmosphere, but the great enhancement in the rate of ion-ion recombination during the SPE inhibits the formation of NO_3^- and leads to an increase in the fraction of CO_3^-. The initial ions O^- and O_2^- are the main components at the top of the mesosphere.

The ultimate loss of ions in the middle atmosphere takes place mainly through recombination with electrons or with ions of the opposite charge. In the case of negative ions, photodetachment by sunlight provides another loss mechanism about which little quantitative information is available for the main atmospheric species. Loss to aerosol particles is presumably also a significant sink, especially in the vicinity of the stratospheric aerosol layer and possibly of the polar aerosol layers near the summer mesopause. Neither photodetachment nor aerosol attachment were included in the calculations represented by Figure 13.6.

MESOSPHERIC AEROSOLS

Stratospheric aerosols were discussed in Chapter 12 (this volume). The aerosol content of the mesosphere and the effect of these aerosols on the electrical properties are much more speculative. Noctilucent clouds provide direct evidence that particulate material does exist at mesospheric heights, at least on some occasions. These silvery-blue translucent clouds appear sporadically at high latitudes during the summer months, when their great height allows them to be seen by scattered sunlight long after the Sun has set at the surface. Their phenomenology has been reviewed by Fogle and Haurwitz (1966). Optical measurements suggest that the particle concentrations are 1 to 50 cm^{-3} in these clouds, with an individual particle radius of the order of 0.1 μm. Satellite measurements of backscattered sunlight reveal the existence near the mesopause of a denser semipermanent particle layer over the summer polar

cap (Donahue et al., 1972; Thomas et al., 1982), which is probably related to the noctilucent cloud layer.

There is general agreement that the particles forming these layers are ice crystals formed in the extremely low temperatures of the summer mesopause region by condensation from the low background concentration of water vapor (Hesstvedt, 1961). Several model calculations have shown that it is reasonable to expect ice crystals to form under these conditions (e.g., Charlson, 1965; Reid, 1975; Turco et al., 1982) provided that suitable nucleation centers exist. Mass-spectrometer results suggest that nucleation does occur on positive ions (Goldberg and Witt, 1977; Björn and Arnold, 1981), and the particles themselves could then act as surfaces for ion capture. There is evidence for abnormally low electron concentrations in the high-latitude summer mesopause region, perhaps indicating enhanced electron loss through attachment to particles (e.g., Pedersen et al., 1970).

Below the mesopause, the evidence is much less direct. Volz and Goody (1962) found evidence from twilight measurements of low concentrations of dust particles throughout the mesosphere. Hunten et al. (1980) calculated the flux of particles produced by condensation of meteor ablation products. Depending on the model conditions, this calculation predicted mesospheric concentrations of 10^2 to 10^3 cm^{-3} and individual particle radii of a few nanometers. The influence of such a particle distribution on the ion and electron concentrations would be small. Chesworth and Hale (1974) proposed the existence of mesospheric particle concentrations of 10^3 to 10^4 cm^{-3} to explain certain discrepancies in the electrical parameters. The evidence for such large concentrations was indirect, and further work is needed in this area. In particular, the relationship, if any, between the meteor-ablation particles of Hunten et al. (1980) and the particles suggested by Chesworth and Hale (1974) should be studied.

CONDUCTIVITY IN THE MIDDLE ATMOSPHERE

The current density, j, and the electric field, E, are related by the familiar Ohm's law expression

$$j = \sigma E, \qquad (13.2)$$

where σ is the conductivity. In the lower atmosphere and most of the middle atmosphere, σ is a scalar, and the electric field and the current lie in the same direction. Above about 70 km, however, collisions between electrons and air molecules become infrequent enough that the bending of the electron path by the Earth's magnetic field becomes appreciable between collisions, and motion perpendicular to the field becomes less easy than motion along the field. In these circumstances, σ becomes a tensor, and an applied electric field with a component perpendicular to the magnetic field drives a current in a different direction. This anisotropy of the conductivity becomes a dominant influence on the electrical properties of the thermosphere (see Chapter 14, this volume).

Conductivity can be measured directly by rocket-borne probes, and a substantial number of such conductivity values are reported in the literature, especially using the so-called blunt-probe technique (Hale *et al.*, 1968). The problem of shock-wave effects associated with a supersonic rocket are usually avoided in these experiments by deploying the probe with a parachute at the top of the trajectory and making the measurements during the subsonic descent.

In the presence of a mixture of ions, the conductivity can be expressed in terms of the mobilities of the individual species as

$$\sigma = e\Sigma n_i^+ k_i^+ + e\Sigma n_j^- k_j^- + e n_e k_e, \qquad (13.3)$$

where n and k are the concentrations and mobilities of the positive ions, negative ions, and electrons. [Equation (13.3) is identical to Eq. (12.1) in Chapter 12, this volume, except that in the middle atmosphere electrons come into consideration.] An experimentally derived relationship between reduced mobility and ion mass in nitrogen is shown in Figure 13.7 (Meyerott *et al.*, 1980).

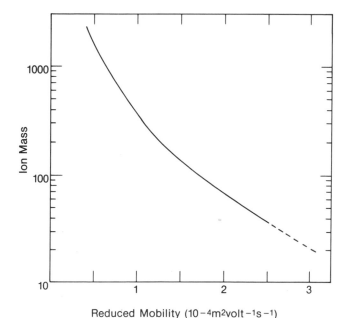

FIGURE 13.7   Reduced mobility as a function of ion mass.

The reduced mobility, k_0, is the mobility in a standard atmosphere and is related to the actual mobility, k, by

$$k = k_0 \left(\frac{T}{273}\right)\left(\frac{1000}{p}\right), \qquad (13.4)$$

where T is absolute temperature and p is pressure in millibars.

The ion mobility can be measured by the Gerdien condenser technique (e.g., Conley, 1974), and a number of measurements using rockets have been reported. As discussed by Meyerott *et al.* (1980), there is no general agreement among the various measurements, although the data of Conley (1974) and Widdel *et al.* (1976) suggest a single reduced mobility of about 2.7×10^{-4} m^2 V^{-1} sec^{-1} for the entire middle atmosphere. This corresponds to an ion mass of about 30 amu according to Figure 13.7, which is clearly in disagreement with both the mass-spectrometer measurements and the model studies discussed above. Meyerott *et al.* (1980) suggested that the Gerdien condenser measurements were affected by the breakup of cluster ions by both shock-wave and instrumental electric-field effects. Much heavier ions have been seen in some flights. Rose and Widdel (1972), for example, reported a group of positive ions above 60 km whose altitude dependence gives a reduced mobility of about 3.7×10^{-5} m^2 V^{-1} sec^{-1}, corresponding to an ion mass of several thousand amu. The origin of these ions is unknown.

Ion concentrations are also measured by the Gerdien condenser technique, and here again there are puzzling differences between observations and predictions. Above about 60 km, ionization of nitric oxide by solar Lyman-alpha radiation becomes an important source, and a considerable amount of variability in ion concentration is expected (even in quiet conditions) owing to the variability in NO concentration (Solomon *et al.*, 1982b). At lower altitudes, however, the only significant source is cosmic-ray ionization, and the ion-production rate due to cosmic rays can be calculated with a fair degree of certainty. The loss rate through ion-ion recombination is also well known (Smith and Church, 1977; Smith and Adams, 1982), and the steady-state ion concentration can thus be calculated with corresponding accuracy. Figure 13.8 shows a typical set of results. The points are taken from experimental data reported by Widdel *et al.* (1976), and the solid line is the positive-ion concentration calculated from the same model used to produce the ion-composition profiles shown in Figure 13.6. The overall shape of the altitude profile shows reasonable agreement, but the measured values are generally lower than the calculated values by about a factor of 3. A similar discrepancy was found by Meyerott *et al.*

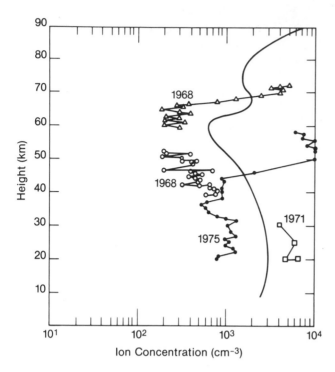

FIGURE 13.8   Results of measurements of ion concentration (Widdel *et al.*, 1976) and of model calculations for quiet conditions.

(1980), who suggested that it might be due to the lack of the experimental technique to measure low-mobility ions. Further measurements are clearly needed.

Another puzzling feature shown in Figure 13.8 is the abrupt increase in concentration occurring at about 65 km in the 1968 data and 45 km in the 1975 data. A similar abrupt increase in conductivity at about 65 km is seen in blunt-probe measurements and has been attributed to the transition from a cosmic-ray ionization source below to a Lyman-alpha source above (Mitchell and Hale, 1973). The model results show such a transition, but it is much less abrupt and smaller in magnitude than the one observed. The same explanation is certainly ruled out for the 45-km transition, since solar Lyman-alpha radiation is almost totally extinguished below 50 km.

Figure 13.9 shows the theoretical profiles of positive-ion, negative-ion, and electron contributions to the daytime conductivity, using the same model as before. The dotted curve shows an average of several blunt-probe measurements of positive conductivity in quiet conditions (Mitchell and Hale, 1973) and is taken as representative of the current experimental situation. Clearly the overall shape of the theoretical positive-conductivity profile matches the observations reasonably well—particularly below 60 km where cosmic rays are the principal source of ionization. At higher levels, the expected

variability of ion concentration should lead to a corresponding variability in positive conductivity.

The role of electrons is noteworthy. The theoretical profiles show that electrons make a negligible contribution to the total conductivity below 50 km but completely dominate the conductivity above 60 km, where they give rise to an extremely steep upward gradient in conductivity [the "equalizing" layer (Dolezalek, 1972)]. The electron mobility is so large that the electron contribution to the conductivity becomes equal to the ion contribution at a level where the electron concentration is less than 1 cm^{-3}. In this region the model predictions of the electron-ion balance are not trustworthy. In particular, the model used here does not include negative-ion photodetachment as a source of electrons, as photodetachment cross sections of the principal atmospheric negative ions are not known. Even small amounts of photodetachment, however, will have an important effect on electron concentrations in the region of the stratopause, and hence on the model conductivity profiles.

The conductivity is greatly reduced at night at all heights above 50 km. The decrease is partly due to the absence of solar ionizing radiation and partly to changes in neutral chemistry, notably the conversion of atomic oxygen to ozone in the mesosphere. Figure 13.10 shows model profiles of daytime and nighttime conductivity for quiet mid-latitude conditions, in which the nighttime sources of ionization are mostly solar Lyman-alpha radiation scattered from the geocorona and the zonally averaged flux of energetic electrons at 45° magnetic latitude given by Vampola and Gorney (1983). The night-

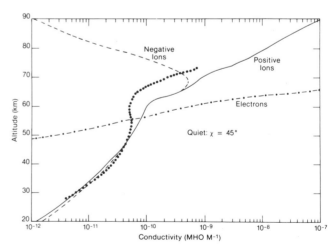

FIGURE 13.9   Middle-atmosphere daytime conductivity. The dotted curve shows direct measurements (Mitchell and Hale, 1973) of positive conductivity, and the other curves show the result of model calculations of the contributions of the electrons, positive ions, and negative ions.

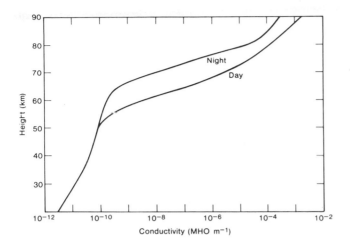

FIGURE 13.10   Model calculations of the daytime and nighttime conductivity of the middle atmosphere during quiet conditions.

time equalizing layer lies nearly 10 km higher than the daytime layer, and the transition between the two should take place rapidly during twilight.

During a solar-proton event, the conductivity profile of the middle atmosphere is greatly changed. Individual events are so variable in ionization rate, however, that a representative SPE-conductivity profile would not be meaningful. Figure 13.11 is simply intended as an illustration of the conductivity profile calculated from the above model for a fairly intense SPE during daytime. The largest fractional increase from the quiet-time conductivity occurs between 50 and 60 km, and the greatly enhanced ionization in this region tends to eliminate, or

at least substantially reduce, the steep gradient immediately above.

To summarize, neither direct measurements nor modeling studies of the bulk electrical parameters of the middle atmosphere are in a satisfactory condition. The measurements are difficult to make and tend to be beset by problems of interpretation. Models are based on deficient knowledge of several of the important mechanisms and cannot yet give more than estimates of the altitude profiles of such important parameters as the ion concentration and the conductivity. Much remains to be done to place our knowledge on a secure footing.

ELECTRIC FIELDS IN THE MIDDLE ATMOSPHERE

The normal electric field in the middle atmosphere is a superposition of fields mapped upward from thunderstorm generators in the lower atmosphere and fields mapped downward from magnetospheric and ionospheric dynamo generators. The possible existence of local electric-field generators within the middle atmosphere is a controversial topic and will be discussed briefly later.

The mapping of the fair-weather field in the vertical direction has been studied by a number of authors (e.g., Mozer and Serlin, 1969; Park and Dejnakarintra, 1973), and a full three-dimensional model that includes a realistic thunderstorm distribution and surface topography has been constructed by Hays and Roble (1979). The principal component of the middle-atmosphere electric field provided by this tropospheric source is vertically directed and arises from the necessity for continuity of the vertical current. The vertical electric field is thus roughly inversely proportional to the conductivity, and its order of magnitude varies from 10^{-1} V/m at balloon altitudes to 10^{-6} V/m at the base of the thermosphere. The horizontal component of the electric field arises from the nonuniform distribution of thunderstorm generators over the Earth and is largely removed by the short-circuiting effect of the equalizing layer in the mesosphere, where the conductivity increases sharply (see Figure 13.10). The attenuation is not complete, however, and the model of Hays and Roble (1979) predicts horizontal electric fields of magnitude up to a few tenths of a millivolt per meter in the lower thermosphere arising from the fair-weather source.

The corresponding problem of mapping the ionospheric and magnetospheric electric fields downward through the middle atmosphere to the ground has also been examined by several authors (e.g., Mozer and Serlin, 1969; Volland, 1972; Chiu, 1974; Park, 1976) using simple one-dimensional models and by Roble and

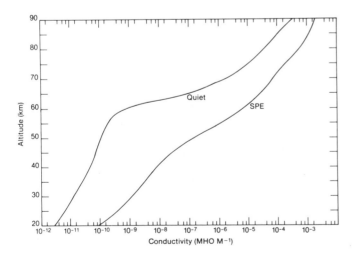

FIGURE 13.11   Model calculations of the daytime conductivity of the middle atmosphere during quiet conditions and during an intense solar-proton event.

Hays (1979) with a full three-dimensional model. The generation of these fields and their mapping into the global electrical circuit will be discussed in Chapters 14 and 15 (this volume).

The dramatic increase in middle-atmosphere conductivity during a major solar-proton event causes large changes in the local electric fields at high latitudes. Holzworth and Mozer (1979) carried out balloon measurements of the stratospheric electric field over northern Canada during the intense event of August 1972 and reported a decrease in the vertical field by more than an order of magnitude. The decrease closely paralleled the increase in solar-proton flux and could be explained qualitatively by conservation of the fair-weather current in the presence of the greatly enhanced conductivity. The upward mapping of the thunderstorm-generated fields of the lower atmosphere is sensitive to changes in middle-atmosphere conductivity, since the middle atmosphere represents a low-resistance load to the generator even in quiet conditions. The downward mapping of electric fields generated in the ionosphere and magnetosphere, however, is much less sensitive, since the conductivity of the middle atmosphere remains much less than that of the lower ionosphere even during a major solar-proton event (see Figure 13.11).

The changes in the global electric circuit arising from the August 1972 SPE have been examined in detail by Reagan et al. (1983) and Tzur and Roble (1983), all of whom pointed out the importance in estimating the changes in middle-atmosphere electrical parameters of including the current carried by the precipitating protons themselves in the polar-cap region. Changes in the global circuit, however, arose mainly from the Forbush decrease in galactic cosmic-ray flux that accompanied the event rather than from the solar-proton flux.

Measurements of the electric field in the mesosphere have been carried out with rocketborne techniques and have yielded conflicting and unexpected results. Most startling of these is the measurement of strong electric fields in the lower mesosphere (Tyutin, 1976; Hale and Croskey, 1979; Maynard et al., 1981) with intensities that can be orders of magnitude larger than those required to maintain continuity of the fair-weather current. The reality of these fields has been questioned (Kelley et al., 1983), and the possibility that they are instrumental artifacts has not been entirely laid to rest. No satisfactory explanation of their existence, either as a genuine atmospheric phenomenon or as an instrumental effect, has yet been proposed. However, they remain an intriguing feature of the middle atmosphere. The anomalous fields, if they are real, cannot be mapped from above or below, since they are present only in relatively well-defined height ranges. Any plausible explanation

must involve either a local mesospheric generation mechanism or a dramatic local decrease in conductivity, in which case the strong electric field would be needed to maintain current continuity. The latter possibility appears to be ruled out by simultaneous conductivity measurements made on the same flight (Maynard et al., 1981). These measurements show that the conductivity is indeed low in the region of the strong electric fields but is still large enough to provide a vertical current density about 200 times larger than the fair-weather value.

The fact that the strong fields are usually seen near the 60- to 65-km height region, where the equalizing layer exists, is a possible clue. In this region the dominant negatively charged current carriers change from the slow-moving negative ions below to the highly mobile electrons above, and a fairly sharp change in electric field must result from the need to conserve vertical current alone. However, as mentioned above, the fields measured are much larger than those associated with this upward mapping process, and there is no obvious reason why strong fields should be generated in this neighborhood. The observations challenge our picture of the middle atmosphere as a passive element in the global electrical circuit and suggest that there may be field-generating mechanisms that we do not yet understand. Even if the electric fields do turn out to be instrumental artifacts, their explanation will contribute to our understanding of the limitations of in situ electric-field measurements in the terrestrial environment.

CONCLUSION

In this brief review we have summarized our present understanding of the electrical structure of the atmosphere in the 30- to 100-km height range. The sources of ionization in this region are reasonably well known, and their variations in time and space are at least qualitatively understood. The complexities of the ion chemistry that connects the ionization sources to the ambient ion composition still require a great deal of unraveling. We are still quite ignorant of many aspects, including photodetachment of negative ions and the role of reactive neutral species with extremely low concentrations. These uncertainties lead to corresponding uncertainties in ion concentration and mobility and in such bulk electrical parameters as the conductivity. Direct experimental measurements have led to considerable progress, but they are beset by difficulties of interpretation and by inconsistencies among themselves. Finally, the recent observations of large mesospheric electric fields have raised first-order questions about our understanding of atmospheric field-generating mechanisms or of the instrumental techniques used in these measurements. Significant challenges for the future exist in almost every aspect of middle-atmospheric electricity.

REFERENCES

Arnold, F., and G. Henschen (1978). First mass analysis of stratospheric negative ions, *Nature* 275, 521.

Arnold, F., J. Kissel, H. Wieder, and J. Zähringer (1971). Negative ions in the lower ionosphere: A mass spectrometric measurement, *J. Atmos. Terrest. Phys. 33*, 1969.

Arnold, F., H. Böhringer, and G. Henschen (1978). Composition measurements of stratospheric positive ions, *Geophys. Res. Lett. 5*, 653.

Arnold, F., A. A. Viggiano, and E. E. Ferguson (1982). Combined mass spectrometric composition measurements of positive and negative ions in the lower ionosphere. II. Negative ions, *Planet. Space Sci. 30*, 1307.

Becker, K. H., and A. Ionescu (1982). Acetonitrile in the lower troposphere, *Geophys. Res. Lett. 9*, 1349.

Björn, L. G., and F. Arnold (1981). Mass spectrometric detection of precondensation nuclei at the Arctic summer mesopause, *Geophys. Res. Lett. 8*, 1167.

Charlson, R. J. (1965). Noctilucent clouds: A steady-state model, *Q. J. R.Meteorol. Soc. 91*, 517.

Chesworth, E. T., and L. C. Hale (1974). Ice particulates in the mesosphere, *Geophys. Res. Lett. 1*, 347.

Chiu, Y. T. (1974). Self-consistent electrostatic field mapping in the high-latitude ionosphere, *J. Geophys. Res. 79*, 2790.

Conley, T. D. (1974). Mesospheric positive ion concentrations, mobilities, and loss rates obtained from rocketborne Gerdien condenser measurements, *Radio Sci. 9*, 575.

Cook, J. W., G. E. Brueckner, and M. E. VanHoosier (1980). Variability of the solar flux in the far ultraviolet 1175-2100 Å, *J. Geophys. Res. 85*, 2257.

Dolezalek, H. (1972). Discussion of the fundamental problem of atmospheric electricity, *Pure Appl. Geophys. 100*, 8-43.

Donahue, T. M., B. Guenther, and J. E. Blamont (1972). Noctilucent clouds in daytime: Circumpolar particulate layers near the summer mesopause, *J. Atmos. Sci. 29*, 1205.

Fehsenfeld, F. C., and E. E. Ferguson (1969). Origin of water-cluster ions in the D region, *J. Geophys. Res. 74*, 2217.

Ferguson, E. E. (1974). Laboratory measurements of ionospheric ion-molecule reaction rates, *Rev. Geophys. Space Phys. 12*, 703.

Ferguson, E. E., and F. Arnold (1981). Ion chemistry of the stratosphere, *Acc. Chem. Res. 14*, 327.

Ferguson, E. E., D. B. Dunkin, and F. C. Fehsenfeld (1972). Reactions of NO_2^- and NO_3^- with HCl and HBr, *J. Chem. Phys. 57*, 1459.

Ferguson, E. E., F. C. Fehsenfeld, and D. L. Albritton (1979). Ion chemistry of the Earth's atmosphere, in *Gas Phase Ion Chemistry, Vol. 1*, M. T. Bowers, ed., Academic Press, New York, Chapter 2.

Fogle, B., and B. Haurwitz (1966). Noctilucent clouds, *Space Sci. Rev. 6*, 278.

Forbush, S. E. (1938). On cosmic-ray effects associated with magnetic storms, *Terrest. Magn. Atmos. Electr. 43*, 203.

Goldberg, R. A., and G. Witt (1977). Ion composition in a noctilucent cloud region, *J. Geophys. Res. 82*, 2619.

Good, A., D. A. Durden, and P. Kebarle (1970). Mechanism and rate constants of ion-molecule reactions leading to formation of $H^+(H_2O)_n$ in moist oxygen and air, *J. Chem. Phys. 52*, 222.

Hale, L. C., and C. L. Croskey (1979). An auroral effect on the fair weather electric field, *Nature 278*, 239.

Hale, L. C., D. P. Hoult, and D. C. Baker (1968). A summary of blunt probe theory and experimental results, *Space Res. 8*, 320.

Hays, P. B., and R. G. Roble (1979). A quasi-static model of global atmospheric electricity. 1. The lower atmosphere, *J. Geophys. Res. 84*, 3291-3305.

Heaps, M. G. (1978). Parametrization of the cosmic-ray ion-pair production rate above 18 km, *Planet. Space Sci. 26*, 513.

Hesstvedt, E. (1961). Note on the nature of noctilucent clouds, *J. Geophys. Res. 66*, 1985.

Holzworth, R. H., and F. S. Mozer (1979). Direct evidence of solar flare modification of stratospheric electric fields, *J. Geophys. Res. 84*, 363-367.

Hunten, D. M., R. P. Turco, and O. B. Toon (1980). Smoke and dust particles of meteoric origin in the mesosphere and stratosphere, *J. Atmos. Sci. 37*, 1342.

Kelley, M. C., C. L. Siefring, and R. F. Pfaff (1983). Large amplitude middle atmospheric electric fields: Fact or fiction? *Geophys. Res. Lett. 10*, 733.

Luhmann, J. G. (1977). Auroral bremsstrahlung spectra in the atmosphere, *J. Atmos. Terrest. Phys. 39*, 595.

Maynard, N. C., C. L. Croskey, J. D. Mitchell, and L. C. Hale (1981). Measurement of volt/meter vertical electric fields in the middle atmosphere, *Geophys. Res. Lett. 8*, 923.

Meyerott, R. E., J. B. Reagan, and R. G. Joiner (1980). The mobility and concentration of ions and the ionic conductivity in the lower stratosphere, *J. Geophys. Res. 85*, 1273.

Mitchell, J. D., and L. C. Hale (1973). Observations of the lowest ionosphere, *Space Res. 13*, 471.

Mozer, F. S., and R. Serlin (1969). Magnetospheric electric field measurements with balloons, *J. Geophys. Res. 74*, 4739-4754.

Narcisi, R. S., and A. D. Bailey (1965). Mass spectrometric measurements of positive ions at altitudes from 64 to 112 kilometers, *J. Geophys. Res. 70*, 3687.

Narcisi, R. S., A. D. Bailey, L. Della Lucca, C. Sherman, and D. M. Thomas (1971). Mass spectrometric measurements of negative ions in the D and lower E regions, *J. Atmos. Terrest. Phys. 33*, 1147.

Park, C. G. (1976). Downward mapping of high-latitude ionospheric electric fields to the ground, *J. Geophys. Res. 81*, 168-174.

Park, C. G., and M. Dejnakarintra (1973). Penetration of thundercloud electric fields into the ionosphere and magnetosphere. 1. Middle and subauroral latitudes, *J. Geophys. Res. 78*, 6623.

Pedersen, A., J. Troim, and J. A. Kane (1970). Rocket measurements showing removal of electrons above the mesopause in summer at high latitude, *Planet. Space Sci. 18*, 945.

Pomerantz, M. A., and S. P. Duggal (1974). The Sun and cosmic rays, *Rev. Geophys. Space Phys. 12*, 343-361.

Reagan, J. B. (1977). Ionization processes, in *Dynamical and Chemical Coupling Between the Neutral and Ionized Atmosphere*, B. Grandal and J. A. Holtet, eds., Reidel, Dordrecht, Holland, p. 145.

Reagan, J. B., R. E. Meyerott, J. E. Evans, W. L. Imhof, and R. G. Joiner (1983). The effects of energetic particle precipitation on the atmospheric electric circuit, *J. Geophys. Res. 88*, 3869.

Reid, G. C. (1974). Polar-cap absorption—Observations and theory, *Fundam. Cosmic Phys. 1*, 167-200.

Reid, G. C. (1975). Ice clouds at the summer polar mesopause, *J. Atmos. Sci. 32*, 523.

Reid, G. C. (1977). The production of water-cluster positive ions in the quiet daytime D region, *Planet. Space Sci. 25*, 275.

Roble, R. G., and P. B. Hays (1979). A quasi-static model of global atmospheric electricity. 2. Electrical coupling between the upper and lower atmosphere, *J. Geophys. Res. 84*, 7247-7256.

Rose, G., and H. U. Widdel (1972). Results of concentration and mobility measurements for positively and negatively charged particles taken between 85 and 22 km in sounding rocket experiments, *Radio Sci. 7*, 81.

Smith, D., and N. G. Adams (1982). Ionic recombination in the stratosphere, *Geophys. Res. Lett. 9*, 1085.

Smith, D., and M. J. Church (1977). Ion-ion recombination rates in the Earth's atmosphere, *Planet. Space Sci. 25*, 433.

Solomon, S., P. J. Crutzen, and R. G. Roble (1982a). Photochemical coupling between the thermosphere and the lower atmosphere. 1. Odd nitrogen from 50 to 120 km, *J. Geophys. Res. 87*, 7206.

Solomon, S., G. C. Reid, R. G. Roble, and P. J. Crutzen (1982b). Photochemical coupling between the thermosphere and the lower atmosphere. 2. D region ion chemistry and the winter anomaly, *J. Geophys. Res. 87*, 7221.

Strobel, D. F., T. R. Young, R. R. Meier, T. P. Coffey, and A. W. Ali (1974). The nighttime ionosphere: E region and lower F region, *J. Geophys. Res. 79*, 3171.

Thomas, G. E., G. H. Mount, C. P. McKay, and L. Sitongia (1982). Satellite observations of noctilucent clouds, *EOS 63*, 1049.

Turco, R. P., O. B. Toon, R. C. Whitten, R. G. Keesee, and D. Hollenbach (1982). Noctilucent clouds: Simulation studies of their genesis, properties and global influences, *Planet. Space Sci. 30*, 1147.

Tyutin, A. A. (1976). Mesospheric maximum of the electric-field strength, *Cosmic Res. 14*, 132.

Tzur, I., and R. G. Roble (1983). Ambipolar diffusion in the middle atmosphere, *J. Geophys. Res. 88*, 338-344.

Vampola, A. L., and D. J. Gorney (1983). Electron energy deposition in the middle atmosphere, *J. Geophys. Res. 88*, 6267.

Viggiano, A. A., R. A. Perry, D. L. Albritton, E. E. Ferguson, and F. C. Fehsenfeld (1980). The role of H_2SO_4 in stratospheric negative-ion chemistry, *J. Geophys. Res. 85*, 4551.

Viggiano, A. A., F. Arnold, D. W. Fahey, F. C. Fehsenfeld, and E. E. Ferguson (1982). Silicon negative ion chemistry in the atmosphere—in situ and laboratory measurements, *Planet. Space Sci. 30*, 499.

Volland, H. (1972). Mapping of the electric field of the Sq current into the lower atmosphere, *J. Geophys. Res. 77*, 1961-1965.

Volz, F. E., and R. M. Goody (1962). The intensity of the twilight and upper atmosphere dust, *J. Atmos. Sci. 19*, 385.

Widdel, H. U., G. Rose, and R. Borchers (1976). Experimental results on the variation of electric conductivity and ion mobility in the mesosphere, *J. Geophys. Res. 81*, 6217.

Upper-Atmosphere Electric-Field Sources

14

ARTHUR D. RICHMOND
National Center for Atmospheric Research

The Earth's space environment is filled with electrons and positive ions, comprising a plasma of very low density. These charged particles collide only infrequently and are strongly influenced by magnetic and electric fields. In turn, the charged particles affect the distributions of the magnetic and electric fields in space. The space plasma environment, therefore, is dominated by electrodynamic processes. The regions of space involved in creating upper atmosphere electric fields are illustrated in Figure 14.1 and are described below.

The solar wind is a plasma with electron and ion number densities of order 5×10^6 m^{-3} flowing continually outward from the Sun at a speed of 300-1000 km/sec. Imbedded within it is the interplanetary magnetic field (IMF), which is maintained by electric currents flowing throughout the solar-wind plasma. The IMF strength at the orbit of the Earth is roughly a factor of 10^{-4} smaller than the strength of the surface geomagnetic field. Most of the time, an interplanetary field line near the Earth can be traced back to the surface of the Sun, where magnetic fields are ubiquitous. As the Sun rotates (once every 27 days) different magnetic regions influence the IMF near the Earth. The combination of solar-rotation and outward solar-wind flow produce a roughly spiral IMF pattern. In addition, the solar-wind velocity can change dramatically and produce both large-scale and small-scale distortions of the IMF so that the field direction and strength vary greatly. These changes have been found to influence the electrical state of the magnetosphere and ionosphere.

The magnetosphere is the region of space where the geomagnetic field has a dominant influence on plasma properties. As the charged particles of the solar wind are deflected by the geomagnetic field, an electric current layer is formed at the boundary between the solar wind and the magnetosphere, called the magnetopause. This current layer distorts the geomagnetic field from the dipole-like configuration that it would otherwise have and helps to create a long magnetized tail trailing the Earth. Although the full extent of this tail has not yet been determined, it is known to be more than 500 Earth radii. The magnetosphere contains the radiation belt, composed of energetic charged particles trapped in the magnetic field. The number density of electron-ion pairs in the magnetosphere is highly variable, ranging in order of magnitude from a low of 10^6 m^{-3} in parts of the tail up to 10^{12} m^{-3} in the densest portions of the dayside ionosphere.

The ionosphere is the ionized component of the Earth's upper atmosphere. It is not distinct from the magnetosphere, but rather forms the base of the magnetosphere in terms of electrodynamic processes. The lower boundary of the ionosphere is not well defined but can be taken as about 90 km altitude for the present pur-

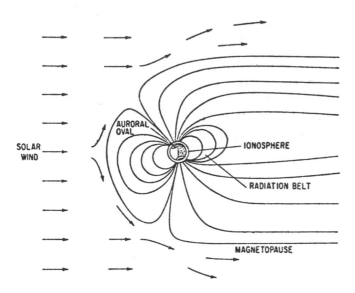

FIGURE 14.1 Configuration of the magnetosphere. The magnetic field is shown by continuous lines, and locations of important magnetospheric features are pointed out.

poses, representing the level where the density of electron-ion pairs falls to roughly 10^{10} m^{-3} and below which electric currents become relatively small. The ionization is formed largely by the effect on the upper atmosphere of solar extreme-ultraviolet and x-ray radiation at wavelengths shorter than 102.6 nm, but energetic particles impacting the upper atmosphere from the magnetosphere also create important enhancements. The ionospheric plasma has a temperature on the order of 1000 K, which is much cooler than the energetic plasma farther out in the magnetosphere. Collisions between charged particles and neutral atmospheric molecules become important below 200 km altitude and strongly affect the electrodynamic characteristics of the ionosphere.

At high latitudes, where magnetic-field lines connect the ionosphere with the outer magnetosphere, the ionospheric features are quite complex. Ionospheric phenomena become better organized in a coordinate system based on the geomagnetic field than in geographic coordinates, with the difference arising mainly from the 11° tilt of the dipolar field from the Earth's axis. Different magnetic coordinate systems exist, but for descriptive purposes the differences are not crucial, and the simple terms "magnetic latitude" and "magnetic local time" will suffice here.

At high magnetic latitudes, aurora are produced as energetic charged particles, mainly electrons, precipitate into the upper atmosphere from the outer magnetosphere, creating both visible emissions and ionization enhancements. The aurora form a belt around the magnetic pole, called the auroral oval. The oval is in fact

roughly circular, of variable size, and has a wider latitudinal extent on the nightside than on the dayside of the Earth. The entire oval is shifted toward the nightside, so that aurora appear at lower magnetic latitudes at night (roughly 67°) than during the day (roughly 78°). The nighttime particle precipitation also tends to be more intense and widespread than on the dayside. Contained within the auroral oval is the polar cap, where auroras are less frequent but where on occasion very energetic protons from solar flares enter and penetrate relatively deep into the upper atmosphere.

A more detailed description of the Earth's space environment can be found in several of the references listed at the end of this paper, especially in the book of Akasofu and Chapman (1972).

ELECTRODYNAMIC PROCESSES IN SPACE

The charged particles of a plasma react strongly to electric and magnetic fields. There is a strong tendency for the particles to short out any electric fields, so that it is often a good approximation to treat the electric field in the frame of reference of the plasma as vanishing:

$$\mathbf{E}_{\text{plasma}} = 0. \qquad (14.1)$$

This is often called the magnetohydrodynamic (or MHD) approximation (e.g., Roederer, 1979). The frame-of-reference choice is important if the plasma is moving and if a magnetic field is present, because the electric field observed in a different reference frame is not the same. If we let $\mathbf{E}$ be the electric field in an Earth-fixed reference frame, $\mathbf{V}$ be the velocity of the plasma with respect to the Earth, and $\mathbf{B}$ be the magnetic-field vector, then a (nonrelativistic) Lorentz transformation yields

$$\mathbf{E}_{\text{plasma}} = \mathbf{E} + \mathbf{V} \times \mathbf{B}, \qquad (14.2)$$

where the vector product $\mathbf{V} \times \mathbf{B}$ results in a vector directed perpendicular to both $\mathbf{V}$ and $\mathbf{B}$. The electric field is then simply related to the plasma velocity and magnetic field by the approximate relation

$$\mathbf{E} = -\mathbf{V} \times \mathbf{B}. \qquad (14.3)$$

Alternatively, the velocity component perpendicular to $\mathbf{B}$ can be related to $\mathbf{E}$ and $\mathbf{B}$ as

$$\mathbf{V} = (\mathbf{E} \times \mathbf{B})/\mathbf{B}^2. \qquad (14.4)$$

Equations (14.3) and (14.4) express the same fact from two different points of view: the electric field and plasma velocity are closely interrelated and help to determine each other. In some cases, as in the solar wind where plasma momentum is high, the electric field quickly adjusts toward the value given by Eq. (14.3). In other cases, as in the upper ionosphere where electric fields tend to be imposed on the plasma as a result of the dynamo processes to be discussed, the plasma quickly is set into motion at the velocity given by Eq. (14.4).

A further important consequence of the MHD approximation, when combined with the Faraday law of magnetic induction, is the following: all plasma particles lying along a common magnetic-field line at one instant of time will forever remain on a common field line. The magnetic field may vary both temporally and spatially, and the plasma may move from one region of space to another, but plasma ions and electrons will continue to share a field line with the same partners. This result affects both the magnetic-field configuration and the plasma velocity. In the solar wind, the magnetic field is distorted to follow the motions of the plasma. Nearer the Earth, where the magnetic field is so strong that it is not easily distorted, the constraint means that all particles on a dipolar field line must move simultaneously together to another field line. Convection of plasma thus can be mapped between the outer magnetosphere and the ionosphere along magnetic-field lines. The electric field similarly maps along the magnetic field.

The MHD approximation is useful in interrelating plasma motions with electric and magnetic fields, but it breaks down under a number of important circumstances, especially where electric current densities are large. Furthermore, the approximation does not explain the distribution of currents, which are a central element in the processes giving rise to upper-atmospheric electric fields. Current flow across magnetic-field lines exerts a force on the medium, a force that must either be balanced by other forces, like pressure gradients, or else result in acceleration of the medium. Consideration of force and momentum balance thus is an important part of understanding currents in plasmas.

One place where the MHD approximation breaks down is in the lower ionosphere, below 150 km, where collisions between ions and the much more numerous air molecules are sufficiently frequent to prevent the ions from maintaining the velocity given by Eq. (14.4). In this region electric current readily flows across magnetic-field lines. As we shall see, neutral-air winds in the lower ionosphere lead to generation of electric currents and fields, and for this reason the height range of roughly 90-150 km is called the "dynamo region" of the ionosphere.

Unlike the outer magnetosphere, the ionosphere behaves as an Ohmic medium, with the current density linearly related to the electric field under most circumstances. The conductivity, however, is highly anisotropic owing to the presence of the geomagnetic field. The conductivity in the direction of the magnetic field is very large, so that the electric-field component in this

direction is almost entirely shorted out, and magnetic-field lines are nearly electric equipotential lines at all altitudes above 90 km. The conductivity characteristics perpendicular to the magnetic field depend on the rate of ion-neutral collisions, and they change with altitude as the neutral density varies. Figure 14.2 shows typical mid-latitude conductivity profiles for day and night conditions. In the nighttime auroral oval the conductivity is more akin to the "Day" profile in Figure 14.2 than to the "Night" profile because of ionization production associated with the aurora. Although the ionospheric plasma density typically maximizes at around 300 km altitude, the conductivity perpendicular to the magnetic field maximizes at around 110 km, with a large day-night difference owing to the day-night difference in ionospheric density. The current component perpendicular to the magnetic field flows in a direction as much as 88° different from the electric-field direction, as also shown in Figure 14.2. This effect results from the fact that electrons are relatively little influenced by collisions above 80 km and move perpendicular to the electric field as given by Eq. (14.4), while positive ions are strongly affected by collisions below 130 km and are unable to cancel current carried by the drifting electrons.

SOLAR WIND/MAGNETOSPHERE DYNAMO

In an analogy with a dynamo-electric machine that generates electricity by rotating a conducting armature through a magnetic field, the motion of a plasma through a magnetic field produces an electromotive

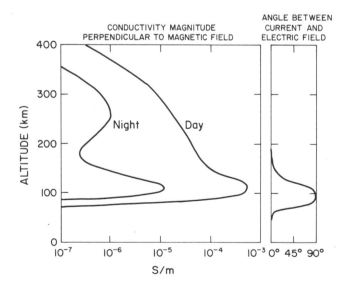

FIGURE 14.2 Typical profiles of the ionospheric conductivity component perpendicular to the geomagnetic field for day and night conditions (left) and angle between the current and electric-field components perpendicular to the geomagnetic field (right).

force and current flow and can also be considered a dynamo. The solar wind/magnetosphere dynamo results from the flow of the solar wind around and partly into the magnetosphere, setting up plasma motion in the magnetosphere as well an electric field and currents (e.g., Stern, 1977; Hill, 1979; Roederer, 1979; Cowley, 1982). All details of this process are not yet understood, but a number of features have become clear.

Strong evidence exists that the magnetosphere is partly open, that is, that some magnetic-field lines traced from the Earth extend indefinitely into interplanetary space (Hill and Wolf, 1977; Stern, 1977; Lyons and Williams, 1984). The amount of magnetic flux that connects to the interplanetary magnetic field may be as great as the entire flux passing through the polar caps. Figure 14.3 shows a schematic figure of the magnetic-field configuration for the simple case where the IMF is directed southward. If the IMF had an east-west component, as it usually does, we would require a three-dimensional representation of the magnetic-field interconnection. Details of the magnetic-field configuration are not yet resolved, so Figure 14.3 should be treated more as a conceptual tool than as a true representation of the magnetosphere. The essential features are the four classes of magnetic-field lines denoted in Figure 14.3: (1) closed field lines connected to the Earth in both northern and southern hemispheres; (2) interplanetary field lines unconnected to the Earth; (3) open field lines connecting the northern polar cap to interplanetary space; and (4) open field lines connecting the southern polar cap to interplanetary space.

The interplanetary electric field, obtained from Eq. (14.3), is directed out of the page in Figure 14.3. To the extent that the MHD approximation is valid and electric fields map along the magnetic field, the polar ionosphere is also subject to an electric field out of the page,

causing ionospheric plasma to convect antisunward. The magnitude of the ionospheric electric field is greater than that of the interplanetary electric field because the bundling of magnetic-field lines at the ionosphere causes electric potential gradients to intensify. On the other hand, the plasma drift velocity in the upper ionosphere, given by Eq. (14.4), is much less than the solar-wind velocity because of the inverse dependence on magnetic-field strength. The polar-cap electric field is typically 20 mV/m, giving an ionospheric convection velocity of roughly 300 m/sec. Other directions of the IMF than shown in Figure 14.3 result in a somewhat altered pattern of polar-cap ionospheric convection, but a usual feature is the general antisunward flow. More about IMF influence on the high-latitude electric field is discussed in a later section.

The physical processes that determine the amount of magnetic flux that interconnects the geomagnetic field and the IMF are not well understood and are the subject of much study (e.g., Cowley, 1982). They clearly involve a violation of the MHD approximation since the ionospheric and solar-wind plasmas coexisting on an open magnetic-field line at one instant of time could not have lain on a common magnetic-field line throughout their entire histories. The violations of this approximation occur to some extent throughout the magnetosphere but are particularly important in at least two regions: at the sunward magnetopause and somewhere in the magnetospheric tail. At the sunward magnetopause plasma flows together through unconnected interplanetary and magnetospheric magnetic fields and flows out northward and southward on interconnected magnetic-fields lines (Figure 14.3). In the tail plasma flows together on interconnected field lines and flows out through unconnected magnetospheric and interplanetary magnetic fields. In the closed portion of the magnetosphere plasma flows generally toward the Sun, passing around the Earth on the morning and evening sides (out of the plane of Figure 14.3). However, some of the outermost portions of the closed-field region convect away from the Sun because of momentum transfer from the nearby solar wind (e.g., Hones, 1983).

Magnetospheric plasma convection has a number of important consequences, one of which is the energization of plasma and particle precipitation into the ionosphere. As plasma flows from the tail toward the Earth it is compressionally heated because the volume occupied by plasma on neighboring magnetic-field lines is reduced as the magnetic-field strength increases and as the length of field lines decreases. Other particle acceleration processes also help to energize the plasma. Some of the energized particles precipitate into the ionosphere and create ionization enhancements, especially in the auroral oval. The energized plasma also has an impor-

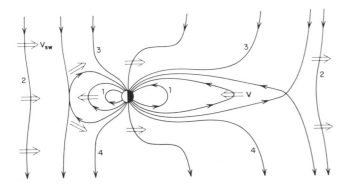

FIGURE 14.3 Schematic diagram of magnetic field and plasma flow in the solar wind/magnetosphere environment (from Lyons and Williams, 1984). Continuous lines show the magnetic field for the case where the IMF is purely southward. Open arrows show the plasma velocity direction. Numbers 1-4 denote magnetic regions of different topology, as discussed in the text.

tant influence on the flow of electric currents and on the distribution of electric fields (e.g., Spiro and Wolf, 1984). Energetic particles drift in the Earth's magnetic field—electrons toward the east and positive ions toward the west—so that a westward current flows within the hot plasma. This westward current, flowing in the geomagnetic field, essentially exerts an electromagnetic force on the plasma directed away from the Earth and thus tending to oppose the earthward convection. Charge separation associated with the current tends to create an eastward electric-field component, opposite to the nightside westward convection electric field, largely canceling the convection electric field in the inner magnetosphere.

The overall pattern of magnetospheric convection tends to map along the magnetic field into the ionosphere even though this mapping is imperfect because of net electric fields that tend to develop within the nonuniform energetic plasma. In the upper ionosphere, where Eq. (14.4) is valid, the general convection pattern looks something like that shown in Figure 14.4. There is antisunward flow over the polar cap and sunward flow in most of the auroral oval. The dayside magnetopause maps perhaps somewhere near the polar cap—auroral oval boundary on the dayside, while the most distant closed field lines in the tail map perhaps somewhere near the polar cap boundary on the nightside (the mapping of these outer magnetospheric regions into the ionosphere is not yet well determined). The flow lines in Figure 14.4 correspond to lines of constant electrostatic potential in a steady state. There is a potential high on the dawn side of the polar cap and a low on the dusk side, with a potential difference of the order of 50,000 V. The electric-field strength in the auroral oval tends to be somewhat larger than the polar-cap electric field.

Currents are an integral part of the electrical circuit associated with the solar wind/magnetosphere dynamo (e.g., Banks, 1979; Roederer, 1979; Stern, 1983; Akasofu, 1984). Figure 14.5 shows schematically the current flow near the Earth. Currents flowing along the direction of the magnetic field (field-aligned currents) couple the auroral oval with outer portions of the magnetosphere. The upward and downward currents are connected by cross-field currents in the ionospheric dynamo region. The anisotropy of the dynamo-region conductivity gives rise to strong current components perpendicular to the electric field in the auroral oval, in the form of eastward and westward auroral electrojets.

IONOSPHERIC WIND DYNAMO

Winds in the dynamo region have the effect of moving an electric conductor (the weakly ionized plasma)

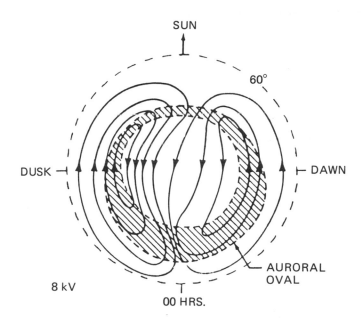

FIGURE 14.4 Schematic diagram of the magnetic north polar region showing the auroral oval and ionospheric convection (from Burch, 1977). The convection contours also represent electric potential contours, with a potential difference of order 8 kV between them.

through a magnetic field (the geomagnetic field), which results in the production of an electromotive force and the generation of electric currents and fields (e.g., Akasofu and Chapman, 1972; Wagner et al., 1980). The effective electric field driving the current, $\mathbf{E}'$, is related to the electric field in the earth frame, $\mathbf{E}$, the wind velocity $\mathbf{u}$, and the geomagnetic field, $\mathbf{B}$, by

$$\mathbf{E}' = \mathbf{E} + \mathbf{u} \times \mathbf{B}. \qquad (14.5)$$

Thus we may consider that two components of current exist, one driven by the "real" (measurable) electric field $\mathbf{E}$ and the other driven by the "dynamo electric field"

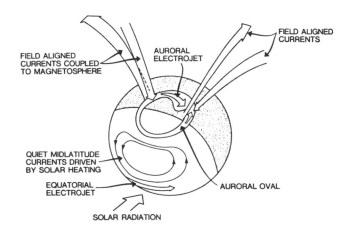

FIGURE 14.5 Schematic diagram of electric currents in the ionosphere and inner magnetosphere.

$\mathbf{u} \times \mathbf{B}$. These two components are not independent, however, because it turns out that the electric field $\mathbf{E}$ itself depends on the dynamo electric field $\mathbf{u} \times \mathbf{B}$.

To see how ionospheric winds cause electric fields to be set up, let us for the moment ignore the effects of the solar wind/magnetosphere dynamo. The dynamo electric field associated with the wind will drive a current. In general, this current would tend to converge in some regions of space and cause an accumulation of positive charge, while in other regions of space it would diverge and cause negative charge to accumulate. These charges would create an electric field directed from the positive toward the negative regions, which would cause current to flow tending to drain the charges. An equilibrium state would be attained when the electric-field-driven current drained charge at precisely the rate it was being accumulated by the wind-driven current. Actually, the time scale for this equilibrium to be achieved is extremely rapid, so that the electric field is effectively always in balance with the wind. The magnitude of the electric field is of order 1 mV/m. A net current flows in the entire ionosphere owing to combined action of the wind and electric field, especially on the sunlit side of the Earth (e.g., Takeda and Maeda, 1980, 1981). Figure 14.5 shows a large-scale current vortex at middle- and low-latitudes flowing counterclockwise in the northern hemisphere. A corresponding clockwise vortex flows in the southern hemisphere. These vortices are known traditionally as the Sq current system because of the nature of the ground-level magnetic variations that they produce: S for solar daily variations and q for quiet levels of magnetic activity. At high latitudes the electric fields and currents produced by the ionospheric wind dynamo are relatively weak in comparison with those of the solar wind/magnetospheric dynamo.

Winds in the upper atmosphere change strongly through the course of the day. They are driven in one form or another by the daily variation in absorption of solar radiation. Atmospheric heating causes expansion and the creation of horizontal pressure gradients, which drive the global-scale upper-atmospheric winds. Solar-ultraviolet radiation absorption at the height of the dynamo region drives a major portion of the winds. Absorption by ozone lower down (30-60 km altitude) also affects the dynamo-region winds by generating atmospheric tides that can propagate upward as global-scale atmospheric waves (e.g., Forbes, 1982a, 1982b). Figure 14.6 gives an example of three days of wind measurements at 100-130 km above Puerto Rico, showing the strong effects of propagating tides. Important contributions to the dynamo also come from altitudes above 130 km, where the conductivity is smaller (Figure 14.2) but where winds tend to be stronger and to vary less in height.

Figure 14.7 shows the average global electrostatic potential generated by the ionospheric wind dynamo, expressed in magnetic coordinates. The zero potential is arbitrarily defined here such that the average ionospheric potential over the Earth is zero. The total potential difference of the average pattern (4.7 kV) is smaller than what can usually be expected on any given day and is much smaller than that associated with the solar wind/magnetospheric dynamo. This pattern was derived from observations of plasma drifts on magnetically quiet days at an altitude of about 300 km. The electric-field maps along magnetic-field lines between hemispheres, providing the symmetry about the magnetic equator even when the wind dynamo action in opposite hemispheres is asymmetric. Magnetic-field lines peaking below 300 km in the equatorial region are not represented in Figure 14.7. Electric fields in the equatorial lower ionosphere have a localized strong enhancement of the vertical component associated with the strong anisotropy of the conductivity in the dynamo re-

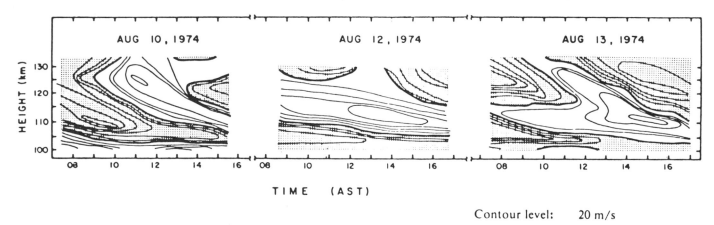

Contour level: 20 m/s

FIGURE 14.6   Observed eastward (unshaded) and westward (shaded) winds above Puerto Rico during three daytime periods in August 1974 (from Harper, 1977). The contour level is 20 m/sec.

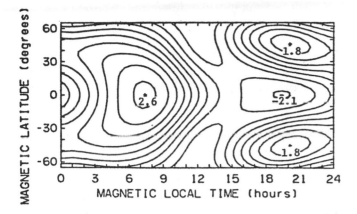

FIGURE 14.7   Average quiet-day ionospheric electrostatic potential at 300 km altitude as a function of magnetic local time (from Richmond *et al.*, 1980). The contour level is 500 V, and extrema relative to the global average are labeled in kilovolts.

gion. This enhanced electric field drives an eastward daytime current along the magnetic equator called the equatorial electrojet, as seen in Figure 14.5 (e.g., Forbes, 1981).

VARIABILITY

The preceding sections discussed the average patterns of ionospheric electric fields. Substantial deviations from these patterns occur, on a global scale as well as on a localized scale, and with a wide range of time scales.

Because the solar wind and particularly the IMF undergo large variations (e.g., Hundhausen, 1979), it is not surprising that the electric fields associated with solar wind/magnetosphere dynamo action similarly show large variations (e.g., Cowley, 1983; Rostoker, 1983). The interconnection of the interplanetary and magnetospheric magnetic fields maximizes when the IMF is southward, as in Figure 14.3. As the IMF direction rotates out of the plane of Figure 14.3 the amount of interconnected magnetic flux appears to lessen. For a northward-directed IMF the pattern of magnetic-field interconnection must be quite different from that shown in Figure 14.3, and it is possible that interconnection becomes insignificant, so that the magnetosphere is closed. The east-west component of the IMF affects the interconnection morphology and consequently also the pattern of high-latitude magnetosphere convection. The third IMF component, in the plane of Figure 14.3 but directed toward or away from the Sun, seems to have a less important role in the solar wind/magnetosphere dynamo than the other two components. However, the toward/away component correlates strongly with the west/east IMF component because of the ten-

dency for the IMF to assume a spiral pattern around the Sun. Some discussion of IMF effects has therefore taken place in reference to the toward/away component rather than the west/east component.

Figure 14.8 shows the average patterns of electric potential above 60° magnetic latitude deduced from ground magnetic variations for four different directions of the IMF. B_Z is the northward component of the IMF; B_Y is the eastward component (toward the east for an observer on the sunward side of the Earth). The electric fields are stronger for a southward IMF ($B_Z < 0$) than for northward IMF ($B_Z > 0$), in accord with the concept that magnetic field interconnection is greater and magnetospheric convection is stronger for a southward IMF. There are also clear differences in the patterns between the westward ($B_Y < 0$) and eastward ($B_Y > 0$) IMF cases, especially on the sunward side of the Earth.

Magnetic storms are dramatic disturbances of the entire magnetosphere lasting a time on the order of 1 day, usually produced by a strong enhancement of the solar-wind velocity, density, and/or southward IMF component (e.g., Akasofu and Chapman, 1972). The enhancements often come from explosive eruptions of plasma

ELECTRIC POTENTIAL

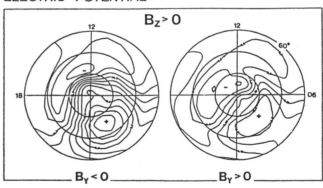

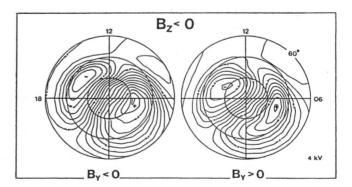

FIGURE 14.8   Average electric potential patterns above 60°N magnetic latitude as a function of magnetic local time deduced from magnetic variations at the ground, for four directions of the IMF (see text). The contour level is 4 kV (from Friis-Christensen *et al.*, 1985).

near the Sun's surface but sometimes are a long-lasting, relatively localized feature of the solar wind that sweeps past the Earth as the Sun rotates. During a magnetic storm magnetospheric convection varies strongly but is generally enhanced; plasma energization and precipitation into the ionosphere are greatly increased; and electric current flow is much stronger, also varying rapidly in time. Figure 14.9 gives an example of ground-level magnetic fluctuations caused by magnetospheric and ionospheric currents during a large storm. Quiet-day variations in the declination (D), vertical component (Z), and horizontal intensity (H) are seen up until 8:30 UT, when a shock wave in the solar wind hit the magnetosphere to produce a storm sudden commencement. Perturbations up to a few percent of the total geomagnetic-field strength occurred during the subsequent hours. Storms are composed of a succession of impulsive disturbances lasting 1-3 hours, called substorms (e.g., Akasofu, 1977; Nishida, 1978; McPherron, 1979). Impulsive disturbances with similar characteristics also occur on the average a few times a day even when no storm is in progress, and these are also called substorms. The characteristics of substorms can vary quite considerably from one to another but are often associated with what appears to be a large-scale plasma instability in the magnetospheric tail. The disturbed electric fields extend beyond the auroral oval and can even be seen at the magnetic equator (e.g., Fejer, 1985).

Magnetic storms are predominantly a phenomenon of the solar wind/magnetosphere dynamo, but they affect the ionospheric wind dynamo as well. In addition to strong auroral conductivity enhancements, conductivities can also be altered at lower latitudes at night by over

an order of magnitude (Rowe and Mathews, 1973), though they still remain well below daytime values. The nighttime ionospheric layer above 200 km altitude can be raised or lowered in response to stormtime electric fields and winds, which changes its conductive properties. During major storms the entire wind system in the dynamo region can be altered by the energy input to the upper atmosphere, so that the pattern of electric-field generation is modified (Blanc and Richmond, 1980). This effect can raise the potential at the equator by several thousand volts with respect to high latitudes.

Regular changes in the electric fields and currents occur over the course of the 11-year solar cycle and with the changing seasons. Ionospheric conductivities change by up to a factor of 2 as the ionizing solar radiation waxes and wanes along with the trend of sunspots. Ionospheric winds also change as the intensity of solar extreme-ultraviolet radiation increases and decreases (e.g., Forbes and Garrett, 1979). We know that ionospheric dynamo currents change by a factor of 2 to 3 with the solar cycle (e.g., Matsushita, 1967), but the variation of electric fields is not yet so extensively documented. At Jicamarca, Peru, however, ionospheric electric fields have been measured on an occasional basis for well over a solar cycle, and the average behavior of the east-west equatorial electric-field component at 300 km altitude is shown in Figure 14.10, as represented by the vertical plasma drift that it produces. The presence of a strong upward drift after sunset at solar sunspot maximum (1968-1971) is not usually present at solar minimum (1975-1976). Clear seasonal variations appear at Jicamarca as well as at higher latitudes. Day-to-day changes in middle- and low-latitude electric fields and

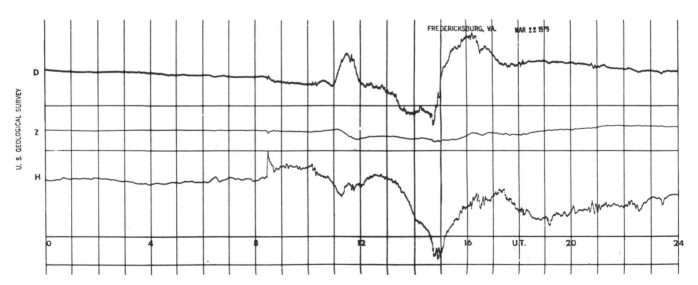

FIGURE 14.9 Magnetogram from Fredericksburg, Virginia, on March 22, 1979, showing variations of the magnetic declination (D), vertical magnetic field (Z), and horizontal magnetic intensity (H) during a magnetic storm. Scale values are not shown.

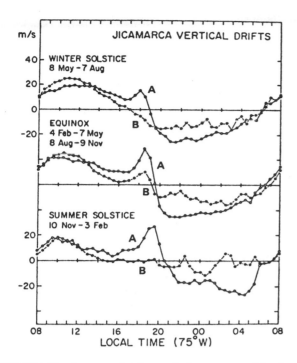

FIGURE 14.10 Average quiet-day vertical component of the plasma drift velocity caused by east-west electric fields over Jicamarca, Peru, as a function of local time (from Fejer *et al.*, 1979). 1968-1971 (a) are sunspot cycle maximum years, while 1975-1976 (b) are sunspot cycle minimum years.

currents also occur, produced by corresponding changes in the ionospheric winds.

Rapid variations with time scales ranging from seconds to hours are a common feature of ionospheric electric fields and currents. The magnitude of the fluctuations is often as large as that of the regular daily variations, generally increasing with magnetic latitude up to the auroral oval. Besides storm and substorm phenomena, global-scale disturbances in the electric fields and currents also occur with the arrival of solar-wind shocks, with fluctuations of the IMF, and with rapid ionosphere conductivity changes during solar flares. Localized ionospheric electric-field fluctuations may be associated with wind or conductivity irregularities or with small-scale magnetospheric processes. For example, localized quasi-periodic oscillations in the electric fields and currents with periods ranging from a fraction of a second to minutes, called pulsations, are often observed at high latitudes (e.g., Nishida, 1978).

EFFECTS OF UPPER-ATMOSPHERIC ELECTRIC FIELDS AND CURRENTS

Electric fields and currents interact strongly with the upper atmosphere and help determine its behavior

(e.g., Banks, 1979). The drifting ions, as they interact collisionally with neutral molecules, exert a force on the air and tend to bring it toward the ion motion. Above 200 km altitude this effect can be important: at high latitudes winds are common that approach the rapid velocity of the convecting plasma (e.g., Meriwether, 1983), while at low latitudes, where plasma drifts are much smaller, the collisional interaction tends to retard the winds driven by pressure gradients. Of even greater importance is the heating of the upper atmosphere caused by currents in the auroral region. The heating can make a significant contribution to the upper-atmospheric energy budget and can even be the dominant heat source above 120 km during magnetic storms. As the temperature increases the upper atmosphere expands, and the drag on near-Earth satellites is increased, changing their orbits (e.g., Joselyn, 1982).

The ionosphere is affected in many ways by electric fields. Above 200 km, where the chemical lifetimes of ions range from several minutes to hours, rapid convection of ionization at high latitudes can bring dense dayside plasma to the nightside of the Earth in some places, cause stagnation and prolonged nighttime decay of ionization at other places, and generally produce highly complex patterns of ionization density (e.g., Sojka *et al.*, 1983). Plasma temperatures and chemical reaction rates are also affected by the rapid ion convection through the air. Even at middle and low latitudes plasma drifts have an important influence on the upper ionosphere, primarily by raising or lowering the layer into regions of lower or higher neutral density, so that chemical decay is retarded or accelerated. During magnetic storms the plasma-drift effects on the ionosphere are not only intensified but are also supplemented by indirect effects through modification of the neutral atmosphere (e.g., Prölss, 1980). Auroral heating induces atmospheric convection that alters the molecular composition of the upper atmosphere and leads to more rapid chemical loss of the ionization, even at middle latitudes. Winds generated by the magnetic storm impart motion to the ionization along the direction of the magnetic field, causing redistribution of the plasma as well as further modification of the loss rate. All these ionospheric phenomena affect radio-wave transmissions that reflect off the ionosphere.

Radio waves at frequencies greater than about 30 MHz do not normally reflect off the ionosphere and are much less influenced by large-scale plasma density variations than are lower frequencies. However, these waves are affected by small-scale plasma irregularities that cause radio signals to scintillate, undergoing substantial amplitude and phase modulations (e.g., Aarons, 1982). The scintillations are bothersome for transmissions between satellites and the ground. Electric fields are involved in producing the ionospheric irregularities, often by plasma instabilities resulting from drifts or from steep density gradients created by nonuniform convection (e.g., Fejer and Kelley, 1980). The irregularities tend to be strongest at high latitudes, where electric fields and their consequences are greatest, but the equatorial region is also subject to strong irregularities, both in the equatorial electrojet and in the nighttime upper ionosphere.

Rapid magnetic fluctuations caused by changing ionospheric currents disturb geophysical surveys of magnetic anomalies in the Earth's crust. In addition, the fluctuations induce electric currents in the Earth. Analysis of the Earth currents can be useful for geophysical studies, but when they enter large man-made structures like electric transmission lines and pipelines they can cause disruptive electrical signals and corrosion. Lanzerotti and Gregori discuss these problems in more detail in Chapter 16, this volume.

Finally, it should be noted that the ionospheric electric field extends down into lower atmospheric regions. In fact, the presence of the field at lower altitudes has been used for many years to measure ionospheric electric fields from stratospheric balloons (e.g., Mozer and Lucht, 1974). Roble and Tzur (Chapter 15, this volume) discuss the relation of the ionospheric electric field to the global atmospheric electric circuit.

SOME OUTSTANDING PROBLEMS

The entire range of subjects discussed in this chapter is undergoing scientific investigation, but there are certain topics that currently present questions of fundamental importance to the understanding of upper-atmospheric electric-field sources.

Since the beginning of the artificial satellite era measurements of space plasmas, energetic particles, and electromagnetic fields have provided a broad picture of the interaction of the solar wind with the magnetosphere. Yet theory still is unable to explain quantitatively, without the use of ad hoc parameterizations, how much magnetic flux interconnects the Earth with the solar wind or what the details of the magnetic- and electric-field configurations are in the vicinity of the magnetopause and in the distant magnetospheric tail. Measurements alone cannot answer these questions because of the impossibility of measuring simultaneously the entire spatial structure of the continuously changing magnetosphere. However, some future observational programs may aid in the solution of the problem by combining continuous monitoring of electrodynamic features of the solar wind and of the entire auroral oval

ionosphere with spot measurements in the poorly explored high-latitude magnetopause and distant tail regions.

Because solar wind/magnetosphere dynamo processes intimately involve magnetospheric plasma energization, ionospheric conductivity alterations, and coupling of magnetospheric and ionospheric electric fields and currents, the dynamo cannot be fully understood without taking account of the magnetosphere-ionosphere interactions. Theoretical models incorporating the mutual interactions have made great strides of progress in recent years (e.g., Spiro and Wolf, 1984), but much remains to be learned. Observational programs that can measure a large number of the important phenomena involved in the interactions are a necessary adjunct to theoretical studies.

The distribution of ionospheric conductivity is important in determining the distributions of both electric fields and currents in the ionosphere and magnetosphere, yet our understanding of nightside conductivities is still rudimentary. We know that the conductivities vary greatly but they are difficult to measure when the ionization density is low, and we do not yet fully understand the nature of nighttime ionization sources. In the auroral oval and polar cap the nightside conductivities can be highly structured, having features correlated with structured features of the electric fields and currents. Because the conductivity features are so variable, useful models are difficult to construct, although a number of encouraging developments in modeling auroral-oval conductivities have been made in the past few years (Reiff, 1984).

Concerning the ionospheric wind dynamo, theoretical models and observations are in general agreement, except that the morphology of nighttime electric fields is not yet well explained, a problem related to the conductivity uncertainties. The nature of day-do-day variability in the electric fields and currents, as well as smaller-scale variations, are only poorly understood at present. Winds in the dynamo region, especially those of a tidal nature, exhibit day-to-day variability that has not been fully explained and is currently unpredictable. It appears that day-to-day variability in the distribution of atmospheric heating as well as changes in atmospheric tidal propagation characteristics and interactions with other wave motions are largely to blame, but these conditions are not now monitored on a global scale. Advances in this area may be forthcoming in the wake of intensified study of the middle atmosphere (10-120 km altitude). In addition, further clarification of upper-atmospheric circulation changes during magnetic storms is needed in order to understand how storms affect the ionospheric wind dynamo.

ACKNOWLEDGMENT

Most of this chapter was prepared while the author was employed at the NOAA Space Environment Laboratory.

REFERENCES

Aarons, J. (1982). Global morphology of ionospheric scintillations, *Proc. ISEE 70*, 360-378.

Akasofu, S.-I. (1977). *Physics of Magnetospheric Substorms*, D. Reidel, Dordrecht, Holland.

Akasofu, S.-I. (1984). The magnetospheric currents: An introduction, in *Magnetospheric Currents*, T. A. Potemra, ed., American Geophysical Union, Washington, D.C., pp. 29-48.

Akasofu, S.-I., and S. Chapman (1972). *Solar-Terrestrial Physics*, Clarendon Press, Oxford.

Banks, P. M. (1979). Magnetosphere, ionosphere and atmosphere interactions, in *Solar System Plasma Physics, Vol. II*, C. F. Kennel, L. J. Lanzerotti, and E. N. Parker, eds., North-Holland Publ. Co., Amsterdam, pp. 57-103.

Blanc, M., and A. D. Richmond (1980). The ionospheric disturbance dynamo, *J. Geophys. Res. 85*, 1669-1686.

Burch, J. L. (1977). The magnetosphere, in *The Upper Atmosphere and Magnetosphere*, NRC Geophysics Study Committee, National Academy of Sciences, Washington, D.C., pp. 42-56.

Cowley, S. W. H. (1982). The causes of convection in the Earth's magnetosphere: A review of developments during the IMS, *Rev. Geophys. Space Phys. 20*, 531-565.

Cowley, S. W. H. (1983). Interpretation of observed relations between solar wind characteristics and effects at ionospheric altitudes, in *High-Latitude Space Plasma Physics*, B. Hultqvist and T. Hagfors, eds., Plenum Press, New York, pp. 225-249.

Fejer, B. G. (1985). Equatorial ionospheric electric fields associated with magnetospheric disturbances, in *Solar Wind-Magnetosphere Coupling*, Y. Kamide, ed., D. Reidel, Dordrecht, Holland.

Fejer, B. G., and M. C. Kelley (1980). Ionosphere irregularities, *Rev. Geophys. Space Phys. 18*, 401-454.

Fejer, B. G., D. T. Farley, R. F. Woodman, and C. Calderon (1979). Dependence of equatorial F region vertical drifts on season and solar cycle, *J. Geophys. Res. 84*, 5792-5796.

Forbes, J. M. (1981). The equatorial electrojet, *Rev. Geophys. Space Phys. 19*, 469-504.

Forbes, J. M. (1982a). Atmospheric tides, 1. Model description and results for the solar diurnal component, *J. Geophys. Res. 87*, 5222-5240.

Forbes, J. M. (1982b). Atmospheric tides, 2. The solar and lunar semidiurnal components, *J. Geophys. Res. 87*, 5241-5252.

Forbes, J. M., and H. B. Garrett (1979). Theoretical studies of atmospheric tides, *Rev. Geophys. Space Phys. 17*, 1951-1981.

Friis-Christensen, E., Y. Kamide, A. D. Richmond, and S. Matsushita (1985). Interplanetary magnetic field control of high-latitude electric fields and currents determined from Greenland magnetometer data, *J. Geophys. Res. 90*, 1325-1338.

Harper, R. M. (1977). Tidal winds in the 100- to 200-km region at Arecibo, *J. Geophys. Res. 82*, 3243-3250.

Hill, T. W. (1979). Generation of the magnetospheric electric field, in *Quantitative Modeling of Magnetospheric Processes*, W. P. Olson, ed., American Geophysical Union, Washington, D.C., pp. 297-315.

Hill, T. W., and R. A. Wolf (1977). Solar-wind interaction, in *The Upper Atmosphere and Magnetosphere*, NRC Geophysics Study Committee, National Academy of Sciences, Washington, D.C., pp. 25-41.

Hones, E. W., Jr. (1983). Magnetic structure of the boundary layer, *Space Sci. Rev. 34*, 201-211.

Hundhausen, A. J. (1979). Solar activity and the solar wind, *Rev. Geophys. Space Phys. 17*, 2034-2048.

Joselyn, J. C., ed. (1982). *Workshop on Satellite Drag*, NOAA Space Environment Laboratory, Boulder, Colo., 251 pp.

Lyons, L. R., and D. J. Williams (1984). *Quantitative Aspects of Magnetospheric Physics*, D. Reidel Publishing Co., Dordrecht, Holland.

Matsushita, S. (1967). Solar quiet and lunar daily variation fields, in *Physics of Geomagnetic Phenomena, Vol. 1*, S. Matsushita and W. H. Campbell, eds., Academic Press, New York, pp. 301-424.

McPherron, R. L. (1979). Magnetospheric substorms, *Rev. Geophys. Space Phys. 17*, 657-681.

Meriwether, J. W., Jr. (1983). Observations of thermospheric dynamics at high latitudes from ground and space, *Radio Sci. 18*, 1035-1052.

Mozer, F. S., and P. Lucht (1974). The average auroral zone electric field, *J. Geophys. Res. 79*, 1001-1006.

Nishida, A. (1978). *Geomagnetic Diagnosis of the Magnetosphere*, Springer-Verlag, New York, 256 pp.

Prölss, G. . (1980). Magnetic storm associated perturbations of the upper atmosphere: Recent results obtained by satelliteborne gas analyzers, *Rev. Geophys. Space Phys. 18*, 183-202.

Reiff, P. H. (1984). Models of auroral-zone conductances, in *Magnetospheric Currents*, T. A. Potemra, ed., American Geophysical Union, Washington, D.C., pp. 180-191.

Richmond, A. D., M. Blanc, B. A. Emery, R. H. Wand, B. G. Fejer, R. F. Woodman, S. Ganguly, P. Amaynec, R. A. Behnke, C. Calderon, and J. V. Evans (1980). An empirical model of quiet-day ionospheric electric fields at middle and low latitudes, *J. Geophys. Res. 85*, 4658-4664.

Roederer, J. G. (1979). Earth's magnetosphere: Global problems in magnetospheric plasma physics, in *Solar System Plasma Physics, Vol. II*, C. F. Kennel, L. J. Lanzerotti, and E. N. Parker, eds., North-Holland Publ. Co., Amsterdam, pp. 1-56.

Rostoker, G. (1983). Dependence of the high-latitude ionospheric fields and plasma characteristics on the properties of the interplanetary medium, in *High-Latitude Space Plasma Physics*, B. Hultqvist and T. Hagfors, eds., Plenum Press, New York, pp. 189-204.

Rowe, J. F., Jr., and J. D. Mathews (1973). Low-latitude nighttime E region conductivities, *J. Geophys. Res. 78*, 7461-7470.

Sojka, J. J., R. W. Schunk, J. V. Evans, and J. M. Holt (1983). Comparison of model high-latitude electron densities with Millstone Hill observations, *J. Geophys. Res. 88*, 7783-7793.

Spiro, R. W., and R. A. Wolf (1984). Electrodynamics of convection in the inner magnetosphere, in *Magnetospheric Currents*, T. A. Potemra, ed., American Geophysical Union, Washington, D.C., pp. 247-259.

Stern, D. P. (1977). Large-scale electric fields in the Earth's magnetosphere, *Rev. Geophys. Space Phys. 15*, 156-194.

Stern, D. P. (1983). The origins of Birkeland currents, *Rev. Geophys. Space Phys. 21*, 125-138.

Takeda, M., and H. Maeda (1980). Three-dimensional structure of ionospheric currents—1. Currents caused by diurnal tidal winds, *J. Geophys. Res. 85*, 6895-6899.

Takeda, M., and H. Maeda (1981). Three dimensional structure of ionospheric currents—2. Currents caused by semidiurnal tidal winds, *J. Geophys. Res. 86*, 5861-5867.

Wagner, C. U., D. Möhlmann, K. Schäfer, V. M. Mishin, and M. I. Matveev (1980). Large-scale electric fields and currents and related geomagnetic variations in the quiet plasmasphere, *Space Sci. Rev. 26*, 391-446.

The Global Atmospheric-Electrical Circuit

15

RAYMOND G. ROBLE *and* ISRAEL TZUR
National Center for Atmospheric Research

Lightning was recognized as a grand manifestation of static electricity within thunderstorm clouds in the eighteenth century. It was also recognized that electrical phenomena are not confined to thunderclouds and that a weak electrification exists as a permanent property of the atmosphere even during fair weather. Further research established that the Earth's surface is charged negatively and the air is charged positively, with a vertical electric field of about 100 V/m existing in the atmosphere near the Earth's surface. An electrostatic explanation for the phenomena was sought at first, and one theory suggested that the electric field of the atmosphere was the result of an intrinsic negative charge on the Earth, probably collected during the Earth's formation. With the discovery of cosmic-ray ionization in the early twentieth century, it was realized that air possesses an electrical conductivity due to its ion content. As a result of the finite electrical conductivity, vertical conduction currents flow from the atmosphere to the Earth, tending to neutralize the charge on the Earth. On the basis of actual conductivity values it was calculated that charge neutralization would take place in less than an hour, and the continued existence of an electric field suggested some generation mechanism to oppose the leakage currents flowing to the Earth. The search for this generation mechanism soon became the main object of research on global atmospheric electricity.

In the early twentieth century the concept of a global circuit of atmospheric electricity slowly began to evolve (Israël, 1973; Pierce, 1977). The net positive space charge in the air between the ground and a height of about 10 km is nearly equal to the negative charge on the surface of the Earth. The electrical conductivity of the air increases rapidly with altitude, and the product of the local vertical electric field and local conductivity at any altitude within an atmospheric column gives a constant air-earth current flowing downward. This constant air-earth current with respect to altitude implies that the current flow is mainly driven by a constant difference in potential between the surface of the Earth and some higher altitude in the atmosphere. The discovery of the highly conducting ionosphere in the 1920s explained the long-range propagation of radio waves and was important for the evolution of the concept of the global electric circuit. The ionosphere, with its large electrical conductivity, provided a means of closing the global circuit. It, however, is not a perfect conductor parallel to the Earth's surface, but it possesses a finite conductivity, and the electric currents and fields within it are driven by the combined action of the ionospheric and magnetospheric dynamo systems as well as by current generation from the lower atmosphere.

Wilson (1920) first demonstrated that a thunderstorm supplies a negative charge to the Earth. In the 1920s, it was also known that over the oceans and in polar areas

206

the diurnal maximum of the fair-weather potential gradient at the Earth's surface occurred at the same universal time (about 1900 UT). Furthermore, radio measurements of atmospherics showed that global thunderstorm activity also peaked near 1900 UT, with the main thunderstorm centers being in Africa and South America. Scientists studying meteorological statistics of thunderstorm activity found similar diurnal variations. The diurnal UT variation of potential gradient over the oceans was similar to the diurnal UT variation of thunderstorm occurrence frequency with no phase delay. These experimental facts all contributed to the concept of the Earth's global electrical circuit and furthermore suggested that thunderstorms were the generators within the circuit. This concept of the global electrical circuit persists today, although there are still basic problems and details that need to be resolved (Dolezalek, 1971, 1972; Kasemir, 1979).

THUNDERSTORMS AS GENERATORS IN THE GLOBAL CIRCUIT

The vast majority of clouds in the atmosphere form and dissipate without ever producing precipitation or lightning. A cloud, however, interacts with atmospheric ions and becomes electrified to a certain degree. The small, fast ions in the atmosphere almost exclusively provide the electrical conductivity. The intermediate and large ions do not greatly contribute because of their lower mobility. The fast ions within clouds become attached to more massive cloud particles and thereby decrease the electrical conductivity within the cloud relative to the surrounding clear air. As a result of this electrical conductivity change alone, clouds act as an electrical obstacle; space charge develops on the surface of the cloud, and the distribution of fair-weather conduction currents and fields flowing in the vicinity of the cloud are altered. As convective activity intensifies, electrification increases. Strong electrification generally begins with the rapid vertical and horizontal development of a fair-weather cumulus cloud into a cumulonimbus type. Most of the lightning on Earth is produced by strongly convective cumulonimbus clouds with a vigorous system of updrafts and downdrafts.

The updrafts and downdrafts associated with convection and the interactions between cloud and precipitation particles (Tzur and Levin, 1981; Rawlins, 1982) act in some manner that eventually separates charges within the thundercloud. Charge-separation processes usually fill the upper portion of the thundercloud with the main positive charge and the lower portion with the main negative charge. The measured charge structure within individual clouds is complex (Krehbiel et al.,

1979), and a simplifying assumption that is usually made for modeling purposes considers the charge to be distributed in a spherical fashion within some finite volume. A finite spherical distribution of charge has the same distant electric-field structure in space as an equivalent point charge. Therefore, for simplicity in studies of the electrical interaction of a thunderstorm with its immediate environment, the lightning charges, which are lowered to ground during a lightning stroke, and the main thunderstorm charges are usually represented as a series of point charges.

The main negative charge in the lower part of a thunderstorm (Krehbiel et al., 1979) occurs at a height where the atmospheric temperature is between $-10°C$ and $-20°C$. This temperature range is typically between 6 and 8 km for summer thunderstorms and around 2 km for winter thunderstorms. The positive charge at the top of the storm does not have so clear a relationship with temperature as the negative charge but can typically occur between $-25°C$ and $-60°C$ depending on the size of the storm. This temperature range usually lies between 8 and 16 km in altitude. The ensemble of thunderstorms occurring over the globe at any given time will have positive and negative charge centers located over a large altitude range depending on the atmospheric structure, the size of the thunderstorm, its type (e.g., air mass, frontal), its location (e.g., ocean, plains, or mountains), its latitude, its stage of development, and other factors.

There is no generally accepted model of thunderstorm electrification that can be used to calculate the current that storms release into the global electrical circuit. Measurements by Gish and Wait (1950), Stergis et al. (1957a), Vonnegut et al. (1966, 1973), and Kasemir (1979) showed that the total current flowing upward from thunderstorms areas ranges from 0.1 to 6 A with an average of about 0.7 A per thunderstorm cell. It is, thus, possible to use this value in a theoretical model without referring to the details of the charge-separation mechanisms. The simplest model used to investigate the electrical interactions of a thunderstorm with its immediate environment assumes a quasi-static dipolar charge distribution embedded within the thundercloud, which is immersed in a conducting atmosphere whose electrical conductivity increases exponentially from the surface of the Earth to a highly conducting region somewhere within the ionosphere above about 60 km.

In the quasi-static state these charges are maintained in equilibrium against discharge currents by assuming that a steady convection current acts between the two charge centers in the updraft and downdraft regions of the storm. In earlier studies only conduction currents were assumed to flow in the environment, and although this assumption is probably valid in the highly conducting region above a storm it is not valid within and below the storm, where corona, lightning, precipitation, convection, and displacement currents all contribute to the charge exchange between charge centers and between the storm and the surface of the Earth. Holzer and Saxon (1952) and Kasemir (1959) presented analytic solutions for the potential distribution around thunderstorms whose charge distributions are represented as point dipole current sources. A calculated normalized potential distribution around a quasi-static dipolar charge structure within the atmosphere (Kasemir, 1959) is shown in Figure 15.1(a). The lines represent streamlines of current flow between the two charge centers and between the charge centers and the ionosphere and the ground. The highly conducting Earth's surface is considered by including image point sources within the Earth for the solution of the electrostatic potential distribution around the thunderstorm. The currents flow upward from the positive charge center toward the ionosphere and from the ground toward the negative charge. There is also a current flow between the charges. All currents are calculated as conduction cur-

rents in this simple model and have considerably more complexity in actual storms.

Holzer and Saxon (1952) showed that the current output from such a thunderstorm model is sensitive to the charge-separation distance, with a greater current output from larger storms having intense vertical velocities and charge-separation capability. They also showed that storms with frequent cloud-to-ground lightning strokes move negative charge to the Earth's surface and leave the cloud with a net positive charge that has a greater current output than a dipole. Thus, according to their calculations, even a weak storm with charge separation could supply a current to the ionosphere and the global circuit.

Kasemir (1959) showed that the exponential conductivity increase with altitude in the atmosphere is important for the current flow from thunderstorms toward the ionosphere. He pointed out that if the atmosphere had a constant conductivity at all altitudes the primary thunderstorm source current would flow downward to the image source, with no current flow toward the ionosphere and into the global circuit. Anderson and Freier (1969) considered both the limiting fast and slow time variations within thunderstorms. Their dipole quasi-static solutions for slow time variation show that the potential distribution about thunderstorms is affected by both the conductivity and the charge-separation distance within storms. Their calculations also suggest that the thunderstorm, in a quasi-static sense, may be electrically closed and only the pumping action of lightning discharges can provide the current necessary to maintain the ionospheric potential.

Park and Dejnakarintra (1973, 1977a) considered a dipolar thunderstorm model and analytically solved for the current output and the mapping of thundercloud electric fields into the ionosphere. This model considered the anisotropy of the electrical conductivity above about 60 km and assumed that the Earth's geomagnetic-field lines were vertical. The largest ionospheric fields were calculated to occur at night over giant thunderstorms with values on the order of 10^{-4} V/m at 100 km. During the day the calculated dc electric fields are 1 to 2 orders of magnitude smaller because of the increased ionospheric electrical conductivity caused by solar extreme ultraviolet (EUV) radiation ionization. Dejnakarintra and Park (1974) examined the penetration of lightning-induced ac fields into the ionosphere and found that the lightning electric-field signal recovery time decreases rapidly with increasing altitude until at 100 km the electric-field wave form appears as a sharp pulse. The ac fields are also larger at night than during the day, when ionospheric conductivities are larger.

Wait (1960) pointed out that there is a constant radial

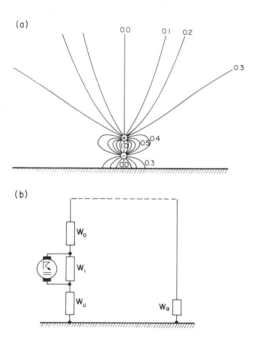

FIGURE 15.1 (a) Schematic giving the current streamlines of dipolar point current sources embedded within an atmosphere with exponentially increasing conductivity over a perfectly conducting Earth, (b) schematic of a lumped parameter representation of the global atmospheric electrical circuit where the thunderstorm is represented as a current generator with internal resistance W_i, W_0 represents the resistance between cloud top and the ionosphere, W_a represents the fair-weather load resistance, and W_u represents the resistance between the thunderstorm and the ground (Kasemir, 1959).

potential associated with the lower-order ($n = 0$) mode in a concentric spherical cavity excited by a radial current-moment element. He estimated that the potential across the earth-ionosphere gap caused by a single 3-km-length lightning discharge of 1000 A would be on the order of 1 V, and he suggested that the omnipresent constant voltage during fair weather might result from the accumulated action of a number of lightning strokes. Hill (1971) further examined the resonant hypothesis and showed that the atmospheric field is generated through electrostatic induction by equivalent charge dipoles from thunderclouds. Using parameters of lightning frequency and the magnitude of the current moment within storms available at that time, he found an ionospheric potential only about one third of the observed intensity.

Schumann resonances in the earth-ionosphere cavity are also excited by worldwide lightning activity, and this subject has been reviewed by Polk (1982).

Willett (1979) pointed out that the total current supplied to the global electric circuit from a thunderstorm differs depending on whether the storm is considered to be a current or a voltage generator. For a current thunderstorm generator, where the current supplied to the global circuit is independent of the load, his results showed that current output is proportional to the assumed source current strength and to a proportionality factor that depends on the ratio of the height of the ionosphere to the electrical conductivity scale height. For a voltage-source thunderstorm generator, where the charging current increases with time until limited by some voltage-sensitive dissipative mechanism such as lightning or corona, the current output may be somewhat smaller depending on assumptions made concerning the storm's internal resistance. Not enough is known about the thunderstorm generator to distinguish between the current and voltage source models.

Freier (1979) presented a thunderstorm model with more details than point-charge sources and conduction currents. He considered the atmosphere between the Earth and the ionosphere divided into three separate regions as shown in Figure 15.2. In region 1, below the negative layer of charge in a thunderstorm, he considered conduction, displacement, and precipitation current densities, allowing all to vary with altitude. In region 2, between the bottom and top of the thunderstorm cloud, he considered charging, conduction, displacement, and precipitation current densities that also vary in space and time. In region 3, above the storm, only conduction and displacement are considered. All lightning currents are considered to be discontinuous charge transfers, and in the fair-weather regions far to the side of the thunderstorm only conduction currents flow.

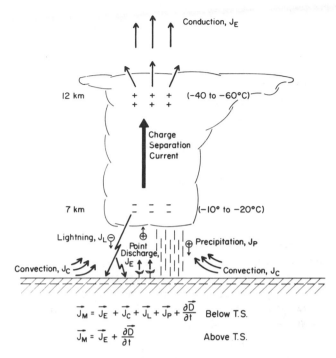

FIGURE 15.2 Schematic illustrating the various currents that flow within and in the vicinity of thunderstorms: $\mathbf{J}_E$ is the conduction current, $\mathbf{J}_c$ is a convection current, $\mathbf{J}_L$ is the lightning current, $\mathbf{J}_p$ is the precipitation current, $\partial D/\partial t$ is the displacement current, and $\mathbf{J}_M$ is the total Maxwell current.

This model is an improvement over previous thunderstorm models, which considered lumped circuit parameters of columnar resistance as shown in Figure 15.1(b) for the dipolar conduction current generator. Freier (1979) used it to describe a charge-transfer mechanism that may occur during large severe storms when precipitation is heavy and when the storms are, in certain instances, accompanied by tornadoes. The precipitation current during such storms may be so large that it removes much of the negative charge in the lower portion of the storm and allows the generator to operate between the positive charge and the ground. During such storms much more electrical energy is produced, with a possibly greater current output that must be considered in any global electric model.

Krider and Musser (1982) pointed out that the time variations in thunderstorm electric fields, both aloft and at the ground, can be interpreted as a total Maxwell current density that varies slowly in intervals between lightning discharges. The total Maxwell current, $\mathbf{J}_M$, consists of field-dependent currents, both linear and nonlinear (corona), $\mathbf{J}_E$; convection currents, $\mathbf{J}_C$; lightning currents, $\mathbf{J}_L$; and the displacement current density, $\partial D/\partial t$.

$$J_M = J_E + J_C + J_L + \partial D/\partial t. \qquad (15.1)$$

The field-dependent currents and convection currents are distributed over large areas and, in the absence of lightning, tend to vary rather slowly with time. The lightning current occurs impulsively and represents a discontinuous transfer of charge in both space and time. Corona and lightning usually transfer negative charge to the ground, and precipitation and turbulence transfer positive charge.

Krider and Musser (1982) showed that the average Maxwell current density is usually not affected by lightning discharges and varies slowly throughout the evolution of the storm. Since the Maxwell current is steady at times when the electric field both at the ground and aloft undergoes large changes in amplitude, and sometimes even polarity, they inferred that the cloud electrification processes are substantially independent of the electric field. Also, since the Maxwell current varies slowly throughout the storm, it probably represents an electrical quantity that is coupled directly to the meteorological structure of the storm. Thus, the thunderstorm appears to be a current source that produces a quasi-static Maxwell current density.

Grenet (1947, 1959) and Vonnegut (1953, 1965) proposed a convective thunderstorm generator in which the electrification process is related to the updrafts and downdrafts acting within the thunderstorm. In this generation process positive space charge is convected to the top of the thunderstorm cloud by updrafts within it, and the negative charge attracted to the top of the cloud is swept downward by convective downdrafts at the edges of the cloud. For this generator, the magnitude of the cloud electrification is related to the strength of thunderstorm convective activity. The convection mechanism also involves conduction currents flowing from the upper atmosphere to the top of the cloud. The magnitude of the electrification process can, therefore, increase with an increase in the electrical conductivity of the atmosphere over the cloud.

The above discussion reveals the complexity and difficulty involved in modeling the thunderstorm as a generator in the global circuit. Significant progress has been made in recent years, but there is still a long way to go before a generally accepted thunderstorm model will be available for use in a global model of atmospheric electricity. Yet such a model is of prime importance, and research in this area should be pursued vigorously.

SOME PROPERTIES OF THE GLOBAL CIRCUIT

According to the classical picture of atmospheric electricity (Dolezalek, 1972; Israël, 1973), the totality of thunderstorms acting together at any time charges the ionosphere to a potential of several hundred thousand volts with respect to the Earth's surface. This potential difference drives a vertical electric conduction current downward from the ionosphere to the ground in all fair-weather regions on the globe. The fair-weather electric conduction current varies according to the ionospheric potential difference and the columnar resistance between the ionosphere and the ground. Horizontal currents flow freely along the highly conducting Earth's surface and in the ionosphere. A current flows upward from a thunderstorm cloud top toward the ionosphere and also from the ground into the thunderstorm generator, closing the circuit. A lumped parameter schematic of the global circuit is shown in Figure 15.1(b). This schematic does not represent the real circuit but rather illustrates its basic concepts. The global fair-weather load resistance is given as W_a and is about 250 Ω. The thunderstorm generator source is shown along with its equivalent internal resistance, W_i, which is not well known. The total resistance between the thunderstorm and ionosphere is represented as W_0 and is about 10^5-10^6 Ω, and the total resistance between the thunderstorm and ground is represented by W_u and is also not well known. Markson (1978), however, suggests that W_u is small because of corona discharge beneath the storm, having a value of about 10^4-10^5 Ω. Some of the overall properties of the global circuit are summarized in Table 15.1. There are many additional elements that complicate this simplified classical picture, and these are discussed in the following subsections.

Global Thunderstorms

The thunderstorm generator hypothesis proposed by Wilson (1920) was based on his observations that beneath the thundercloud negative charge is transferred to the Earth and above the thundercloud positive charge is transferred to the conductive upper atmosphere. A subsequent discovery was the close correlation between the diurnal universal time variation of the thunderstorm generator current (represented by the frequency of thunderstorm occurrence) and the load current (represented by the fair-weather ground electric field or air-earth current density), integrated over the surface of the Earth.

In about the 1920s the electric field over the oceans was found to vary diurnally in accordance with universal time (Parkinson and Torreson, 1931), as shown in the upper frame of Figure 15.3. The diurnal change of electrical conductivity over the oceans is relatively small, and therefore Ohm's law requires that the air-earth current density also follow the diurnal variation of electric field. The maximum value of both the average

TABLE 15.1   Some Properties of the Global Circuit

Number of Thunderstorms Acting at One Time	1500-2000
Currents above Thunderstorms (A)	
(a) Range	0.1 to 6
(b) Average	0.5 to 1
Global Current (A)	750-2000
Ionospheric Potential (kV)	
(a) Range	150-600
(b) Mean	280
Columnar Resistance at Sea Level (Ω/m^2)	
(a) Low latitude	1.3×10^{17}
(b) High latitude	3×10^{17}
(c) Tibet and Antarctic plateau	2×10^{16}
Total Resistance (Ω)	230
(including resistance decrease by mountains)	200
Current Density (A/m^2)	
(a) Inhabited and industrialized areas	1×10^{-12}
(b) Vegetated ground and deserts	2.4×10^{-12}
(c) South Pole Station	2.5×10^{-12}
Potential Gradient (V/m)	
(a) Equator	120
(b) 60° latitude	155
(c) South Pole	71
(d) Industrial areas	300-400
Average Charge Transfer over the Entire World	
($C\,km^{-2}\,yr^{-1}$)	+90 C
Total Charge on the Earth (C)	500,000
Electrical Relaxation Times	
(a) 70 km	10^{-4} sec
(b) 18 km	4 sec
(c) 0.01 km	5-40 min
(d) Earth's surface	10^{-5} sec
Electrical Conductivity (mho/m)	
Sea level	10^{-14}
Tropopause	10^{-13}
Stratopause	10^{-10}
Ionosphere	
(a) Pedersen conductivity	10^{-4}-10^{-5}
(b) Parallel conductivity	10

electric field and the current over the oceans occurs near 1900 UT, and minimum values near 0400 UT.

Whipple and Scrase (1936) obtained the average thunderstorm probability as a function of local time at Kew, England, from corona current records. Assuming that the same thunderstorm probability curve also exists on other continents as a function of local time, they combined this curve with the world thunderstorm day statistics of Brooks (1925) and obtained the diurnal variation of worldwide thunderstorm activity as a function of universal time, shown in the bottom frame of Figure 15.3. The three major component curves with their maxima at 0800, 1400, and 2000 UT represent the contributions of the major thunderstorm regions of Asia and Australia, Africa and Europe, and America, respectively. The summation of the component curves represents the diurnal variation of worldwide thunderstorm activity. The similarity of the diurnal variation of electric field over

the ocean and the diurnal variation of worldwide thunderstorm activity supports the hypothesis that thunderstorms are the electrical generator in the global circuit. The maxima and minima of both curves occur at about the same universal times. The amplitudes of modulation for the two curves, however, are different. The amplitude of the electric-field curve is about 20 percent, and the amplitude of the thunderstorm curve is about 45 percent. Whipple and Scrase (1936) suggested that this difference might be resolved if a steady supply current from worldwide ocean thunderstorms is added to the worldwide continental distribution of storms. This suggestion, however, is not supported by current data. The difference in amplitudes probably arises from the enormous variability in thunderstorm electrification.

Although the similarity between the diurnal UT worldwide thunderstorm frequency curve and the diurnal UT electric-field curve suggests that thunderstorms

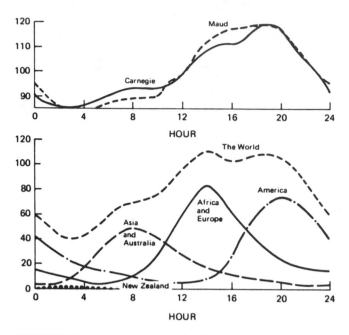

FIGURE 15.3 (a) Annual curve of the diurnal variation of the atmospheric electric field on the oceans (volts per meter) as measured by the *Carnegie* and *Maud* expeditions (Parkinson and Torreson, 1931) and (b) annual curve of the diurnal variations of global thunderstorm activity according to Whipple and Scrase (1936).

are the generators in the global electrical circuit, there is still considerable uncertainty concerning the details, as discussed by Dolezalek (1972) and Kasemir (1979). Moreover, the data on which the thunderstorm activity is based are only of a qualitative nature: "a thunderstorm day is a day when thunder has been heard." The commonly quoted UT diurnal patterns, shown in Figure 15.3, are averages over a long time that were made to reduce the influence of various disturbing factors. When shorter time averages are used—and even single diurnal variations—the correlations show great departures from the average curves. The variability of thunderstorm frequency can be large, with significant departures from the average; for example, whole continents may be cloudless for a long time. The measured electric field on the ground is also highly variable as a result of local influences, and it generally takes a week's worth of averaging or more to bring out the diurnal UT pattern. Most measurements are made on continents, where the electric field displays variations with local time, and these measurements do not fit into a daily worldwide pattern but must be averaged to determine worldwide characteristics. Paramanov (1950) suggested that local influences might cancel out if the electric fields measured on many continents were synchronized to universal time and averaged.

Dolezalek (1972), in examining the data available at the time, concluded that a globally controlled current does flow vertically through the atmosphere but that its connection to thunderstorm activity is tenuous and, in fact, is often contradicted by the proper interpretation of available measurements.

More recently, Orville and Spencer (1979) examined lightning flashes recorded in photographs by two satellites in the Defense Meteorological Satellite Program (DMSP) and found that most of the lightning is confined to land areas and that the ratio of global lightning frequency during northern summer to that of southern summer is about 1.4 for both the dusk and midnight satellite data. They pointed out that this summer-winter difference in global lightning frequency is opposite to the electric-field measurements. The hypothesized relation of the global atmospheric electric current to thunderstorms is still an unsettled question and clearly needs to be resolved to make further progress in understanding the Earth's global atmospheric electric circuit.

Current above Thunderstorms

A few measurements have been made that give the magnitude of the current flowing upward over the whole area of a thundercloud (Gish and Wait, 1950; Stergis *et al.*, 1957a; Vonnegut *et al.*, 1966; Imyanitov *et al.*, 1969; Kasemir, 1979). The currents range from 0.1 up to 6 A, with an average between about 0.5 and 1 A per thunderstorm cell. Gish and Wait (1950) flew an aircraft over a thunderstorm in the central United States and found an average upward current of 0.8 A at an altitude of about 12 km. They measured electric fields of up to 70 kV/m. Stergis *et al.* (1957b), in a series of balloon flights at altitudes of about 25 km in central Florida, measured an average upward current of 1.3 A. The electric fields that they measured at this altitude were on the order of a few hundred volts per meter. Holzworth (1981), with a balloon near 20 km over a large thunderstorm at Fort Simpson, NWT, Canada, on August 15, 1977, measured a vertical upward electric field of more than 6.7 V/m (the instrumentation threshold) for longer than 2 hours. These few data indicate that a positive current flows toward the ionosphere above thunderstorm regions that is of sufficient magnitude to account for fair-weather conduction current. Is it possible to relate current output to such factors as frequency of cloud-to-ground strokes, charge structure and separation distances, and cloud-top height? Both Pierce (1970) and Prentice and Macherras (1977) presented relationships for a latitudinal variation in the ratio between cloud-to-ground and cloud-to-cloud flashes. This ratio is about 0.1 in the equatorial region, increasing to about 0.4 near

50° latitude. Do lightning-intensive tropical thunderstorms deliver more or less current than thunderstorms at higher latitudes, even though the ratio of cloud-to-ground to cloud-to-cloud flashes is smaller? A considerable amount of research is necessary to determine the processes responsible for regulating the current flow from thunderstorms into the global circuit. This research is important for understanding the role of worldwide thunderstorm activity as the generator for the global circuit.

Electrical Conductivity, Columnar Resistance, and Global Resistance

Galactic cosmic rays are the main source of ionization that maintains the electrical conductivity of the atmosphere from the ground to about 60 km in altitude. Near the ground, however, there is additional ionization due to release of radioactive gases from the soil, and above about 60 km solar ultraviolet radiation becomes important. During geomagnetic storms, ionization due to energetic auroral electron precipitation and to auroral x-ray bremsstrahlung radiation and proton bombardment during solar proton events can all be significant sources for the high-latitude middle atmosphere.

The galactic cosmic rays that bombard the Earth's atmosphere are influenced by the Earth's geomagnetic field, which produces a magnetic latitudinal effect in the incoming cosmic-ray flux. The full cosmic-ray spectrum is only capable of reaching the Earth at geomagnetic latitudes higher than about 60°. At lower geomagnetic latitudes, the lower-energy particles are successively excluded by the Earth's geomagnetic field, and only particles with energies greater than about 15 GeV reach the equator. The cosmic-ray spectrum thus hardens with a decrease in geomagnetic latitude, with the height of the maximum ion production rate decreasing from about 20 km at high latitudes to about 10 km near the equator. The cosmic-ray ion production rate profile for various geomagnetic latitudes during solar-cycle minimum using the data from Neher (1967) is shown in Figure 15.4. The ionization rate increases with geomagnetic latitude, and both the height of the peak and the slope of the ionization rate above the peak also increase. Near the ground there is about a 20 percent variation between the equatorial region and higher latitudes (Israël, 1973).

The cosmic-ray ionization undergoes a regular solar-cycle variation with maximum values near solar minimum. At high latitudes the ionization rate may vary by about 50 percent at 15 km and about 75 percent at 20 km. At 20 km the ionization-rate variation through the solar cycle is about 40 percent in mid-latitudes and

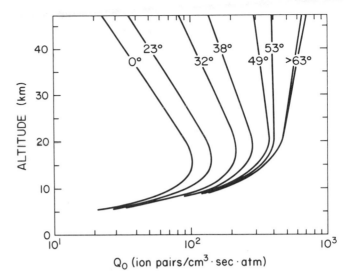

FIGURE 15.4 Cosmic-ray ion-production rate vertical profiles at various geomagnetic latitudes (Neher, 1967; R. Williamson, Stanford University, personal communications, 1981).

about 20 percent near the equator. In addition to the solar-cycle variation of galactic cosmic-ray flux, there are shorter-term variations that are associated with some magnetic storms, a 27-day quasi-periodic variation, variations due to solar flares, a diurnal variation, and a variation in the amplitude of the diurnal variation with time. All these variations have been reviewed by Forbush (1966). The diurnal variations are generally small. The solar-flare and magnetic-storm variations of cosmic-ray fluxes are larger (about 2 to 20 percent) and are more important for understanding solar-terrestrial electrical coupling mechanisms (Roble and Hays, 1982). The full impact of all these variations on the electrical conductivity and the properties of the global circuit has not yet been evaluated.

Over land, the natural radioactivity of the solid ground adds to the cosmic-ray ion-production rate, not by direct radiation from the solid surface but in the release of gaseous intermediaries from rocks and soil on the surface and from soil capillaries. These gaseous intermediaries are then carried upward by vertical mass transfer, and radiation from them can affect the ion-production rate within the first kilometer above the Earth's surface (Israël, 1973).

Above about 60 km the solar ultraviolet radiation ionization exceeds the ion production owing to galactic cosmic rays. The major source of ionization within the mesosphere is nitric oxide, which can be ionized to NO^+ by solar Lyman-alpha radiation at 121.6 nm. Above about 80 km, the EUV and soft x-ray radiation from the Sun produces ionization in the ionospheric E and F regions. These EUV radiations all have a large day-to-day variability as well as periodic 27-day solar rotation and solar-cycle variations. During solar flares the solar EUV and x-ray radiation can be greatly enhanced, thereby increasing the ionization source and electrical conductivity of the ionosphere.

The ionization sources mentioned above all contribute to the electrical conductivity of the atmosphere through the production of positive and negative ions and electrons. The electrical conductivity is governed by the number of positive and negative ions and electrons as well as by their respective mobility. The number densities of various ionic species are controlled by complicated chemical reactions between ions and neutral species. Many of the neutral constituents important for ion chemistry, such as water vapor and NO, are furthermore transported by atmospheric motions, thereby increasing the complexity of determining the global distribution of electrical conductivity. All these various processes are discussed by Reid (Chapter 14, this volume).

Above about 80 km, the electrical conductivity is governed by collisions between neutrals, ions, and electrons and by the influence of the geomagnetic field on the mobility of the plasma components. The electron gyrofrequency is greater than the electron-neutral collision frequency above about 80 km, and therefore the electrons are restricted by the geomagnetic-field line. For the ions this occurs above about 140 km. The differential motion of the ions and the electrons in the dynamo region between about 80 and 200 km in altitude gives rise to an anisotropic behavior of the conductivity. The conductivity parallel to the geomagnetic-field lines is not affected by the field, and it increases rapidly with altitude, limited only by collisions between electrons and neutrals and ions. The Pedersen conductivity is parallel to an applied electric field and orthogonal to the magnetic field. It is smaller than the parallel conductivity, and it has a maximum near 140 km, where the ion-neutral collision rate equals the gyrofrequency of ions. The conductivity orthogonal to both an applied electric field and the magnetic field is the Hall conductivity, which maximizes near 105 km. The Pedersen conductivity is carried by electrons below 105 km and by ions above that altitude, whereas the Hall current is mainly due to electrons.

A typical profile of electrical conductivity through the daytime atmosphere is shown in Figure 15.5. Near the Earth's surface the electrical conductivity is about 10^{-14} mho/m. It increases exponentially with altitude, having a scale height of about 6 km until about 60 km, where the effects of free electrons become important and there is an abrupt increase in the conductivity.

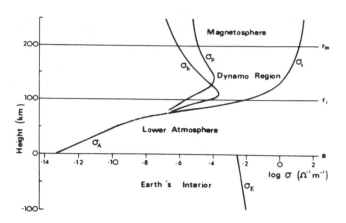

FIGURE 15.5 Altitude variation of the electrical conductivity in the Earth's atmosphere and ionosphere from the ground to 200 km. The electrical conductivity within the Earth is shown for comparison (Volland, 1982).

Above about 80 km the geomagnetic field introduces anisotropic conductivity components. The parallel conductivity continues to increase with altitude, whereas the Hall and Pedersen conductivities peak near 105 and 140 km, respectively, before decreasing with altitude. The electrical conductivity of the Earth is about 10^{-3} mho/m, and therefore the Earth's atmosphere can be considered a leaky dielectric sandwiched between two highly conducting regions—the Earth's surface and the ionosphere.

A calculation of the latitudinal variation of the vertical and horizontal electrical conductivity components in the daytime atmosphere during equinox by Tzur and Roble (1983) is shown in Figure 15.6. The electron densities above about 10 km are calculated using the model of Reid (1976, 1977) and the properties of the neutral atmosphere are specified from the model of Solomon *et al.* (1982a, 1982b). Below 10 km, the electrical conductivity is represented by the Gish formula, with a latitudinal variation as determined from Israël (1973). The electrical conductivity increases abruptly above about 60 km because free electrons are present in the daytime ionosphere, and the effect of the geomagnetic-field line becomes apparent above 80 km. The electrical conductivity parallel to the geomagnetic-field line is larger than either the Hall or the Pedersen conductivity, so the vertical component of the electrical conductivity is small in the equatorial region, where the magnetic-field line is horizontal; and the horizontal component of the electrical conductivity is small in the polar region, where the magnetic-field line is vertical.

This calculation for the global distribution of electrical conductivity represents an idealized case considering fair-weather conditions. In reality, the conductivity is

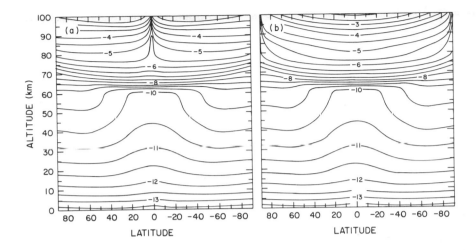

FIGURE 15.6 Latitudinal distribution of the (a) vertical and (b) horizontal components of $\log_{10} \sigma$ (mhos/m), where σ is the electrical conductivity (Tzur and Roble, 1983).

probably quite variable depending on variations of ionization rates, ion and neutral chemical reactions, aerosol and cloud interactions, and a host of other meteorological factors. The electrical columnar resistance that determines the local air-earth current flow and electric field is determined by the height integral of the reciprocal of the electrical conductivity distribution. The bulk of the columnar resistance resides in the troposphere, which can be strongly controlled by electrical conductivity variation owing to factors such as clouds, fog, aerosols, and pollution. Boeck (1976) showed that changes in the global conductivity may result in from [85]Kr being released into the atmosphere. The low-frequency radiation from power lines (Vampola, 1977) may also affect the precipitation of electrons and consequently affect ionization in the stratosphere. The columnar resistance derived from the Gish formula is about 1.3×10^{17} Ω/m^2. This fair-weather value probably varies considerably in place and time as determined by aerosol and weather conditions. The global variation of the columnar resistance is not well known, yet it is an important property of the global electrical circuit. The global resistance is the parallel circuit resistance obtained by adding the various columnar resistance values. Mühleisen (1977) estimated a global resistance of 230 Ω without mountains and 200 Ω when the Earth's orography is considered. The magnitude of the variability of global resistance is poorly known, and it is an important parameter that needs to be determined in order to improve our understanding of the Earth's global electrical circuit.

Fair-Weather Vertical Current Density and Total Current

The total current flowing in the global circuit is not well known, and only crude estimates have been made.

Its value is generally estimated by integrating the measurable fair-weather vertical current density over the fair-weather area of the globe. A recent estimate by Mühleisen (1977) used 10^{-12} A/m^2 for inhabited and industrialized areas, 2-4 $\times$ 10^{-12} A/m^2 for vegetated ground and for deserts, 2.5×10^{-12} A/m^2 over the Atlantic Ocean to derive a total global current of about 1000 A. About 750 A is derived for the current flow over oceans and 250 A for the current flow over the continents. Mühleisen (1977) pointed out that the columnar resistance over mountains is much smaller than over flat land near sea level, and he estimated that as much as 20 percent of the global vertical current streams toward mountains. Gathman and Anderson (1977) showed that the air-earth current also has a latitudinal variation due to the effect of the Earth's geomagnetic field on the cosmic-ray ionization rate throughout the troposphere.

Another means of estimating the total current flow in the circuit is to estimate the total number of thunderstorms working simultaneously and multiply that value by the average current output determined from measurements over thunderstorms, as was discussed previously. Estimates of the number of global thunderstorms range from 1500 to 2000. If these numbers are multiplied by the average current output of thunderstorms (0.5 to 1.0 A), the total current is about 750 to 2000 A, which is nearly the same magnitude as the total current derived by air-earth current estimates, about 1000 A. Blanchard (1963) indicated that currents as large as hundreds of amperes flow from the ocean surface into the atmosphere as a result of electrified droplets that are ejected from the bursting of small bubbles. It also should be mentioned that intense electrification is associated with some volcanic eruptions. When such phenomena occur, they may have a significant input into the global electrical circuit. There is considerable uncertainty with these estimates that needs to be resolved. It is unlikely that the global current can be accurately determined from only a few localized measurements, and some other means must be used to determine its value. Localized measurements, however, are needed to determine the range of variability, especially from globally representative stations such as on a mountaintop (Reiter, 1977a).

Ionosphere Potential

An important parameter for determining the electrical state of the global circuit is the ionospheric potential, which specifies the potential difference between the ground and the ionosphere assuming that the ground is arbitrarily referenced to zero. This quantity can be estimated by using the values of the global resistance, about 200 Ω, and the values of the total current flowing in the circuit, 750 to 2000 A, to get 150 to 400 kV.

The ionospheric potential is of fundamental importance for the global circuit because it is one of the few measurable global parameters. It can be determined by integrating the altitude profile of the electric field measured from ascending or descending balloons or aircraft. This technique uses the facts that the electrical conductivity of the ionosphere is so large that any horizontal potential difference is small and the entire ionosphere has a uniform potential difference with respect to the Earth. At high latitudes, however, it is necessary to consider the horizontal potential differences that are gener-

ated by the solar-wind/magnetospheric generator discussed later. Another factor is that the electric field decreases exponentially with altitude in such a manner that the product of the electrical conductivity and electric field remains constant with altitude. Therefore, the bulk of the ionospheric potential drop occurs within the first few kilometers above the Earth's surface, a region that is easily attainable by balloons and aircraft.

The ionosphere potential can also be determined from measurements of the air-earth current density and electrical-conductivity profiles. Mühleisen (1977) summarized the distribution of his ionospheric potential measurements made from balloons during the period 1959 to 1970. The minimum measured potential was 145 kV, the mean 278 kV, and the maximum 608 kV. The measurements also showed that the diurnal UT variation of ionospheric potential is similar to the *Carnegie* curve, shown in Figure 15.3(top), and that there is an 11-year variation in ionospheric potential that is out of phase with the solar sunspot cycle. Markson (1976, 1977) made aircraft measurements of the ionospheric potential that are shown in Figure 15.7. These measurements have a diurnal UT variation similar to the *Carnegie* curve and also show the magnitude of day-to-day variability that is associated with the inherent variability of the worldwide thunderstorm generator and, to a smaller extent, the global resistance. Measurements of the ionospheric potential are important for understanding the global electric circuit and should be continued.

FIGURE 15.7   Summary of ionospheric potential measurements as a function of time (UT) of aircraft soundings made by Markson (1976, 1977).

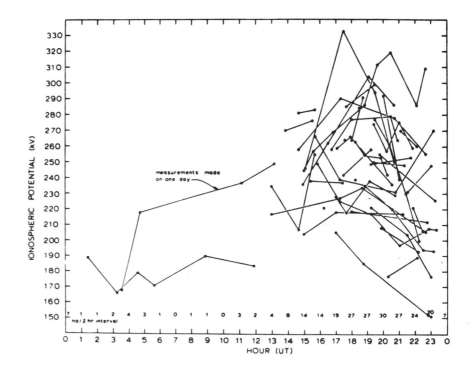

Electric Fields

The electric field at the ground is a readily measurable quantity that has been determined at various stations for nearly a century. The main characteristics of these measurements are summarized in detail by Israël (1973). In fair weather, the electric field is directed downward perpendicular to the ground and is typically 100-150 V/m. Field reversals are rare during undisturbed periods, but they are frequent during stormy weather and in conditions of dust, smoke, and fog. The electric field in densely populated areas is usually much larger than 150 V/m, whereas in small towns and far away from cities it can be smaller. The main reason for the variability of the electric field near the ground is that it is a complex quantity subject to universal time variations of the global circuit as well as the local influences of turbulence, weather, smoke, aerosols, and other anthropogenic factors.

Israël (1973) summarized electric-field observations that define (a) a latitudinal variation; (b) altitude variations including effects of clouds and aerosol layers; (c) diurnal variation giving continental and oceanic types; (d) annual variation with maximum values during southern hemisphere summer; and (e) possible variation due to solar influences. In fair weather the ground electric field varies owing to changes in columnar resistance, ionospheric potential, and the local electrical conductivity at the ground. During disturbed periods it fluctuates rapidly due to the space-charge variation associated with turbulence, lightning, thunderstorm charge location, precipitation in the form of snow or rain, cloud passages, fog, blowing snow, dust or aerosols, and other meteorological properties. The relationship of the ground electric field to these processes has been studied extensively. More recently, the rapid variation of electric field observed at a number of ground stations has been used to derive the charge structure within clouds, as discussed by Krehbiel (Chapter 8, this volume).

Electrical Relaxation Time

A fundamental property of the global atmospheric electrical circuit is the electrical relaxation time at various altitudes, which is defined as the time the electric current takes to adjust to $1/e$ of its final value after an electric field is suddenly applied, assuming that the conductivity remains constant. At high altitudes, near 70 km, the relaxation time is about 10^{-4} sec, increasing with decreasing altitude to about 4 s near 18 km and to about 5-40 min near the Earth's surface. The electrical relaxation time of the land surface of the Earth is about

10^{-5} sec. The maximum value of about 40 min in the atmosphere near the Earth's surface is the characteristic time that the global circuit would take to discharge if all thunderstorm activity suddenly ceased. Measurements have never shown a complete absence of a fair-weather electric field for any length of time, thereby suggesting a continuous operation of thunderstorms and other generators in maintaining the currents flowing in the global circuit. For time variations longer than about 40 min a quasi-static approximation can be applied when one is modeling the electrical properties of the global circuit.

MATHEMATICAL MODELS OF GLOBAL ATMOSPHERIC ELECTRICITY

Only a few mathematical models of global atmospheric electricity have appeared over the years (Kasemir, 1963, 1977; Hill, 1971; Hays and Roble, 1979; Volland, 1982). Since it is difficult to obtain global measurements to deduce the instantaneous properties of the global circuit, these models provide a convenient means of examining, through numerical experiments, the various interacting processes operating in the global circuit. The overall success of the models is judged on how well they represent observed properties at any place and time within the circuit.

Some of the elements that need to be considered in any global model of atmospheric electricity are schematically illustrated in Figure 15.8. Thunderstorms are extremely complex, and some simplifying assumptions must be made to represent their properties in a global model. The usual assumption is to consider thunderstorms as dipolar current sources, with a positive source in the cloud top and negative source in the cloud bottom. The storms are smaller than the grid scale for a

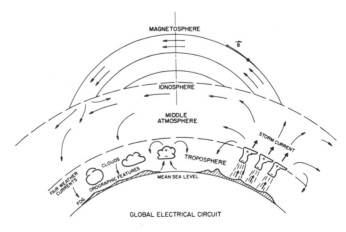

FIGURE 15.8   Schematic of various electrical processes in the global electrical circuit.

global model that has a representative grid size of 5° in latitude and longitude. The storms provide current to the global circuit, which flows upward toward the ionosphere. The rapid increase in electrical conductivity with height confines the current to a vertical column that flows from the storm to the ionosphere where it then is rapidly distributed over the globe at ionospheric heights. Part of the upward-directed current also flows along the Earth's geomagnetic-field line into the magnetic conjugate ionosphere, where it is also redistributed globally. The dipolar magnetic-field-line configuration is important for current closure within the global model. From the ionosphere the current flows downward toward the Earth with a magnitude that is governed by the potential difference between the ionosphere and the Earth's surface and by the local columnar resistance. The Earth's orography is important because of the decreased columnar resistance and consequentially higher current flows over mountains than at sea level. Clouds, fog, aerosols, and other meteorological phenomena must be considered because of their influence on electrical conductivity, columnar and global resistance, and local generation processes.

Perhaps the most difficult region to model is the planetary boundary layer whose electrical characteristics are discussed by Hoppel *et al.* (Chapter 11, this volume). The electric current flows freely along the highly conducting Earth's surface to the region underneath a thunderstorm, where it then flows upward into the storm, thus closing the circuit. The upward current flow into the thunderstorm is extremely complicated, consisting of lightning, precipitation, corona, conduction, convection, and displacement currents as discussed in a previous section.

Simplifying assumptions are usually used to make the mathematical problem of modeling the global circuit more tractable. The model of Hays and Roble (1979) used the quasi-static approximation to simplify the mathematical procedure. They assumed that thunderstorms, on a global scale, can be represented as dipolar current-generator point sources that are randomly distributed in preferred storm regions around the surface of the Earth. In fair-weather regions far away from the storm centers, the problem of the distribution of the electrostatic potential is determined by the current return from the sources to the Earth's surface.

Hays and Roble used the basic Green's function for the problem of the global electric circuit to solve for the electrostatic potential considering only conduction currents flowing in the global circuit. The model included orography in geomagnetic coordinates as shown in Figure 15.9 but neglects clouds, fog, pollution, and aerosols. The electric potential is determined by solving the

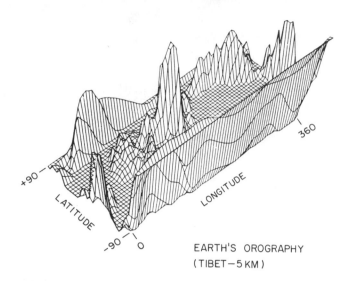

FIGURE 15.9 Perspective illustration of the Earth's orography in geomagnetic coordinates that is used in the global model of atmospheric electricity.

electrostatic equation in spherical coordinates. Variables are separated and the horizontal solution is determined by an expansion in tesseral harmonics. Thirty-seven spherical harmonic components were used, giving an effective 5° grid in latitude and longitude. The model of Hays and Roble (1979) included an exponentially increasing conductivity variation with altitude and a latitudinal distribution that follows the latitudinal distribution of cosmic-ray ion production. It also included the coupling of thunderstorm currents flowing along geomagnetic-field lines into magnetic conjugate hemispheres. The model is on geomagnetic coordinates to account for the cosmic-ray variation and geomagnetic-field-line coupling.

We use the model of Hays and Roble (1979) and Roble and Hays (1979) to present a model simulation that illustrates various properties of the global circuit. In Figure 15.10 the boxed-in areas show regions assumed to have enhanced thunderstorm activity. The distribution is taken from the maps of thunderstorm frequency generated by Crichlow *et al.* (1971) and representing northern hemisphere summer conditions at 1900 UT. It is assumed that 500 point dipolar current sources are randomly distributed in the Gulf of Mexico area (region 3), 300 in Africa (region 1), 100 off the coast of Argentina (region 4), and 50 in Southeast Asia (region 2). The storms are randomly distributed in latitude, longitude, and altitude, and some fall over land areas, plains, and mountains and some over oceans.

The combined current output from the ensemble of thunderstorms generates an ionospheric potential of 280 kV. The calculated global resistance including orogra-

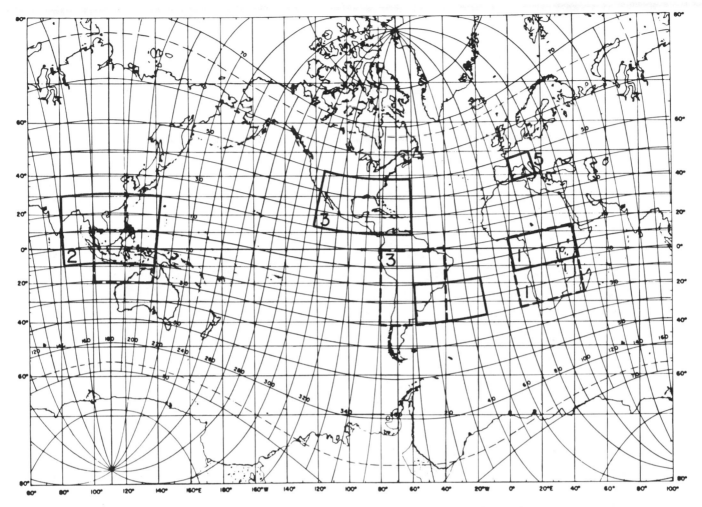

FIGURE 15.10 Regions of assumed thunderstorm distribution for model calculations. The five solid boxes indicate regions of thunderstorm occurrence for 1900 UT during northern hemisphere summer.

phy is 275 Ω, and the total current flowing in the circuit is 1010 A. To illustrate regions of upward and downward current flow, the results are presented as differences from the ionospheric potential $[\phi(\sigma) - \phi_\infty]$ in volts along constant-conductivity surfaces near 105, 50, 25, 8, 4, and 2 km as shown in Figure 15.11. In the middle atmosphere the model calculates the largest positive potential difference over the thunderstorm regions, indicating that currents are flowing upward toward the ionosphere. Negative potential regions indicate regions of downward current flow toward the Earth's surface.

In the equatorial region the geomagnetic-field lines are horizontal; therefore, any upward current flow into the equatorial ionosphere is redistributed only by horizontal conduction currents, and a single positive potential difference develops over the main thunderstorm region in equatorial Africa. The Gulf of Mexico thunderstorm region is off the geomagnetic equator,

and upward-flowing currents can flow freely along the geomagnetic-field line into the conjugate hemisphere. Since the electrical conductivity along the geomagnetic-field line is large, the magnetic potential is symmetric about the equator with no potential differences existing in the ionosphere between the feet of conjugate geomagnetic-field lines. The calculated potential differences along the 105-km constant-conductivity surfaces shown in Figure 15.11(a) are about 1.2 kV, which is relatively small compared with the calculated 280-kV constant ionospheric potential.

The contours of the calculated potential difference along the constant-conductivity surface near 50 km are shown in Figure 15.11(b). The overall pattern is similar to that calculated at 105 km, indicating very little attenuation between the two altitudes. There is, however, a small attenuation in the magnetic conjugate region of the Gulf of Mexico thunderstorm region. Along the con-

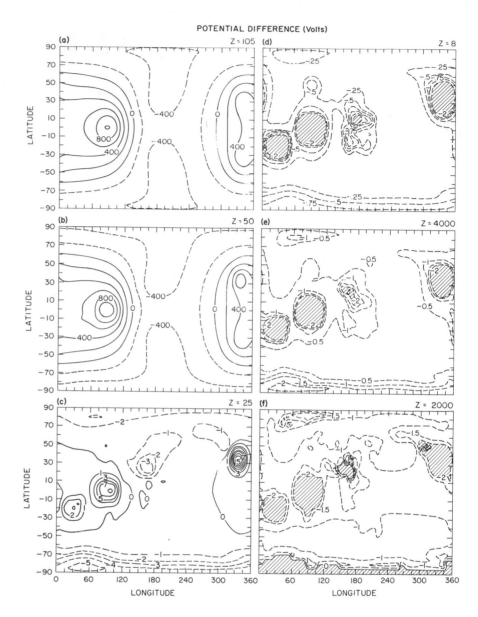

FIGURE 15.11 Contours of calculated potential difference $[\phi(\sigma) - \phi_\infty]$ in kilovolts along a constant-conductivity surface σ. Here, is the ionospheric potential, globally averaged at ionospheric heights (approximately 105 km). (a) Potential difference along the $\sigma_m = 4.5 \times 10^{-6}$ mho/m surface at about 105 km over the equator; (b) potential difference along the $\sigma_m = 4.7 \times 10^{-10}$ mho/m surface, which is 50 km over the equator; (c) potential difference along the $\sigma_m = 7.3 \times 10^{-12}$ mho/m surface, which is at an altitude of 25 km over the equator; (d) potential difference along the $\sigma_m = 4.3 \times 10^{-13}$ mho/m surface, which is at an altitude of 8 km over the equator; (e) and (f) give the potential difference along the constant-height surface at 4 km and 2 km, respectively.

stant-conductivity surface near 25 km, the potential difference pattern is different from that at 50 and 105 km, illustrating how the rapid increase of electrical conductivity alters the potential distribution in the upper atmosphere. Maximum potential variations occur over the main thunderstorm region, indicating a strong current flowing toward the ionosphere. In other regions the current flow is downward, with maximum values over the mountainous areas, primarily Tibet, Antarctica, and Greenland. The potential differences along constant-conductivity surfaces at 8, 4, and 2 km are shown in Figures 15.11(c), 15.11(d), and 15.11(e), respectively. In plotting these contours, the intense negative potentials generated under the thunderstorm areas are arbi-

trarily suppressed for contouring to illustrate the potential distortion due to orography. The hatched areas indicate these intensely disturbed regions. Also, the potential is zero whenever the height surface cuts a mountainous area that lies above it. These figures illustrate the effect of orography in altering the potential surfaces within the troposphere. These same potential surfaces are presented as perspective illustrations in Figure 15.12 for ease in visualizing the global circuit.

The calculated vertical ground potential gradient and current along the Earth's orographic surface are shown in Figures 15.13(a) and 15.13(b), respectively. The electric field in the fair-weather areas of the equatorial region is typically 130 V/m and increases with in-

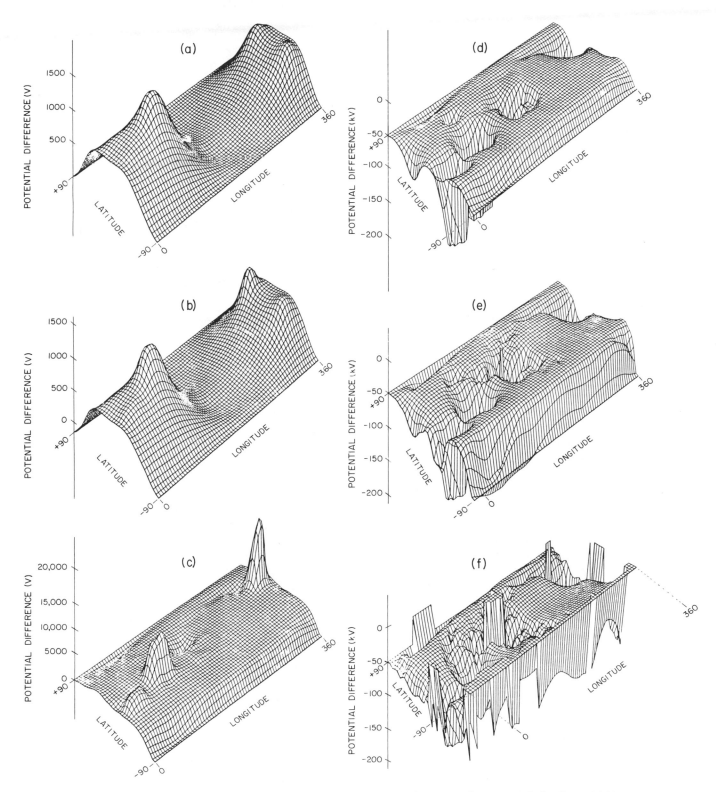

FIGURE 15.12 Perspective illustration of the calculated potential difference along the same surfaces as specified in Figure 15.11.

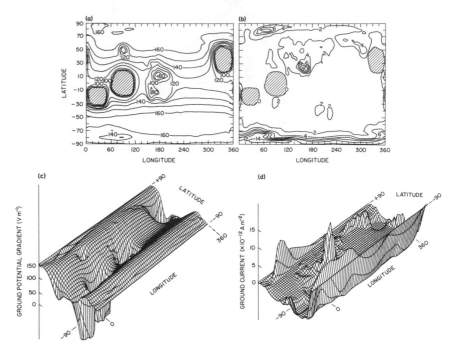

FIGURE 15.13 Contours of calculated (a) ground potential gradient in volts per meter along the Earth's surface and (b) ground current density in amperes per square meter when multiplied by 10^{-12}. (c) and (d) are perspective illustrations of the ground potential gradient and ground current density, respectively.

creasing geomagnetic latitude to about 160 V/m above about 60°. The electric field is greatly disturbed in regions under thunderstorms, indicating an upward-directed field. The electric field is not greatly modified by the mountains because of the large grid spacing (about 5° in latitude and longitude). The ground electric current, however, is strongly influenced by the mountains, as shown in Figure 15.13(b). The contours of the enhanced fair-weather current flow nearly outline the continental regions, with the largest current flowing into the high mountain areas (e.g., Tibet, Andes, Antarctica, Rocky Mountains). This is primarily due to the larger electrical conductivity, with respect to sea level, that exists on the high mountain peaks and to the decreased columnar resistance over mountains. A comparison with a similar calculation made without mountains reveals that about 20 percent of the total current flows into the high mountain areas. Other features of the model calculations are described in detail by Hays and Roble (1979) and Roble and Hays (1979).

Model Improvements

The global models of atmospheric electricity that have been constructed are primarily analytical models that have considerably simplified mathematical prescriptions. These models, nonetheless, provide considerable insight into the electrodynamics of the global circuit. There is a clear need to develop numerical models that allow a more realistic prescription of physical pro-

cesses. For example, the analytic model of Hays and Roble (1979) assumed an ionosphere with a uniform conductivity and electrical vertical profiles that are represented by two exponential functions and simulates the latitudinal variations of cosmic-ray ion production rates only crudely.

A numerical model could adopt a more realistic calculation with latitude and longitude, for example, such as that shown in Figure 15.6, and also allow for a day-night variation of electrical conductivity. Realistic perturbations to the global pattern due to solar-terrestrial influences could then be modeled to determine the magnitude of the global response to such events (Tzur *et al.*, 1983). A numerical model could also be expanded to include the total Maxwell current instead of only electrical conduction currents. Such modifications are important, especially for the middle atmosphere where ambipolar diffusion can alter the distribution of currents and fields as discussed by Tzur and Roble (1983), and for the troposphere, where convection, precipitation, conduction, lightning, and displacement currents can be important in disturbed regions.

The main improvements in any global model will come primarily by more accurate parameterizations of electrical processes within the troposphere. The numerical model should include the electrical charge structures and conductivity modifications due to clouds and fog, a prescription of turbulent convective processes within the planetary boundary layer, physical charge-transfer process due to the nature of the Earth's surface (e.g., forests, deserts, glaciers), aerosol and smoke generation and dispersion, and anthropogenic processes (Anderson, 1977). A global-scale representation of these lower-atmospheric processes is necessary to evaluate their impact on the global circuit. One means of obtaining information on such global processes is to use output from the various general circulation models (GCMs) that have been developed to study the dynamic meteorology of the Earth's atmosphere. The GCM-calculated winds, temperature, humidity, cloudiness, turbulence, and other meteorological phenomena could be used to develop the electrical parameterizations for use in global models of atmospheric electricity. A coupled interactive electrical-dynamic GCM may greatly improve our understanding of various electrical processes within the global circuit.

Regional Modeling

Global models of atmospheric electricity are generally constrained by computer size to a grid that is on the order of 5° in latitude and longitude (about 500 km). The electrical processes need to be parameterized on that scale for insertion into the global models. There is a clear need for the development of regional and local electrical models that use appropriate boundary conditions provided by a global model to resolve subgrid-scale phenomena and to investigate electrical phenomena in a more limited area. Considerably more physics can be incorporated into such models, and the results in turn can then be used to provide appropriate parameterizations of these processes for inclusion into global models.

The calculated electrostatic potential contours and vectors of current flow over a mountain plateau and mountain peak (Tzur and Roble, 1985a) are shown in Figures 15.14(a) and 15.14(b), respectively. The regional model employed for these calculations used a detailed representation of electrical conductivity throughout the atmosphere, with boundary conditions that allow free current exchange between the global and regional models. It is assumed that the plateau and mountain perturb electrical quantities locally but are small enough with respect to the globe that their feedback into the global circuit is small. Over the plateau the vertical electric current flow is about three times larger than over sea level, primarily because of the reduced columnar resistance over the elevated surface. The results also show considerable horizontal current flow in the upper atmosphere, because of the presence of the plateau, indicating a local readjustment of the current system. A similar calculation for a mountain peak is shown in Figure 15.14(a). The mountain is seen to distort significantly the potential pattern in such a manner as to cause

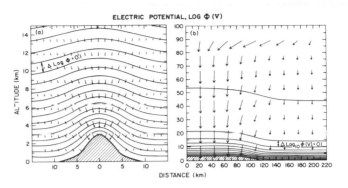

FIGURE 15.14 Regional calculation of the potential distribution and current density flow over a mountain peak (a) and a mountain plateau (b). The length of the maximum arrow indicates a current density of 2×10^{-12} A m^{-2} in (a) and 1.8×10^{-11} A/m^2 in (b). (Tzur and Roble, 1985b.)

an enhanced current to flow into the peak. These calculations are numerical extensions of the analytic mathematical procedure that Kasemir (1977) used to calculate the electric current and field distributions around mountains. Such calculations are important to interpret various measurements of the fair-weather electric field and current in the vicinity of mountaintops (Cobb *et al.*, 1967; Cobb, 1968) and provide a quantitative framework to evaluate the extent of the electrical disturbance source by mountains of various shapes and also to determine the important characteristics that need to be incorporated into global models.

Another important regional problem is to investigate the electrical interaction of a thunderstorm with its immediate environment. Thunderstorms are considered point current sources in the global model, and regional calculations on a much smaller scale are needed to examine such problems as the magnitude of the current output from thunderstorm models and its relationship to the characteristics of its electrical environment. For example, the effect of the ionospheric magnetic-field-line configuration on the vertical current output from a thunderstorm that is represented as a dipolar current source (Tzur and Roble, 1985b) is shown in Figures 15.15(a) and 15.15(b). When the geomagnetic-field lines are assumed to be vertical [Figure 15.15(a)], the upward current flow in the middle atmosphere is confined to the immediate vicinity of the storm, whereas when the geomagnetic-field lines are assumed to be horizontal there is considerably more horizontal current flow in the middle atmosphere. The calculations suggest differences between the current output from thunderstorms in equatorial regions and in high latitudes.

These calculations illustrate the types of problems that need to be addressed with regional models, not only

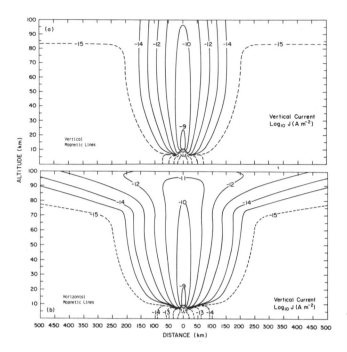

FIGURE 15.15   Contours of $\log_{10} J$ (A/m²), the regional vertical current flow toward the ionosphere from a thunderstorm model considering an ionosphere where (a) the magnetic-field lines are horizontal and (b) the magnetic-field lines are vertical. (Tzur and Roble, 1985b.)

to understand local phenomena but also to provide the appropriate guidance for incorporating these effects into global models.

ELECTRICAL COUPLING BETWEEN THE UPPER AND LOWER ATMOSPHERE

The main generators operating within the Earth's global atmospheric circuit are summarized in Table 15.2. As discussed in the previous section, thunderstorms are generators whose current output maintains a vertical potential difference of about 300 kV between the ground and ionosphere, with a total current flow of about 10^3 A. The classical picture of atmospheric electricity assumes that the ionosphere is at a uniform potential, and it does not account for either ionospheric or

TABLE 15.2   Generators in the Global Electric Circuit

THUNDERSTORMS—current output maintains a *vertical* potential difference of 300,000 V between ground and ionosphere. Current $\sim 10^3$ A.

IONOSPHERIC DYNAMO—tides at ionospheric heights maintain *horizontal* potential differences of 5000-15,000 V between high and low latitudes. Current $\sim 10^5$ A.

MAGNETOSPHERIC DYNAMO—interaction of solar wind with Earth's geomagnetic field maintains a *horizontal* dawn-to-dusk potential drop of 40,000-100,000 V across polar caps. Current $\sim 10^6$ A.

magnetospheric dynamos. The physics of both dynamos are discussed in the chapter by Richmond (Chapter 14, this volume).

The ionospheric dynamo is driven by both tides generated in situ and tides propagating upward from the lower atmosphere. These tides generate horizontal potential differences of 5-10 kV within the ionosphere, with a total current flow on the order of 10^5 A. The magnetospheric dynamo, on the other hand, is driven by the interaction of the solar wind with the Earth's geomagnetic field and generates a horizontal dawn-to-dusk potential drop of typically 40-100 kV across the magnetic conjugate polar cap and a total current flow of 10^6 A. The magnetospheric convection pattern is Sun-aligned relative to the geomagnetic poles (north geomagnetic pole 78.3° N and 291° E, south geomagnetic pole 74.5° S and 127° E), and therefore the pattern remains fixed relative to the Sun but moves in a complex fashion over the Earth's surface as the Earth rotates about its geographic pole.

Several empirical models that describe the horizontal ionospheric potential distribution about the magnetic polar cap have been constructed (Volland, 1975, 1978; Heppner, 1977; Sojka *et al.*, 1979, 1980; Heelis *et al.*, 1983). The magnetospheric convective potential distributions are all similar, with a positive perturbation on the dawn side and a negative perturbation on the dusk side of the magnetic polar cap. The main differences between the various models are due to small horizontal-scale structure variations for various levels of geomagnetic conditions. The calculated average potential patterns over the southern hemisphere polar cap for four different universal times (0000, 0600, 1200, and 1800 UT) using the model of Sojka *et al.* (1980) are shown in Figures 15.16(a)-15.16(d). Satellite observations have shown that the instantaneous magnetospheric convection pattern is highly variable with considerable small-scale deviations from the mean structure, indicating a turbulent plasma flow. The dawn-to-dusk potential drop across the polar cap varies from about 30 kV for quiet geomagnetic activity, to about 60 kV for average geomagnetic activity, and to about 150-200 kV during geomagnetic storms. In addition, for greater geomagnetic activity the convection pattern expands equatorward by about 5° from its normal quiet-time position. The magnetospheric convective electric field is generally confined to the vicinity of the polar cap by shielding charges in the Alfvén layer of the magnetosphere. However, during rapid changes of magnetospheric convection a temporary imbalance in these shielding charges can occur, and the high-altitude electric fields can cause immediate effects at the magnetic equator at all longitudes (Gonzalez *et al.*, 1979). The observed propagation

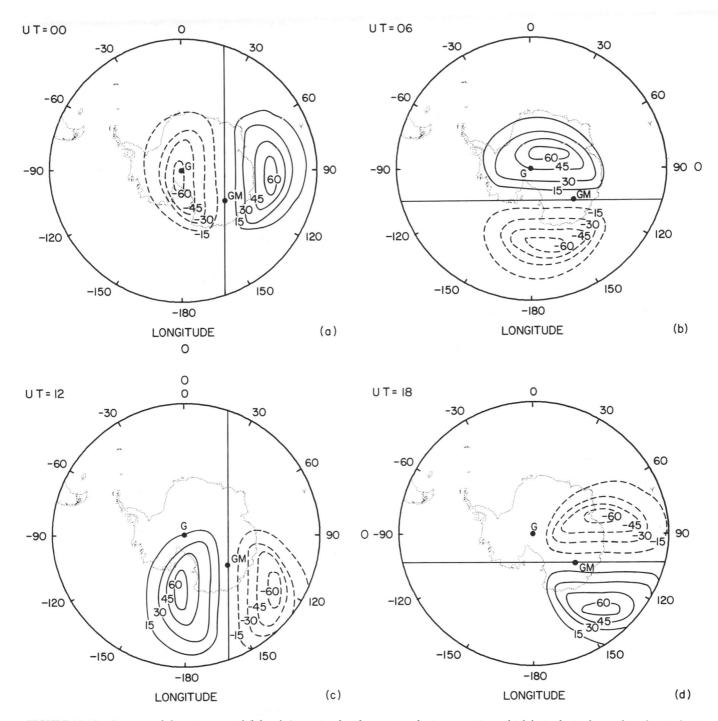

FIGURE 15.16 Contours of electric potential (kilovolts) associated with magnetospheric convection at high latitudes in the southern hemisphere for four universal times (a) 0000 UT, (b) 0600 UT, (c) 1200 UT, and (d) 1800 UT. The symbol G represents the geographic pole and GM the geomagnetic pole.

may be due to currents either within the magnetosphere or ionosphere (Nopper and Carovillano, 1978) or through the lower-atmospheric electrical waveguide (Kikuchi *et al.*, 1978).

The downward mapping of the ionospheric electric fields toward the lower atmosphere has been considered previously by a number of authors (Mozer and Serlin, 1969; Mozer, 1971; Atkinson *et al.*, 1971; Volland, 1972, 1977; Chiu, 1974; Park, 1976, 1979). These studies have all shown that large horizontal-scale electric fields within the ionosphere map efficiently downward in the direction of decreasing electrical conductivity, and that downward electric-field mapping is much more efficient than upward mapping. Both Chiu (1974) and Park and Dejnakarintra (1977b) showed that the anisotropy of the electrical conductivity can have an important influence on the mapping properties of electric fields. Horizontal electric fields of small-scale size (~ 1-10 km) are rapidly attenuated as they map downward into the atmosphere from ionospheric heights, but electric fields of larger horizontal scales (~ 500-1000 km) map effectively right down to the Earth's surface, as shown in Figure 15.17.

Since the electrical conductivity of the Earth's surface is large, horizontal electric fields can usually be neglected, and a vertical electric-field variation results to accommodate horizontal variations of ionospheric potential. Calculations by Park (1976, 1979) and Roble and Hays (1979) showed that the magnetospheric generator can produce perturbations of ± 20 percent in the

air-earth current and ground electric field at high latitudes during quiet geomagnetic periods and larger variations during geomagnetic storms and substorms. A calculation that shows the downward mapping of a 100-kV dawn-to-dusk potential drop superimposed on a 300-kV ionospheric potential is given in Figure 15.18(a). The potential perturbation penetrates from the ionosphere down to the tropopause with little attenuation but rapidly decreases within the troposphere to zero at the Earth's surface. The calculated electric field at the Earth's surface is 190 V/m on the dusk side of the polar cap, as shown in Figure 15.18(b).

Changes in electrical conductivity caused by variations in cosmic-ray ionization during solar-terrestrial events can also change the downward mapping characteristics as discussed by Roble and Hays (1979). In addition, they also have shown that because the magnetospheric potential pattern is Sun-aligned in geomagnetic coordinates, a ground station, balloon, or aircraft at a given geographic location should detect variations organized in magnetic local time. For early magnetic local times the ionospheric potential perturbations of the Earth's potential gradient are positive, and for later magnetic local times the perturbations are negative. At high geomagnetic latitudes these variations are superimposed on the diurnal UT variation of potential gradient maintained by worldwide thunderstorm activity.

Kasemir (1972), using data obtained at the South Pole and Thule, Greenland, noted a departure of the diurnal UT variation measured at these stations from the oceanic diurnal electric-field variation measured during the cruises of the ship *Carnegie*, which is generally accepted as the UT variation due to worldwide thunderstorm activity. The polar curves have a similar shape to the curve derived from the *Carnegie* cruises but at a much reduced amplitude. From these results Kasemir concluded that another agent besides worldwide thunderstorm activity may modulate the global circuit at high latitudes.

The position of the magnetospheric potential pattern over the Earth's surface is shown in Figure 15.16 for four different UTs. It can be seen that the downward mapping of this potential pattern to the Earth's surface gives rise to a complex UT variation due to the displacement of the geographic and geomagnetic poles. The calculated UT variation of the ground electric field at South Pole Station due to the downward mapping of magnetospheric potential pattern is shown in Figure 15.19; the *Carnegie* UT variation and the Kasemir (1972) measurements are also shown. It is seen that the positive potential perturbation maps down over South Pole Station from about 0200 to 1400 UT, and the negative potential perturbation from 1400 UT to 0200 UT. When this pat-

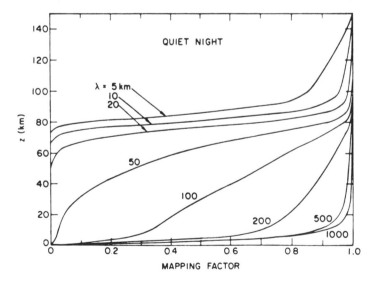

FIGURE 15.17   Downward mapping factor of the horizontal ionospheric electric field as a function of altitude indicating the magnitude of the attenuation of the electric-field strength for various horizontal scale sizes, λ, in kilometers.

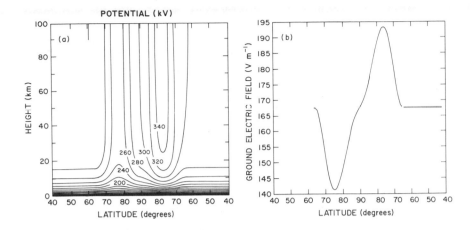

FIGURE 15.18 (a) Calculated contours of potential (kilovolts) in a dawn-to-dusk cross section across the magnetic polar cap. A 100-kV dawn-to-dusk potential drop is superimposed upon a 300-kV ionospheric potential background. (b) Calculated latitudinal variation of the ground potential gradient (volts per meter) over the magnetic polar cap.

tern is superimposed upon the *Carnegie* UT variation, it can be seen that the magnetospheric potential tends to supress the amplitude of the *Carnegie* variation, a result similar to Kasemir's (1972) observations. A similar situation exists at Thule, Greenland, although the predicted amplitude of the magnetospheric potential variation is reduced somewhat because the station is so near the northern geomagnetic pole. There is considerable day-to-day variability associated with the magnetospheric convection potential pattern; however, in the time-average sense the positive and negative variations are both out of phase with the *Carnegie* UT variation at these stations and may be responsible for the suppressed amplitude of the UT diurnal variation observed by Kasemir (1972).

Stratospheric balloon measurements of magnetospheric convection electric fields have been made for a number of years (e.g., Mozer and Serlin, 1969; Mozer, 1971; Holzworth and Mozer, 1979; Holzworth, 1981). Recently D'Angelo *et al.* (1982) processed over 1200 hours of stratospheric balloon data and correlated the vertical electric field with magnetic activity parameters. During quiet geomagnetic conditions the classical *Carnegie* curve was reproduced. During more active geomagnetic conditions, however, the dawn-dusk potential difference of the magnetospheric convection pattern was shown clearly to influence the fair-weather field. The above measurements suggest an electrical coupling between the magnetospheric dynamo and the global electrical circuit and indicate a need for more measurements. With the move of the incoherent-scatter radar from Chatanika, Alaska, to Sonderstrom Fjord, Greenland, and the operation of the EISCAT radar from Tromso, Norway, a unique opportunity exists to examine the high-latitude electrical coupling between the upper and lower atmosphere.

SOME OUTSTANDING PROBLEMS

Component processes within the global circuit have been studied for nearly a century. The ground electric field is the most common measurement, although numerous measurements of the electrical conductivity and air-earth current have also been made. These measurements form the basis of much of our knowledge concerning the influence of local processes on the electrical structure of the global electrical circuit. The measurements of currents and fields over land stations are highly variable, being subject not only to variations of the global generator but also to local meteorological and anthropogenic influences that at times dominate the global electrical variations. Over the oceans the local influences can be less, but considerable averaging is still necessary to derive the global variations.

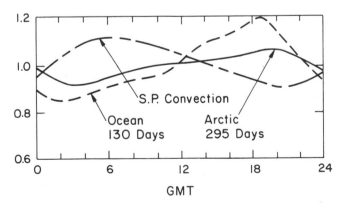

FIGURE 15.19 Normalized diurnal variation of the ground potential gradient as measured in the Arctic and Antarctic by Kasemir (1972) (solid curve), the diurnal potential gradient variation from the *Carnegie* cruise (dashed curve), and the calculated potential gradient at South Pole Station due to the downward mapping of the ionospheric potential pattern (long/short dashed curve) shown in Figure 15.16.

It is unlikely that enough ground-based measurements of electric currents and fields can be made simultaneously to define the instantaneous properties of the global circuit and its temporal variations. A globally representative single measurement is needed to define the characteristics of the global circuit. A measurement that has been used by a number of investigators to define the state of the global circuit is the ionospheric potential. This value is derived from the height integral of the vertical electric-field profile that is measured either by ascending and descending balloons or by aircraft. An inherent property of the measurement is that the ionospheric potential is nearly uniform over the globe because of the highly conducting Earth's surface and ionosphere. The ionospheric potential is equivalent to the product of the worldwide thunderstorm current output and the total global electrical resistance. It is generally assumed that changes in the global electrical resistance are small and that variations in ionospheric potential reflect the variations in worldwide thunderstorm current output. Only occasional electric-field soundings have been made over the years, and there is a clear need to increase the frequency of such measurements. Another means of obtaining the ionospheric potential is from tethered balloons, as described by Vonnegut *et al.* (1966) and Holzworth (1984), that have the capability of providing high-time-resolution measurements for global studies. Markson (1978) called for the establishment of the ionospheric potential as a geoelectric index that gives an indication of the state of the global circuit. This index would be the electrical equivalent of the geomagnetic index that has been used over the years to study geomagnetic phenomena within the Earth's atmosphere.

The main problem impeding progress in understanding the global circuit is the determination of the current output from thunderstorm generators. There are only a few measurements of the total current flow from storms, and there is a clear need for more measurements to define the current output properties in terms of thunderstorm size, duration, lightning flash frequency, charge separation distance, and other parameters. Recent balloon measurements at stratospheric heights over thunderstorms (Holzworth, 1981) showed prolonged periods (1-3 hours) of dc electric fields reversed from fair-weather conditions, indicating a current flow toward the ionosphere.

In addition to single thunderstorm measurements, it is important to be able to obtain information on the global distribution of thunderstorm occurrence. Previous information has been derived from weather stations, when thunder was heard from an observing site (Crichlow *et al.*, 1971), from Schumann resonance (Polk, 1982), and from radio and spheric measurements (Volland, 1984). More recently, lightning detection from satellites has been used to derive information on the global distribution and flash-rate frequency from space, as discussed in the chapter by Orville (Chapter 1, this volume). Krider *et al.* (1980) designed a ground detection network that detects cloud-to-ground lightning flashes and deployed the detectors in regions that cover large parts of North America. And finally, Davis *et al.* (1983) examined the feasibility of detecting lightning from a satellite in synchronous orbit. They estimated that three such satellites could provide worldwide coverage.

These measurement techniques provide a capability of making progress in the century-old problem of understanding the role of thunderstorms as generators within the global circuit. Simultaneous measurements of the ionospheric potential (either from vertical electric-field soundings or from a tethered balloon), along with measurements of lightning flash frequency (either cloud-to-ground flashes from a ground network or total flash rate from satellites), may determine the degree of synchronization between the two phenomena. Such measurements can be used to monitor the electrical state of the global circuit and also can provide a global indicator to help in understanding various local and regional measurements. These measurements would also be useful for improving our understanding of the role of solar-terrestrial perturbations in altering the properties of the global circuit (Reiter, 1969, 1971, 1972; Markson, 1971, 1978; Cobb, 1978; Herman and Goldberg, 1978; Roble and Hays, 1982).

There is also a need to determine the electrodynamic processes operating within the middle atmosphere. According to the classical picture of atmospheric electricity, the middle atmosphere should be passive, yet certain rocket measurements indicate the existence of large electric fields of unknown origin (Bragin *et al.*, 1974; Tyutin, 1976; Hale and Croskey, 1979; Hale *et al.*, 1981; Maynard *et al.*, 1981; Gonzalez *et al.*, 1982). These large electric fields are not understood, and it has been suggested that instrumental effects may be involved (Kelley, 1983; Kelley *et al.*, 1983). It is important to resolve this issue because of the fundamental implications involved in understanding electrodynamic processes within the middle atmosphere.

Finally, progress in understanding the global circuit and possible solar-terrestrial coupling mechanisms requires a collaborative effort between observations and theoretical modeling. The measurements are needed to verify model predictions and guide model development, and the modeling results provide a physical constraint for understanding measurements and suggesting various key experiments. The technology and models are currently available to make progress in resolving the fundamental problem of global atmospheric electricity (Dolezalek, 1972).

REFERENCES

Anderson, F. J., and G. D. Freier (1969). Interactions of the thunderstorm with a conducting atmosphere, *J. Geophys. Res. 74*, 5390-5396.

Anderson, R. V. (1977). Atmospheric electricity in the real world (useful applications of observations which are perturbed by local effects), in *Electrical Processes in Atmospheres*, H. Holezalek and R. Reiter, eds., Steinkopff, Darmstadt, pp. 87-99.

Atkinson, W., S. Sundquist, and U. Fakleson (1971). The electric field existing at stratospheric elevations as determined by tropospheric and ionospheric boundary conditions, *Pure Appl. Geophys. 84*, 46-56.

Blanchard, D. C. (1963). Electrification of the atmosphere by particles from bubbles in the sea, *Prog. Oceanog. 1*, 71.

Boeck, W. L. (1976). Meteorological consequences of atmospheric krypton-85, *Science 193*, 195-198.

Bragin, Yu. A., A. A. Tyutin, A. A. Kocheev, and A. A. Tyutin (1974). Direct measurement of the atmospheric vertical electric field intensity up to 80 km, *Cosmic Res. 12*, 279-282.

Brooks, C. E. P. (1925). The distribution of thunderstorms over the globe, *Geophys. Mem. 24*, Air Ministry, Meteorological Office, London, pp. 147-164.

Chiu, Y. T. (1974). Self-consistent electrostatic field mapping in the high-latitude ionosphere, *J. Geophys. Res. 79*, 2790-2802.

Cobb, W. E. (1968). The atmospheric electric climate at Mauna Loa Observatory, Hawaii, *J. Atmos. Sci. 25*. 470-480.

Cobb, W. E. (1978). Balloon measurements of the air-earth current density at the south pole before and after a solar flare, in preprint volume, *Conference on Cloud Physics and Atmospheric Electricity*, July 31-August 4, Issaquah, Washington, American Meteorological Society, Boston, Mass.

Cobb, W. E., B. B. Phillips, and P. A. Allee (1967). Note on mountaintop measurements of atmospheric electricity in northern United States, *Mon. Weather Rev. 95*, 912-916.

Crichlow, W. Q., R. C. Favis, R. T. Disney, and M. W. Clark (1971). Hourly probability of world-wide thunderstorm occurrence, Res. Rep. 12, Office of Telecommun. Int. Telecommun. Serv., Boulder, Colo., April.

D'Angelo, N., I. B. Iversen, and M. M. Madsen (1982). Influence of the dawn-dusk potential drop across the polar cap on the high-latitude atmospheric vertical current, *Geophys. Res. Lett. 9*, 773-776.

Davis, M. H., M. Brook, H. Christian, B. G. Heikes, R. E. Orville, C. G. Park, R. G. Roble, and B. Vonnegut (1983). Some scientific objectives of a satelliteborne lightning mapper, *Bull. Am. Meteorol. Soc. 64*, 114-119.

Dejnakarintra, M., and C. G. Park (1974). Lightning-induced electric fields in the ionosphere, *J. Geophys. Res. 79*, 1903-1910.

Dolezalek, H. (1971). Introductory remarks on the classical picture of atmospheric electricity, *Pure Appl. Geophys. 84*, 9-12.

Dolezalek, H. (1972). Discussion of the fundamental problem of atmospheric electricity, *Pure Appl. Geophys. 100*, 8-42.

Forbush, S. E. (1966). Time variations of cosmic rays, *Handbuch der Physik, Geophysik III*, J. Bartels, ed., Springer-Verlag, New York.

Freier, G. D. (1979). Time-dependent fields and a new mode of charge generation in severe thunderstorms, *J. Atmos. Sci. 36*, 1967-1975.

Gathman, S. G., and R. V. Anderson (1977). Aircraft measurements of the geomagnetic latitude effect on air-earth current density, *J. Atmos. Terrest. Phys. 39*, 313-316.

Gish, O. H., and G. R. Wait (1950). Thunderstorms and the Earth's general electrification, *J. Geophys. Res. 55*, 473-484.

Gonzalez, C. A., M. C. Kelley, B. G. Fejer, J. F. Vickery, and R. F. Woodman (1979). Equatorial electric fields during magnetically disturbed conditions. 2. Implications of simultaneous auroral and equatorial measurements, *J. Geophys. Res. 84*, 5803-5812.

Gonzalez, W. D. A. E. C. Pereira, A. L. C. Gonzalez, I. M. Martin, S. L. G. Dutra, O. Pinto, Jr., J. Wygant, and F. S. Mozer (1982). Large horizontal electric fields measured at balloon heights of the Brazilian magnetic anomaly and association to local energetic particle precipitation, *Geophys. Res. Lett. 9*, 567-570.

Grenet, G. (1947). Essai d'explication de la charge électrique des nuages nuages d'orages, *Ann. Geophys. 3*, 306-307.

Grenet, G. (1959). Le Nuage d'orage: Machine électrostatique, *Météorologie I-53*, 45-47.

Hale, L. C., and C. L. Croskey (1979). An auroral effect of the fair weather electric field, *Nature 278*, 239-241.

Hale, L. C., C. L. Croskey, and J. D. Mitchell (1981). Measurements of middle-atmosphere electric fields and associated electrical conductivities, *Geophys. Res. Lett. 8*, 927-930.

Hays, P. B., and R. G. Roble (1979). A quasi-static model of global atmospheric electricity 1. The lower atmosphere, *J. Geophys. Res. 84*, 3291-3305.

Heelis, R. A., J. K. Lowell, and R. W. Spiro (1983). A model of the high-latitude ionospheric convection pattern, *J. Geophys. Res. 87*, 6339-6345.

Heppner, J. P. (1977). Empirical models of high-latitude electric fields, *J. Geophys. Res. 82*, 1115-1125.

Herman, J. R., and R. A. Goldberg (1978). Initiation of non-tropical thunderstorms by solar activity, *J. Atmos. Terrest. Phys. 40*, 121-134.

Hill, R. D. (1971). Spherical capacitor hypothesis of the Earth's electric field, *Pure Appl. Geophys. 84*, 67-75.

Holzer, R. E., and D. S. Saxon (1952). Distribution of electrical conduction currents in the vicinity of thunderstorms, *J. Geophys. Res. 57*, 207-216.

Holzworth, R. H. (1981). High latitude stratospheric electrical measurements in fair and foul weather under various solar conditions, *J. Atmos. Terrest. Phys. 43*, 1115-1126.

Holzworth, R. H. (1984). Hy-wire measurements of atmospheric potential, *J. Geophys. Res. 89*, 1395-1401.

Holzworth, R. H., and F. S. Mozer (1979). Direct evidence of solar flare modification of stratospheric electric fields, *J. Geophys. Res. 84*, 363-367.

Imyanitov, I. M., B. F. Evteev, and I. I. Kamaldina (1969). A thunderstorm cloud, in *Planetary Electrodynamics*, S. C. Coroniti and J. Hughes, eds., Gordon and Breach Science Publisher, New York, pp. 401-425.

Israël, H. (1973). *Atmospheric Electricity*, vol. 1 and vol. 2, translated from German, Israel Program for Scientific Translations, Jerusalem.

Kasemir, H. W. (1959). The thunderstorm as a generator in the global electric circuit (in German), *Z. Geophys. 25*, 33-64.

Kasemir, H. W. (1963). On the theory of the atmospheric electric current flow, IV, Tech. Rep. 2394, U.S. Army Electronics Research and Development Laboratories Fort Monmouth, N.J., October.

Kasemir, H. W. (1972). Atmospheric electric measurements in the Arctic and Antarctic, *Pure Appl. Geophys. 100*, 70.

Kasemir, H. W. (1977). Theoretical problems of the global atmospheric electric circuit, in *Electrical Processes in Atmospheres*, H. Dolezalek and R. Reiter, eds., Steinkopff, Darmstadt, pp. 423-438.

Kasemir, H. W. (1979). The atmospheric electric global circuit, in *Proceeding Workshop on the Need for Lightning Observations from Space*, NASA CP-2095, pp. 136-147.

Kelley, M. C. (1983). Middle atmospheric electrodynamics, *Rev. Geophys. Space Phys. 21*, 273-275.

Kelley, M. C., C. L. Siefring, and R. F. Pfaff, Jr. (1983). Large amplitude middle atmospheric electric fields: Fact or fiction, *Geophys. Res. Lett. 10*, 733-736.

Kikuchi, T., T. Araki, H. Maeda, and K. Maekawa (1978). Transmission of ionospheric electric fields to the equator, *Nature 273*, 650-651.

Krehbiel, P. R., M. Brook, and R. A. McCrosy (1979). An analysis of the charge structure of lightning discharges to ground, *J. Geophys. Res. 84*, 2432-2456.

Krider, E. P., and J. A. Musser (1982). The Maxwell current density under thunderstorms, *J. Geophys. Res. 87*, 11171-11176.

Krider, E. P., R. C. Noggle, A. E. Pifer, and D. L. Vance (1980). Lightning direction-finding systems for forest fire detection, *Bull. Am. Meteorol. Soc. 61*, 980-986.

Markson, R. (1971). Considerations regarding solar and lunar modulation of geophysical parameters, atmospheric electricity and thunderstorms, *Pure Appl. Geophys. 84*, 161.

Markson, R. (1976). Ionospheric potential variation obtained from aircraft measurements of potential gradient, *J. Geophys. Res. 81*, 1980-1990.

Markson, R. (1977). Airborne atmospheric electrical measurements of the variation of ionospheric potential and electrical structure in the exchange layer over the ocean, in *Electrical Processes in Atmospheres*, H. Dolezalek and R. Reiter, eds., Steinkopff, Darmstadt, pp. 450-459.

Markson, R. (1978). Solar modulation of atmospheric electrification and possible implications for the sun-weather relationship, *Nature 273*, 103-109.

Maynard, N. C., C. L. Croskey, J. D. Mitchell, and L. C. Hale (1981). Measurement of volt/meter vertical electric fields in the middle atmosphere, *Geophys. Res. Lett. 8*, 923-926.

Mozer, F. S. (1971). Balloon measurement of vertical and horizontal atmospheric electric fields, *Pure Appl. Geophys. 84*, 32-45.

Mozer, F. S., and R. Serlin (1969). Magnetospheric electric field measurements with balloons, *J. Geophys. Res. 74*, 4739-4754.

Mühleisen, R. (1977). The global circuit and its parameters, in *Electrical Processes in Atmospheres*, H. Dolezalek and R. Reiter, eds., Steinkopff, Darmstadt, pp. 467-476.

Neher, H. V. (1967). Cosmic-ray particles that changed from 1954 to 1958 to 1965, *J. Geophys. Res. 72*, 1527-1539.

Nopper, R. E., and R. L. Carovillano (1978). Polar equatorial coupling during magnetically active periods, *Geophys. Res. Lett. 5*, 699-703.

Orville, R. E., and D. W. Spencer (1979). Global lightning flash frequency, *Mon. Weather Rev. 107*, 934-943.

Paramanov, N. A. (1950). To the world time period of the atmospheric electric potential gradient, *Dokl. Acad. Sci. USSR 70*, 37-38.

Park, C. G. (1976). Downward mapping of high-latitude electric fields to the ground, *J. Geophys. Res. 81*, 168-174.

Park, C. G. (1979). Comparison of two-dimensional and three-dimensional mapping of ionospheric electric field, *J. Geophys. Res. 84*, 960-964.

Park, C. G., and M. Dejnakarintra (1973). Penetration of thundercloud electric fields into the ionosphere and magnetosphere, 1. Middle and subauroral latitudes, *J. Geophys. Res. 78*, 6623-6633.

Park, C. G., and M. Dejnakarintra (1977a). Thundercloud electric fields in the ionosphere, in *Electrical Processes in Atmospheres*, H. Dolezalek and R. Reiter, eds., Steinkopff, Darmstadt, pp. 544-551.

Park, C. G., and M. Dejnakarintra (1977b). The effects of magnetospheric convection on atmospheric electric fields in the Polar Cap, in *Electrical Processes in Atmospheres*, H. Dolezalek and R. Reiter, eds., Steinkopff, Darmstadt, pp. 536-542.

Parkinson, W. C., and O. W. Torrenson (1931). The diurnal variation of the electrical potential of the atmosphere over the oceans, Compt. Rend. de l'Assemblée de Stockholm, 1930; *IUGG (sect. Terrest. Magn. Electr. Bull. 8*, 340-345.

Pierce, E. T. (1970). Latitudinal variation of lightning parameters, *J. Appl. Meteorol. 9*, 194-195.

Pierce, E. T. (1977). Stratospheric electricity and the global circuit, in *Electrical Processes in Atmospheres*, H. Dolezalek and R. Reiter, eds., Steinkopff, Darmstadt, pp. 582-586.

Polk, C. (1982). Schumann resonances, in *Handbook of Atmospherics*, Vol. 1, H. Volland, ed., CRC Press, Boca Raton, Fla., pp. 112-178.

Prentice, S. A., and D. Macherras (1977). The ratio of cloud to cloud-ground lightning flashes in thunderstorms, *J. Appl. Meteorol. 16*, 545-550

Rawlins, F. (1982). A numerical study of thunderstorm electrification using a three-dimensional model incorporating the ice phase, *Q. J. R. Meteorol. Soc. 108*, 778-880.

Reid, G. C. (1976). Ion chemistry in the D-region, *Adv. Atom. Mol. Phys. 12*, 375-411.

Reid, G. C. (1977). The production of water-cluster positive ions in the quiet daytime D-region, *Planet. Space Sci. 25*, 275-290.

Reiter, R. (1969). Solar flares and their impact on potential gradient and air-earth current characteristics at high mountain stations, *Pure Appl. Geophys. 72*, 259.

Reiter, R. (1971). Further evidence for impact of solar flares or potential gradient and air-earth current characteristics at high mountain stations, *Pure Appl. Geophys. 86*, 142.

Reiter, R. (1972). Case study concerning the impact of solar activity upon potential gradient and air-earth current in the lower troposphere, *Pure Appl. Geophys. 94*, 218-225.

Reiter, R. (1973). Increased influx of stratospheric air into the lower troposphere after solar H_α and x-ray flares, *J. Geophys. Res. 78*, 6167.

Reiter, R. (1977a). Atmospheric electricity activities of the Institute for Atmospheric Environmental Research, in *Electrical Processes in Atmospheres*, H. Dolezalek and R. Reiter, eds., Steinkopff, Darmstadt, pp. 759-796.

Reiter, R. (1977b). The electric potential of the ionosphere as controlled by the solar magnetic sector structure, result of a study over the period of a solar cycle, *J. Atmos. Terrest. Phys. 39*, 95-99.

Roble, R. G., and P. B. Hays (1979). A quasi-static model of global atmospheric electricity, 2. Electrical coupling between the upper and lower atmosphere, *J. Geophys. Res. 84*, 7247-7256.

Roble, R. G., and P. B. Hays (1982). Solar-terrestrial effects on the global electrical circuit, in *Solar Variability, Weather, and Climate*, NRC Geophysics Study Committee, National Academy Press, Washington, D.C., pp. 92-106.

Sojka, J. J., W. J. Raitt, and R. W. Schunk (1979). Effect of displaced geomagnetic and geographic poles on high-latitude plasma convection and ionospheric depletions, *J. Geophys. Res. 84*, 5943-5951.

Sojka, J. J., W. J. Raitt, and R. W. Schunk (1980). A comparison of model predictions for plasma convection in the northern and southern polar regions, *J. Geophys. Res. 85*, 1762-1768.

Solomon, S., P. J. Crutzen, and R. G. Roble (1982a). Photochemical coupling between the thermosphere and lower atmosphere, I, Odd nitrogen from 50 to 120 km, *J. Geophys. Res. 87*, 7206-7220.

Solomon, S., G. C. Reid, R. G. Roble, and P. J. Crutzen (1982b). Photochemical coupling between the thermosphere and lower atmosphere II, D-region ion chemistry and winter anomaly, *J. Geophys. Res. 87*, 7221-7227.

Stergis, C. G. G. C. Rein, and T. Kangas (1957a). Electric field measurements above thunderstorms, *J. Atmos. Terrest. Phys. 11*, 83-90.

Stergis, C. G., G. C. Rein, and T. Kangas (1957b). Electric field measurements in the stratosphere, *J. Atmos. Terrest. Phys. 11*, 77-82.

Tyutin, A. A. (1976). Mesospheric maximum of the electric field strength, *Cosmic Res. 14*, 132-133.

Tzur, I, and Z. Levin (1981). Ions and precipitation charging in warm and cold clouds as simulated in one dimensional time-dependent models, *J. Atmos. Sci. 38*, 2444-2461.

Tzur, I., and R. G. Roble (1983). Ambipolar diffusion in the middle atmosphere, *J. Geophys. Res. 88*, 338-344.

Tzur, I., and R. G. Roble (1985a). Atmospheric electric field and current configurations in the vicinity of mountains, *J. Geophys. Res. 90*, 5979-5988.

Tzur, I., and R. G. Roble (1985b). The interaction of a dipolar thunderstorm with its global electrical environment, *J. Geophys. Res. 90*, 5989-5999.

Tzur, I., R. G. Roble, H. C. Zhuang, and R. C. Reid (1983). The response of the Earth's global electrical circuit to a solar proton event, in *Solar-Terrestrial Influences on Weather and Climate*, B. McCormac, ed., Colorado Associated University Press, Boulder, Colo., pp. 427-435.

Vampola, A. L. (1977). VLF transmission induced slot electron precipitation, *Geophys. Res. Lett. 4*, 569-572.

Volland, H. (1972). Mapping of the electric field of the Sq current into the lower atmosphere, *J. Geophys. Res. 77*, 1961-1965.

Volland, H. (1975). Models of global electric fields within the magnetosphere, *Ann. Geophys. 31*, 154-173.

Volland, H. (1977). Global quasi-static electric fields in the Earth's environment, in *Electrical Processes in Atmospheres*, H. Dolezalek and R. Reiter, eds., Steinkopff, Darmstadt, pp. 509-527.

Volland, H. (1978). A model of magnetospheric electric convection field, *J. Geophys. Res. 83*, 2695-2699.

Volland, H. (1982). Quasi-electrostatic fields within the atmosphere, in *CRC Handbook of Atmospherics*, Vol. 1, H. Volland, ed., CRC Press, Boca Raton, Fla., pp. 65-109.

Volland, H. (1984). Atmospheric electrodynamics, in *Physics and Chemistry in Space*, Vol. *II*, Springer-Verlag, Berlin, 203 pp.

Vonnegut, B. (1953). Possible mechanism for the formation of thunderstorm electricity, *Bull. Am. Meteorol. Soc. 34*, 378.

Vonnegut, B. (1065). Thunderstorm theory, in *Problems of Atmospheric and Space Electricity*, Proc. 3rd Int. Conf. Atmos. Space Elec., Montreux, Switzerland, May 1963, S. C. Coroniti, ed., Elsevier, New York.

Vonnegut, B., C. B. Moore, R. P. Espinola, and H. H. Blau, Jr. (1966). Electric potential gradients above thunderstorms, *J. Atmos. Sci. 23*, 764-770.

Vonnegut, B., R. Markson, and C. B. Moore (1973). Direct measurement of vertical potential differences in the lower atmosphere, *J. Geophys. Res. 78*, 4526-4528.

Wait, J. R. (1960). Terrestrial propagation of very-low-frequency radio waves, *J. Res. Nat. Bur. Stand. Sec. D 64*, 152-163.

Whipple, F. J. W., and F. J. Scrase (1936). Point discharge in the electric field of the Earth, *Geophys. Memoirs (London) VIII*(68), 20.

Willett, J. C. (1979). Solar modulation of the supply current for atmospheric electricity? *J. Geophys. Res. 84*, 4999-5002.

Wilson, C. T. R. (1920). Investigation on lightning discharges and on the electric field of thunderstorms, *Phil. Trans. A 221*, 73-115.

Telluric Currents: The Natural Environment and Interactions with Man-made Systems

16

LOUIS J. LANZEROTTI
AT&T Bell Laboratories

GIOVANNI P. GREGORI
Istituto di Fisica dell'Atmosfera, Rome

INTRODUCTION

Telluric currents consist of both the natural electric currents flowing within the Earth, including the oceans, and the electric currents originating from man-made systems. Telluric currents could also be considered to include geodynamo currents, i.e., the electric currents that are presumed to flow in the Earth's core and are responsible for the generation of the "permanent" geomagnetic field. This review excludes geodynamo considerations from its purview.

There has been an evolution (see Appendix) in the terminology in the English-language scientific literature related to telluric currents. A common former term used for telluric currents has been "Earth currents," a term that was widely used by Chapman and Bartels (1940) in their classic work, whereas Price (1967) preferred "telluric currents." A difference between the two terms can be recognized in reading historical papers: an impression is obtained that Earth's currents was the name applied to the natural currents (or, more properly, voltages) that are measured between two electrodes which are grounded at some distance apart. Independent of the cause, the observed current was termed an Earth's current. It later became evident that electric currents also flow in seawater. Therefore, the term telluric currents can be interpreted to include currents flowing both

within the solid Earth and within the seas and oceans. However, we note that Earth currents and ocean currents do not form independent electric-current systems. On the contrary, leakage currents exist between continental areas and oceans (see, e.g., references in Gregori and Lanzerotti, 1982; Jones, 1983). In the early French and Italian scientific literature on the subject, however, the term telluric (derived from the Latin *tellus*, for Earth) was always used (e.g., Blavier, 1884; Battelli, 1888; Moureaux, 1896).

The fundamental causes of telluric currents are now believed to be understood. They are produced either through electromagnetic induction by the time-varying, external-origin geomagnetic field or whenever a conducting body (such as seawater) moves (because of tides or other reason) across the Earth's permanent magnetic field. Both causes produce telluric currents, which, in turn, produce magnetic fields of their own—fields that add to the external origin geomagnetic field and produce a feedback on the ionosphere current system (a feedback that, however, is negligible; see, e.g., Malin, 1970).

The complexities associated with telluric currents arise from the complexities in the external sources and in the conductivity structure of the Earth itself. Such complexities have led earlier workers to make statements such as "the simple laws of electromagnetic induction do not fully explain the cause of geoelectric and geomagnetic activity" (Sanders, 1961), while Winckler *et al.*, (1959), in discussing a 2650-V drop across a transatlantic cable produced during a magnetic storm (see below) concluded " . . . either the current circuit [in the Atlantic] is in the horizontal plane or the currents are not the result of the induced emf."

The mathematical modeling of telluric currents, unlike the understanding of their physical causes, is still far from a satisfactory solution. As far as Earth currents are concerned, the investigations have been for the most part carried out on a local or limited regional scale.

In contrast, the understanding of oceanic telluric currents (which cover a considerable fraction of the Earth's surface) has, since the Ashour (1950) estimate of their decay time in an ocean (order of a few hours), undergone substantial progress. The state of the art of ocean-current modeling now takes into account coastlines, although the ocean bottom is usually assumed flat—either nonconducting (although with a conducting mantle; see e.g., Parkinson, 1975; Hobbs and Brignall, 1976; Hobbs and Dawes, 1980; Beamish *et al.*, 1980; Fainberg, 1980, and references therein) or conducting (Hewson-Browne, 1981; Hewson-Browne and Kendall, 1981; Kendall and Quinney, 1983). The ocean currents and their related geomagnetic effects have been investigated by, for example, Barber and Longuet-Higgins (1948), Fraser (1965), Peckover (1973), Klein *et al.*, (1975), and Semevskiy *et al.* (1978). Malin (1970, 1973), in considering the lunar tidal harmonic component M_2 (which is the most important one both in the atmosphere and in the sea, with a period of half a lunar day), succeeded in separating the effect of direct electromagnetic induction from the ionosphere from the currents produced by oceanic tidal flow. He assumed that the geomagnetic variation associated with the tidal component should always be observed, independent of local time, whereas the ionosphere component should be negligible at midnight. In fact, he found that at Irkutsk, the geomagnetic observatory farthest from any ocean, the ocean-produced effect is negligible, unlike the situation at several other observatories closer to a coast, where the ocean component is present.

No equivalently sophisticated modeling, even for long-period geomagnetic variations, can usually be found for Earth currents. This situation exists principally because of the frustrating indeterminacies introduced by local Earth conductivity anomalies. The basic difficulty arises because of the nonuniqueness of the "inversion problem," that is, the nonuniqueness of the evaluation of the underground conductivity structure in terms of the surface geoelectromagnetic recordings. Therefore, the problem is usually tackled in terms of the "forward problem": an external-origin electromagnetic field is assumed to impinge on an underground conducting structure of given geometrical shape, with only the conductivities left to be optimized by a numerical fitting of the model with the actual observational records. The procedure can be worked out only for reasonably simple geometrical shapes (Porstendorfer, 1976) for the conducting bodies, which implies substantial limitations to any attempt to extend such investigations to wider spatial scales (see e.g., Rokityansky, 1982; Hohmann, 1983; Parker, 1983; Varentsov, 1983).

In terms of planetary-scale currents, Gish (1936a, 1936b) presented the results of Figure 16.1, which he deduced for the daytime western hemisphere from diurnal variation (24-h period) recordings of orthogonal Earth current measurements collected during the second International Polar Year (1932-1933) at a number of sites around the world. The directions of current flow were determined by taking the vector sum of the N-S and E-W currents measured at the various sites. Large errors could be expected because of the sparse number of stations and the need, therefore, for large interpolations.

A more recent picture has been provided by Matsushita and Maeda (1965a, 1965b; see also Matsushita, 1967) from analyses of the worldwide geomagnetic field on a planetary scale. These authors performed a standard separation (by means of Gauss' spherical harmonic expansion) of the external- and internal-origin field. Some of the results are presented in Figure 16.2. Notice the obvious, substantial differences compared to the currents of Figure 16.1. However, even the more recent work is obviously unable to recognize the effects, where important, of localized anomalous conductors, such as mid-oceanic ridges, or even the differences between ocean basins and continents. In fact, spherical harmonics vary too smoothly to be able to account for such localized features, particularly if a reasonably limited number of terms is used in the expansion (for recent reviews of spherical harmonic techniques, see Winch, 1981; Fainberg, 1983).

The first 75 to 100 years of Earth current work produced considerable debate as to causes and disagreements among researchers as to the magnitude of the effect at given times. Gish (1936a, 1936b) noted that better agreement between independent measurements often occurred when relatively long lengths of wire were used. Today it is clear that such a situation could easily arise from experimental procedures such as improper grounding of a wire and an insufficiently high impedance in the measuring system (e.g., see Hessler, 1974). For example, in the work of Airy (1868) the wires were grounded to water pipes, which themselves obviously

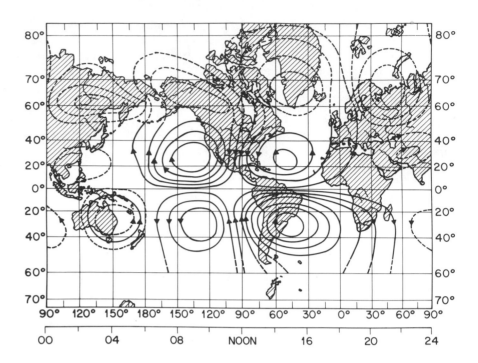

FIGURE 16.1 Planetary-scale distribution
of telluric currents according to Gish (1936a,
1936b) at 1800 GMT.

could carry currents flowing in the entire region over
which the relatively short lengths of wire extended. In
this case, the pipe network was the receiving "antenna,"
even more than the lengths of wire. Recent evidence of
the effect of telluric currents, integrated over a planetary scale, has been provided by analysis of MAGSAT
data. Langel (1982) reported an analysis of the data in
terms of separation, by spherical harmonic expansion,
of the external- and internal-origin geomagnetic field.
The analysis was done for different sets of data, depending on the value of the Dst index (a measure of the particle ring current in the Earth's magnetosphere and,
therefore, of the level of disturbance of the geomagnetic
field). Figures 16.3(a) and 16.3(b) show that the lowest
order and degree terms (i.e., dipole terms), denoted g_1^0
and q_1^0 (external and internal, respectively) change with
the level of Dst. The internal term increases with decreasing Dst, unlike g_1^0, a consequence of the fact that
induced currents must flow in the direction opposite to
the inducing currents.

Summarizing, accurate knowledge of telluric current
patterns in the Earth on a planetary scale still remains a
basically open problem even though the subject has a
long history. In addition to the actual role of ocean water and sediments, largely unknown is the influence of
localized conductivity anomalies (such as fold belts,
mid-oceanic ridges, and trenches and subduction zones)
on such patterns. The current patterns will obviously be
different for different periods of the external inducing
field. The higher-frequency patterns will be highly time

variable because of the temporal and spatial variability
of the external-origin fields, variabilities that are not yet
amenable to accurate predictive modeling. Nevertheless, given all the foregoing caveats, we present in the
following sections additional discussions of many of the
relevant issues, as well as some implications for practical
concerns.

THE NATURAL ENVIRONMENT

The Physical Problem: Hydrology, Geology, Geothermics, and Tectonics

Except during a lightning strike to Earth, essentially
negligible electric current flows between the air and the
ground (integrated over the Earth, the fair-weather
current amounts to some 1000-2000 A). Therefore, the
Earth's surface is a natural surface across which electromagnetic coupling occurs via an electromagnetic field.
This implies that it is possible in many cases to treat the
coupling problem in terms of scalar potentials (at least
for frequencies lower than those used in audio magnetotelluric studies). An attempt by Berdichevsky and Fainberg (1972, 1974) to evaluate, on a global scale, possible
currents between ground and air suffered large uncertainties from the approximations used. As noted briefly
in the Introduction, the cause of telluric currents is either electromagnetic induction by the time-varying geomagnetic field produced by the ionosphere and/or magnetosphere or by water movement across the permanent

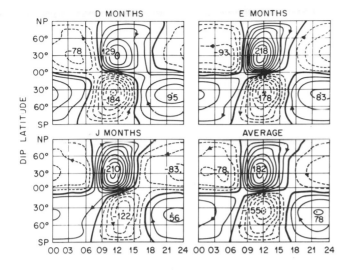

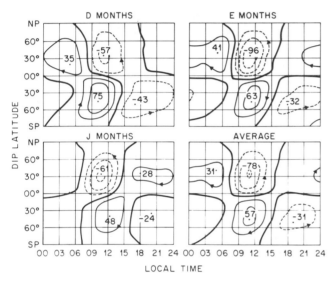

FIGURE 16.2   (a) External *Sq* current systems averaged worldwide for *D* months (northern winter; top left), *E* months (equinox, top right), and *J* months (northern summer, bottom left), and their yearly average (bottom right). The current intensity between two consecutive lines is 25×10^3 A; the thick solid curves indicate the zero-intensity lines. The numbers near the central dots are the total current intensities of these vortices in units of 10 A. (b) Internal *Sq* current systems averaged worldwide for *D* months (top left), *E* months (top right), and *J* months (bottom left), and their yearly average (bottom right). Notice the disagreement with Figure 16.1; the rotational senses of the vortices are opposite. This figure adapted from Matsushita (1967).

geomagnetic field. Considering only the former cause, the longer the period of the time-varying field, the greater the depth in the Earth where the induced currents can be expected to flow. A quantitative criterion can be given in terms of electromagnetic induction in a half-space of uniform conductivity (note that this is a highly idealized case that practically never occurs in re-

ality). The "skin depth" (i.e., the depth at which the external field is damped by a factor $1/e$) is given by $S = 0.5 \, (T/\sigma)^{0.5}$ km, where is the conductivity in mhos/meter and T is the period of the variation in seconds. A signal with period of about 24 h is generally believed to have a skin depth of 600 to 800 km (Hutton, 1976; Gough and Ingham, 1983). (The skin-depth only provides, however, a rough approximation of the depth at which actual telluric currents of a given period are flowing. In fact, the actual conductivity structure underground is most often a matter of considerable indeterminacy.) Saltwater has a conductivity of about 4 mhos/m, hydrated sediments have a conductivity of about 0.1 mho/m, and dry rock has a conductivity of about 0.0001 mho/m. Practically all the materials of the usual geologic environment (see, e.g., ACRES, 1975; Keller, 1966) can be placed between these extremes. Nomograms by which T, σ, and S can be evaluated for different materials and for the "actual" Earth are shown in Figure 16.4. The conductivity of water is largely affected by salinity (and to a minor extent by temperature). The conductivity of soil is largely affected by the state of hydration. Porous materials and sediments can easily be hydrated (see below) by considerable amounts. Hence it might eventually be possible, by electromagnetic means, to distinguish materials of equal density but with different porosities, and hence different hydration (and electrical conductivities), that cannot be distinguished by seismic techniques.

The distributions of sediments, particularly important for shorter-period variations, should be considered on local or regional scales, because minor details in the distributions can be relevant to telluric current flow. A worldwide pattern of sediments has been given by Hopkins (reproduced in Green, 1977, and in Gregori and Lanzerotti, 1982). Fainberg (1980) provided a worldwide model map of the total conductivity of the water shell plus sedimentary cover [Figure 16.5(a)]. Such a map is the result of a more detailed mapping given by Fainberg and Sidorov (1978). For example, Figure 16.5(b) shows the conductivity profile for Europe. Clearly shown are the sedimentary structures responsible for the North German conductivity anomaly and for the channeling in the Seine Basin. The North German anomaly, with a depth-integrated conductivity > 3000 mhos, is equivalent to ≥ 750 m of seawater.

Another physical factor affecting conductivity, and thus telluric currents, is temperature. Since the temperature increases with depth in the Earth, the conductivity is higher with increasing depth. However, the effect is not uniform; the heat flux through the Earth's surface is greater in certain regions than in others, providing thermal anomalies. Whenever a larger geothermal flux

FIGURE 16.3 (a) The spherical harmonic coefficient of lowest degree and order describing the magnetic field originating external to the Earth, as a function of the global Dst index used to describe temporal variations of the equatorial horizontal magnetic field relative to magnetically quiet days. (b) The spherical harmonic coefficient of lowest degree and order describing the field originating within the Earth as a function of the lowest degree and order magnetic-field coefficient describing the magnetic field originating external to the Earth (adapted from Langel, 1982).

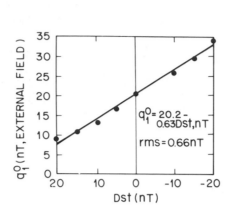

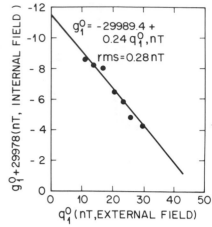

occurs, there is an upward warping of isothermal surfaces. In such a case, telluric currents of a given period will flow in shallower layers. A worldwide mapping of the geothermal flux averaged over a 5° × 5° mesh (Figure 16.6) has been provided by Chapman and Pollack (1975). (This map has largely been obtained using about 5000 direct borehole measurements and, where unavoidable, indirect information. For example, since the heat flow from the ocean floor is a well-defined function of the floor's age, the flow can be approximated even when it has not been directly measured. Analogously, a different function relates the continental heat flow to age.)

Three additional aspects of the conductivity structure of the Earth affect the flow of telluric currents—spatial gradients, temporal variations, and channeling.

The spatial gradients of telluric currents strongly depend, in shallow layers, on geochemical composition,

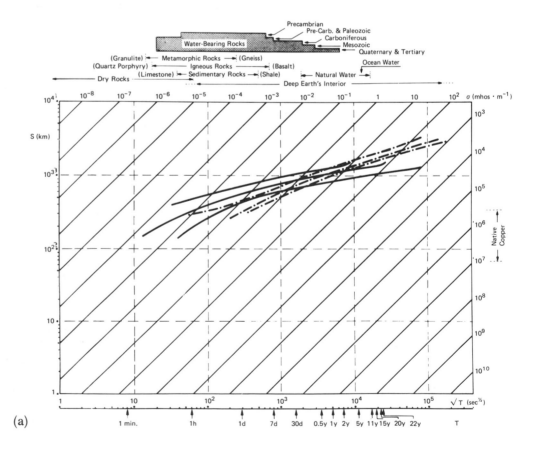

(a)

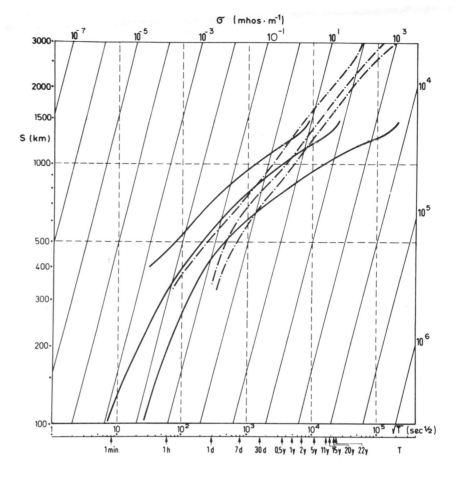

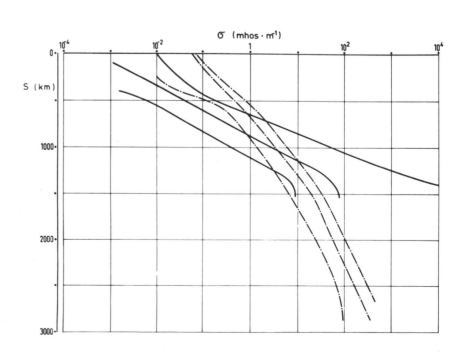

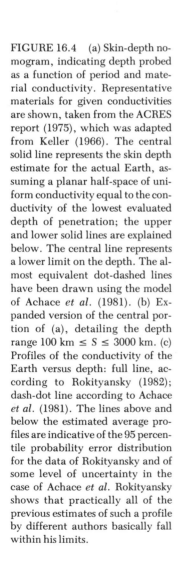

FIGURE 16.4 (a) Skin-depth nomogram, indicating depth probed as a function of period and material conductivity. Representative materials for given conductivities are shown, taken from the ACRES report (1975), which was adapted from Keller (1966). The central solid line represents the skin depth estimate for the actual Earth, assuming a planar half-space of uniform conductivity equal to the conductivity of the lowest evaluated depth of penetration; the upper and lower solid lines are explained below. The central line represents a lower limit on the depth. The almost equivalent dot-dashed lines have been drawn using the model of Achace *et al.* (1981). (b) Expanded version of the central portion of (a), detailing the depth range 100 km ≤ S ≤ 3000 km. (c) Profiles of the conductivity of the Earth versus depth: full line, according to Rokityansky (1982); dash-dot line according to Achace *et al.* (1981). The lines above and below the estimated average profiles are indicative of the 95 percentile probability error distribution for the data of Rokityansky and of some level of uncertainty in the case of Achace *et al.* Rokityansky shows that practically all of the previous estimates of such a profile by different authors basically fall within his limits.

(b)

(c)

238

FIGURE 16.5 (a) Model maps of total conductivity of the water shell plus sedimentary cover. The isolines give the depth-integrated conductivity in units of mhos. The exceptionally solid lines are regions of rapid gradients in conductivity. From Fainberg and Sidorov (1978). (b) Expanded view of the European sector, from Fainberg and Sidorov (1978), where expanded models for other regions of the world are also provided.

(a)

239

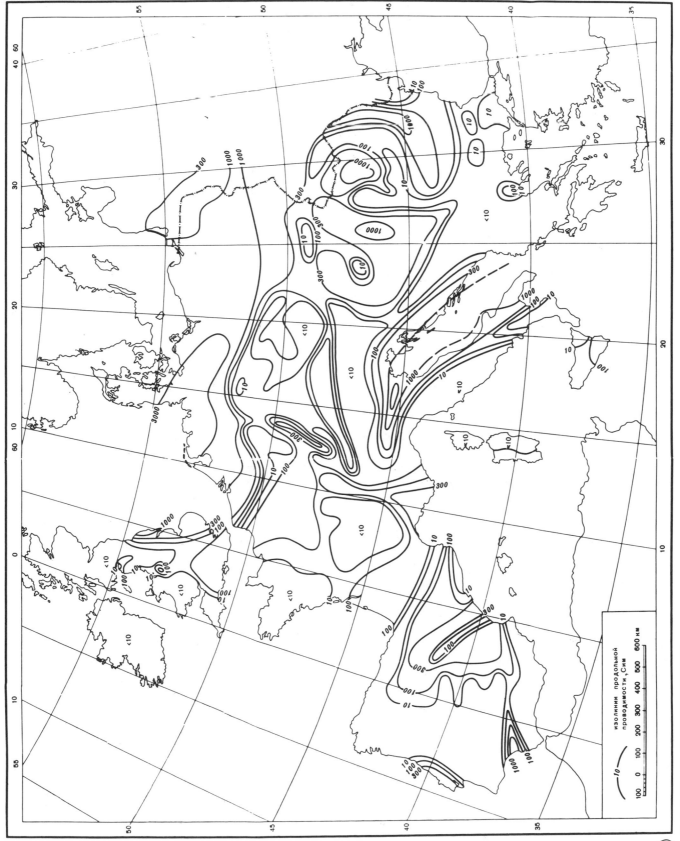

Продольная проводимость осадочного чехла Западной Европы

изолинии продольной
проводимости , См·км

(b)

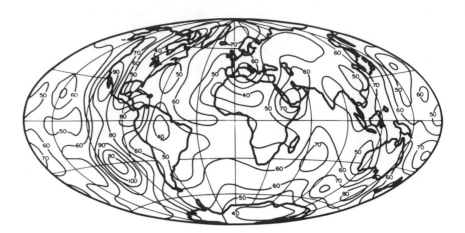

FIGURE 16.6 Spherical harmonic representation (degree 12) of global heat flow from
observations supplemented by predictor.
Heat-flow contour lines are in milliwatts per
square centimeter. Adapted from Chapman
and Pollack (1975).

geological structure, and hydration. [Hydration in this
context can be taken just in terms of water content (producing an increase in conductivity) or in terms of the
formation of particular compounds (clathrate hydrates)
that can decrease the conductivity (although there are
no reports of this in the telluric current literature); see
Miller (1974).] Deeper in the Earth, it is believed that a
more or less thick layer of dry rocks (having reduced
conductivity) is further underlain by layers of increasing
conductivity, which is a function of the increasing temperature with depth. In such deep layers it has generally
been assumed that the Earth becomes increasingly homogeneous with greater depth. More realistically, however, the increasing difficulty (if not impossibility) of
recognizing spatial gradients at greater depths must be
acknowledged. Differently stated, telluric currents as a
means of remote sensing of the underground conductivity provide ever-diminishing spatial (horizontal) resolution with depth.

The problem of spatial gradients of the telluric currents is also related to the state of knowledge of the spatial gradients of the external-origin inducing field. In
fact, the diurnal and the lunar variation fields (Sq and L
fields, respectively) have a planetary scale, albeit showing strong spatial gradients related to the auroral and
equatorial electrojets for quiet conditions (e.g., Schlapp, 1968; Riddihough, 1969; Greener and Schlapp,
1979). For disturbed conditions, the planetary-scale description still plays a relevant, though not singular, role
(e.g., Sato, 1965; Campbell, 1976). Therefore, the ex-
ternal-inducing source at these low frequencies can be
approximately described in terms of a planetary-scale
field, occasionally with strong spatial gradients.

On the contrary, for higher frequencies (magnetic
storms, geomagnetic pulsations) the source can often
appear quite localized (see, e.g., Davidson and
Heirtzler, 1968; Lanzerotti *et al.*, 1977; Southwood and

Hughes, 1978; Reiff, 1983) and is highly time dependent
as well. At the Earth's surface the spatial extent of the
source for pulsations (period of a few to a few hundreds
seconds) is believed to be not smaller than the height of
the ionosphere.

Temporal variations in the Earth's conductivity
structure can be caused by such effects as seasonal cli-
matic changes affecting water salinity and temperature,
ice extension, permafrost and hydration content, and
tectonic processes. The tectonic processes can be either
slow (i.e., those involving the geologic time scale), intermediate (as in earthquake precursors; e.g., Honkura,
1981), or rapid (as in volcanoes).

Channeling of telluric currents in specific, higher-
conductivity regions is an actively debated area at
present. Some recent research papers, without pre-
sumption of completeness, include Lilley and Woods
(1978), Babour and Mosnier (1980), De Laurier *et al.*
(1980), Miyakoshi (1980), Srivastava and Abbas (1980),
Woods and Lilley (1980), Camfield (1981), Chan *et al.*
(1981a, 1981b), Kirkwood *et al.* (1981), Kurtz *et al.*
(1981), Sik *et al.* (1981), Srivastava (1981), Thakur *et al.*
(1981), Booker and Hensel (1982), DeBeer *et al.* (1982),
Le Mouël and Menvielle (1982), Nienaber *et al.* (1982),
and Summers (1982); see also extensive review and discussion by Jones (1983). The issue revolves around the
interpretation of the measured telluric currents. Should
the measurements at some given site be interpreted in
terms of electromagnetic induction on a local (or in any
case on a small-scale) spatial extent, or should they be
considered as the result of a large-scale (i.e., regional,
continental, or planetary scale) induction phenomenon,
whereby telluric currents are channeled from more remote areas within some relevant conducting body not
far away from the recording site? While specific cases
can be discussed (such as the North German anomaly;
see Appendix), a generally valid reply is difficult to give

basically because (1) the planetary-scale response of the actual Earth in terms of telluric currents is poorly known and (2) the temporal and spatial scale of the external-origin inducing field is often poorly known, particularly for shorter-period variations represented by magnetic storms and geomagnetic pulsations.

It is interesting to note that the current channeling was addressed early on in studies of telluric currents. Varley (1873) discussed current channeling from the sea in telegraph wires between the coastal town of Ipswich and London. He also claimed that enhanced currents were seen in the line between Glasgow and Edinburgh, which connected the sea across the British Isle, as compared with a line solely on land.

Summarizing, telluric currents depend on several physical parameters and, if properly interpreted, can be used for studies of the underground electrical structure at both shallow and great depths. It is important for telluric current studies to take into more explicit account the relations of measured currents to the specific tectonic and geomorphological features of the regions under study. Approaches toward such a viewpoint have been presented recently by Hermance (1983). In general, such investigations can best be tackled by means of large arrays of instrumentation (Alabi, 1983).

Shallow Telluric Currents

Effects on shallow telluric currents (generally shorter period) can be found whenever a mineral has some remarkably different electrical conductivity compared with that of the surrounding materials. This gives rise to a localized conductivity anomaly that can be studied by means of a dense network of recording instruments. Shallow currents have also been reported in several sedimentary basins, such as in the Seine Basin and in the northern German anomaly (see Appendix; for other references see, e.g., Gregori and Lanzerotti, 1982). Shallow telluric currents are responsible for a component of the coast effect or magnetic signals, where the geometrical orientation of the magnetic variations at higher frequencies are correlated with the shape of the coast. The coast effect has been reviewed by Fischer (1979), Parkinson and Jones (1979), and Gregori and Lanzerotti (1979b).

The difference between shallow and deep effects (the latter arising from local tectonic features) has been shown by Honkura (1974) for the Japanese islands (Figure 16.7). At shorter periods, when the skin depth is shallower, the coast effect reflects the coast shape. At longer periods, electromagnetic induction evidence suggests a dependence on the downward bending of the lithospheric slab where it approaches the Japanese sub-

duction zone. Similar effects have been reported by Honkura *et al.* (1981) for a small island in the Philippine Sea ("regular" coast effect) and by Beamish (1982) for the island of South Georgia (Scotia Arc, South Atlantic). The threshold period discriminating between shallow and deep effects appears to be about 20 min in the Japanese area, a result obtained from a reinterpretation by Gregori and Lanzerotti (1982) of data published by Yoshimatsu (1964).

Deep Telluric Currents

The best recognized, by seismic waves, underground discontinuity—the Moho (see, e.g., global map presented by Soller *et al.*, 1982)—has no obvious correspondence in geoelectromagnetic phenomena. In fact, the behavior of deep telluric currents is largely controlled by the shape of the isotherms. An idea of the trend of such isotherm surfaces is given by Figure 16.8, which plots isocontours of the thickness of the lithosphere (Chapman and Pollack, 1977), based on the heat-flux results of Figure 16.6. Chapman and Pollack (1977) derived the lithosphere results by determining the depth at which both continental and oceanic geotherms intersect the mantle solidus. They showed this to be a consistent estimator of the depth to the top of the seismic low-velocity channel or of the thickness of the high-velocity lid overlying the channel. They identified the lid as synonymous with the lithosphere.

A similar discussion, limited to the Soviet Union, is given by Cermak (1982). Oxburgh (1981) presented a critical discussion of the method employed for such analyses. For the sake of completeness, however, it should be noted that the concept of the lithosphere is actually more complicated. Depending on the experimental observations used, four different definitions can be distinguished: the elastic or flexural, the thermal, the seismic, and the chemical or mineralogical (U.S. Geodynamics Committee, 1983; Anderson, 1984; Maxwell, 1984). In the context of telluric currents, the thermal structure of the deep Earth is likely the most relevant factor, with the chemical/mineralogical being the second. Hence, in this simple context, Figure 16.8 can provide an idea of the depth where a high electrical conductivity can be expected at a given site. A very general and approximate statement is that the thickness estimates of Figure 16.8 are in reasonable agreement with geomagnetic depth-sounding and magnetotelluric estimates of the depth of the "ultimate conductor": about 200 km below continents (cratons), about 100 km under stable continental areas, about 60-70 under rifts and grabens, and about 10-20 km (or even shallower) under volcanic areas and mid-ocean ridges. [A warning must be given

FIGURE 16.7 (a) The $\Delta Z/\Delta H$ value distribution in Japan for geomagnetic variations corresponding to geomagnetic bays. The profiles AA' and BB' have been investigated in detail, and their results are shown in the subsequent figures. (b) Parkinson vectors along the profile AA' of part (a), for geomagnetic variations with period of 60 min. Contours indicate the sea depth in 10^3 m. The Parkinson vectors are consistent with an interpretation in terms of an asthenosphere bending and deepening in the subduction zone. (c) The same as for profile AA' in part (b), but referring to the profile BB'. The downward bending of the asthenosphere in the subduction zone appears much less pronounced in this region. (d) Parkinson vectors on the Miyakejima island for periods (a) 120, (b) 60, (c) 30, (d) 15, and (e) 5 min, respectively. The coast effect is quite evident at the shorter periods, while at the longer periods the effect of the bending of the asthenosphere is predominant over the coast effect.

The vectors appearing in the (b), (c), and (d) sections of the figure are "Parkinson arrows" or "vectors," defined in the following manner. Consider the deepest surface layer to which the incident electromagnetic wave of a given period can penetrate. Consider a plane (the "Parkinson plane") tangent to such a surface, directly beneath a given recording site. Construct a line perpendicular to this plane and oriented downward. Project this line in

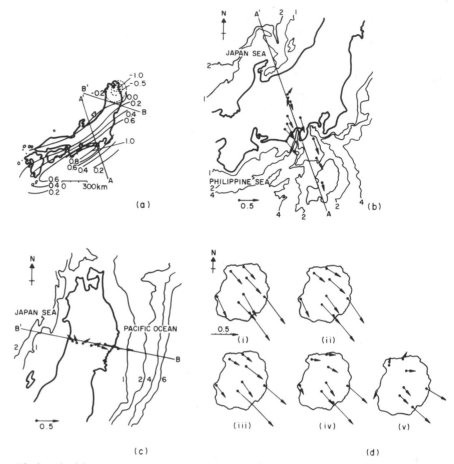

the horizontal plane: this is the direction of the arrow. The length of the arrow is equal to the sine of the tilt of the Parkinson plane with respect to the horizontal plane. Therefore, a vanishing Parkinson arrow implies a horizontal Parkinson plane, a unit length arrow implies a vertical Parkinson plane. A "normal" coast effect on an island shows that Parkinson arrows point outward from the island. (For other details on "induction arrows" refer to the review by Gregori and Lanzerotti, 1980.) Figure is adapted from Honkura (1974).

FIGURE 16.8 Thickness of the lithosphere derived from a spherical harmonic (12 degree) representation of the global heat flow (see Figure 16.6) and continental and oceanic geotherm families. Contours are in kilometers, with variable intervals. From Chapman and Pollack (1977).

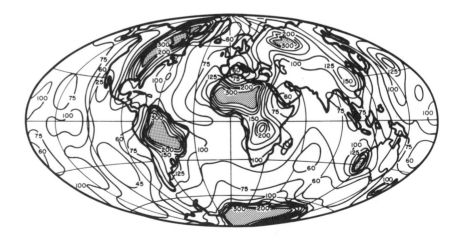

here, however, that these depths are very approximate; the actual structures are generally much more differentiated and complex; see, e.g., Hermance (1983).]

The problem of the deep electrical conductivity structure of the Earth has usually been treated in terms of a concentric spherical shell model of the Earth, where each shell has a uniform electrical conductivity. The interested reader may refer to such classical treatments as Chapter XXII of Chapman and Bartels (1940), Rikitake (1966), Rokityansky (1982), or Parkinson (1982) (a more concise treatment can be found in Price, 1967). A simple but effective procedure was proposed by Schmucker (1970) whereby the Earth is simplified to a two-layer body having an insulating outer layer underlain by a conductor (he considered both flat and spherical Earth models). Using only the ratio of the horizontal to the vertical component of the geomagnetic field (at a prechosen frequency), Schmucker (1970) provided simple formulas by which the thickness of the insulating layer and the electrical conductivity of the underlying conductor can be promptly evaluated. By considering fields of different frequencies it is possible to evaluate different estimates of depth and conductivity. This "Schmucker inversion" technique, which often appears to agree with results obtained by means of other, more involved, methods of handling geomagnetic data, is a simple way of treating the inversion problem—a difficult and much debated problem (e.g., Rokityansky, 1982; Hohmann, 1983; Parker, 1983; Varentsov, 1983; Gough and Ingram, 1983; Berdichevsky and Zhdanov, 1984).

INTERACTIONS OF TELLURIC CURRENTS WITH MAN-MADE SYSTEMS

The natural telluric current environment can significantly affect man-made systems. Conversely, human technology can "pollute" the natural telluric current environment. The mechanisms by which these interactions occur, as well as their modeling, are far from being understood satisfactorily and comprehensively. Geophysicists have often viewed such interactions as an unwanted, unnatural nuisance. Engineers have almost always been concerned with thresholds of system reliability and with a system's capability to react positively to any sudden change in the natural environment, always on a strict basis of yield/cost ratio. Moreover, technological improvements have been progressively introduced within systems to ensure a higher and higher reliability (e.g., Axe, 1968; Anderson, 1979), so that it becomes difficult to compare effects observed on different systems in different years.

Seldomly have man-made systems been viewed as scientific instruments that are useful for studying the natural environment. Often a man-made system can be considered part of the natural environment itself. Geophysicists can then imagine such large man-made tools as similar to specifically designed measuring instruments that, unlike laboratory instruments that are normally presumed to negligibly affect the system, actually interfere with the natural phenomena, often quite seriously. Such huge and expensive man-made systems can allow, in principle, some complex experiments and measurements, which otherwise could not be carried out. For this reason this topic has particular scientific value, much beyond a matter of scientific curiosity or of a more or less minor nuisance affecting the operation of huge engineering systems.

The literature on the subject tends to be rather sparse. However, five principal areas of interest can be considered. These are discussed below; some of this material has been previously reviewed elsewhere (Axe, 1968; Lanzerotti, 1979a, 1979b, 1979c, 1983; Paulikas and Lanzerotti, 1982).

Communication Cables

Historically, this is the best investigated and documented effect of telluric currents on technological systems. In fact, after the lightning rod, the telegraph was essentially the earliest of man-made electromagnetic devices in use. Subsequently, telegraph lines have been progressively supplanted by telephone lines, and submarine cables have supplanted the former radio links between the telephone networks of different continents (e.g., Blackwell, 1928; Bown, 1930, 1937; Schelleng, 1930). Even with the advent of communication satellites, cable systems are still of major economic importance for long-distance communications.

The first detection of effects on a telegraph wire dates back to the years 1847-1852. The first observations appear to be from England by Barlow (1849). As stated by Prescott (1866):

M. Matteucci had the opportunity of observing this magnetic influence under a new and remarkable form. He saw, during the appearance of the aurora borealis of November 17, 1848, the soft iron armatures employed in the electric telegraph between Florence and Pisa remain attached to their electromagnetics, as if the latter were powerfully magnetized, without, however, the apparatus being in action, and without the currents in the battery being set in action. This singular effect ceases with the aurora, and the telegraphs, as well as the batteries, could operate anew, without having suffered any alteration. Mr. Highton also observed in England a very decided action of the aurora borealis, November 17, 1848. The magnetized needle was always driven toward the same side, even with much force. But it is in our own country that the action of the aurora upon the telegraph-wires has been the most remarkable. . . . In September, 1851, . . . there was remarkable aurora, which took complete possession of all the telegraph lines in New England and prevented any business from being transacted during its continuance.

The days between August 28 and September 2, 1859, were also quite remarkable, not only for some wonderful auroral displays (Clement, 1860; Hansteen, 1860; Prescott, 1860, 1866). Clement's (1860) book had a self-explanatory title: *The Great Northern Light on the Night before 29 August 1859 and the Confusion of the Telegraph in North America and Europe.* According to Chapman and Bartels (1940), this aurora was seen in the Atlantic at a latitude as low as 14° N, while in France 800 V were induced on a wire over a distance of 600 km. From Prescott (1866):

We have, however, the second yet more wonderful effects of the aurora upon the wires; namely, *the use of auroral current for transmitting and receiving telegraphic dispatches.* This almost incredible feat was accomplished . . . on the wires of the American Telegraph Company between Boston and Portland, upon the wires of the Old Colony and Fall River Railroad Company between South Braintree and Fall River, and upon other lines in various parts of the country. . . . Such was the state of the line on the September 2nd, 1859, when for more than one hour they held communication over the wires with the aid of celestial batteries alone.

Other studies of historical interest on telluric currents in communication cables are mentioned in the Appendix.

In 1910 work was begun in Norway by Carl Störmer of measuring the height of polar aurorae (Störmer, 1955). Störmer used photographs taken simultaneously from two sites separated by a few tens of kilometers. He was able to send a message of alert to his co-workers about an imminent night of photographic work whenever he measured disturbances in the local telegraph wires.

A geomagnetic storm in Sweden in May 1921 (Germaine, 1942; Sanders, 1961) produced voltages of 6.3 to 20 V/km (i.e., 1 kV or more over 100 to 200 km, with 2.5 A, while the threshold for serious troubles was 15 mA). A large magnetic storm on April 16, 1938, produced potentials of several hundred volts over local wires in Norway (Chapman and Bartels, 1940).

On March 24, 1940 (Germaine, 1942; Harang, 1951; Brooks, 1959; Sanders, 1961), a geomagnetic storm damaged the Norwegian wirelines ($\leq$50-60 V/km, $\sim$600 V, >4 A), while in the United States, more than

500 V were estimated to have occurred along some lines. Reports from two sites near Tromso, Norway, stated

. . . Sparks and permanent arcs were formed in the coupling racks and watch had to be kept during the night to prevent fire breaking out. . . . One line was connected to earth through a 2 mm thick copper wire, which at once got red hot, corresponding to a current more than 10 amps (Harang, 1951).

In the second half of the nineteenth century, Earth currents in submarine cables were rather extensively investigated. Saunders (1880, 1881) and Graves (1873) reported some of their work, which included a cable between Suez and Aden and a cable between Valentia and Newfoundland. Wollaston (1881) concluded that his current measurements on a submarine cable across the English Channel resulted from tidal currents and related an 1851 conversation with Faraday on the matter. The latter was quoted as quite enthused about this confirmation of his earlier predictions.

Axe (1968) listed several geomagnetic storm-induced effects on submarine cables occurring in 1957-1967 (total voltage drops range from 50 V to 2700 V for the different occurrences). The largest voltage drop (Figure 16.9) occurred across a transatlantic cable (equivalent to 0.75 V/km) at the time of the huge storm on February 11, 1958, which produced a well-known spectacular auroral display down to low latitudes (Brooks, 1959; Winckler *et al.*, 1959; Sanders, 1961; Akasofu *et al.*, 1966). It is noteworthy that "the cable to Hawaii which originates about 140 miles north of San Francisco exhibited no major voltage swings" (Winckler *et al.*, 1959).

A major geomagnetic event on August 4, 1972, caused the outage of a continental cable in the midwestern United States. The outage has been investigated (Anderson *et al.*, 1974; Anderson, 1979) by modeling the tel-

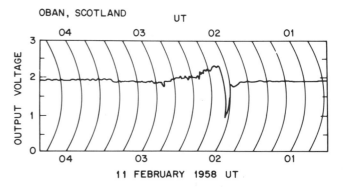

FIGURE 16.9   Output voltage of the power-feed equipment at the Oban, Scotland, end of the Oban-Clareville, Newfoundland, cable. The voltage variation in North America was somewhat larger, leading to a total variation of about 2700 V across the cable. From Axe (1968).

luric currents in terms of a compressed magnetosphere with magnetopause and magnetosphere currents electromagnetically inducing over a three-layer conducting Earth.

Summarizing, shutdowns in both land and sea cables, as well as fires, have been caused by telluric currents induced by geomagnetic storms, and suitable precautions have to be taken (Root, 1979) in order to attempt to avoid them.

A singular example of man-made telluric current "pollution" occurred when a high-altitude nuclear bomb test produced perturbations in the Earth's radiation belts and geomagnetic field. As recounted in Axe (1968):

The disturbance was just detectable on the power-feeding voltage and current recorder charts on the Australia-New Zealand, United Kingdom-Sweden and Bournemouth-Jersey systems. On a circuit originally set up on the Donaghadee-Port Kail No. 3 cable for the measurement of voltage due to water flow, the disturbance was clearly recorded.

The data at the time of the event are shown in Figure 16.10 (Axe, 1968).

All the effects considered above refer to electromagnetic induction from ionospheric and magnetospheric variations. However, there are also effects on submarine communication cables related to water flows (tidal and otherwise). The problem has been extensively reviewed by Meloni *et al.* (1983); see later section. Less dramatic, although relevant, man-induced telluric current perturbations on land cables should be expected in heavily industrialized or populated areas (e.g., Kovalevskiy *et al.*, 1961).

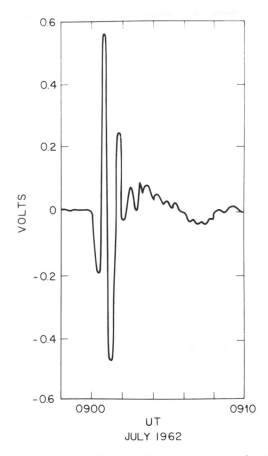

FIGURE 16.10   Effect of the Starfish explosion measured on the center-conductor voltage of the Donaghadee-Kail (Irish Sea) number 3 cable. From Axe (1968).

Powerlines

The historical record of powerlines being greatly distributed or completely disrupted by geomagnetic storms appears somewhat less detailed than that for communications cables. One interruption of service occurred on March 24, 1940, in New England, New York, eastern Pennsylvania, Minnesota, Quebec, and Ontario (Davidson, 1940; Brooks, 1959). As well, during the great geomagnetic storm of February 11, 1958, the Toronto area suffered from a blackout produced by a geomagnetic storm. Currents up to about 100 A were induced in some northern latitude transformers during the great storm of August 4, 1972 (McKinnon, 1972). Induced currents on power systems in the auroral zone have been discussed by Aspnes *et al.* (1981) and Akasofu and Aspnes (1982; see Figure 16.11). Some of the most detailed investigations aimed at establishing engineering relations for power systems have been carried out by

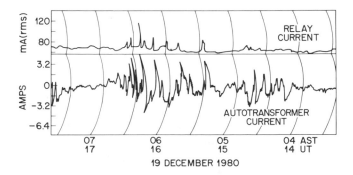

FIGURE 16.11   Simultaneous recordings of geomagnetic induction effects observed as current surges in a positive relay system and an auto transformer in a power substation near Fairbanks, Alaska, on December 19, 1980 (from Akasofu and Aspnes, 1982).

Albertson and Van Baelen (1970), Albertson *et al.* (1970, 1973, 1974), ACRES (1975) (see also references therein), Boerner *et al.* (1983), and Pirjola (1983).

The geomagnetic currents induced in a power system can produce problems of several different types (Albertson *et al.*, 1973, 1974; review by Williams, 1979). First, the arbitrary differential relay operation in power distribution systems during geomagnetic storms can produce a judgmental problem; system operators are unsure of whether the malfunctioning relay indication is an induced-current effect in a transformer or a real transformer malfunction. Second, the currents actually induced in the winding of a power transformer can result in half-cycle saturation of the transformer core. This saturation can produce fluctuations in the transformer operation itself. This local heating can greatly shorten the lifetime of a transformer.

Summarizing, the effects of induced telluric currents on power systems produce outages as well as damages to expensive transformers. Gorely and Uvarov (1981) estimated that in the Norilsk region (Siberia) up to tens of amperes can be expected on powerlines of 100 to 150 km length. Since 500-kV transformers capable of withstanding even 3 to 4 A without saturating appear to cause problems for manufacturing (Sebesta, 1979), a way of avoiding such serious damage is to use powerlines of limited total length (e.g., Akasofu and Merritt, 1979, suggest no more than 500 km for Alaska). Pirjola (1983), from measurements made at four locations in Finland, concluded that currents of the order of 100 A lasting about 1 h should damage transformers.

Pipelines

Varley (1873) reported that large Earth currents on a short length of telegraph cable in London appear to have been related to currents flowing on large, nearby gas pipelines. Studies of induced telluric currents on pipelines took renewed importance when the long, Trans-Alaskan pipeline (1280 km long) was built. The effects of telluric currents appear to be of most importance in affecting electronic equipment related to operational monitoring and corrosion control rather than in producing specific serious corrosion problems.

Viewing a pipeline as a man-made part of the natural environment, it is noteworthy to mention the 30-A current reported by Peabody (1979) to cross the Panama Isthmus, from ocean to ocean, a current that also changes direction. Such specific currents can produce corrosion failures at some ocean terminals of the pipelines, even before the pipeline is in operation. Such problems can be avoided most simply by suitable separate ground connections (Peabody, 1979).

The Alaskan pipeline has been the subject of careful investigations, principally because of its location across the auroral zone (Hessler, 1974). Campbell (1978, 1979, 1980) and Campbell and Zimmerman (1980) provided a comprehensive account of the problem and concluded that the current I expected to flow within the pipeline is related to the geomagnetic index A_p by the linear relationship $I = 5.0 A_p - 0.7$. Based on the statistics of occurrence of the A_p index (larger for greater geomagnetic activity), at least once a year about 600 A should be observed, 800 A should be observed at least once every 2 years, and 1200 A should be observed at least once every 5 years. The dimensions of the Alaskan pipeline (diameter of ~ 1.22 m, a mean wall thickness of ~ 1.30 cm, a resistance per unit length of $\sim 2.81 \times 10^{-6}\ \Omega/\text{m}$, and an end-to-end total resistance of 3.6 Ω; Campbell, 1979) suggest that it is a large man-made conductor that is capable of significantly affecting the local natural regime of telluric currents.

Railways

Pollution by artificially produced telluric currents associated with railway operations have been investigated from several viewpoints. Burbank (1905) reported the effects in 1890 of the South London Electric Railway on the Earth current records being made at Greenwich. The nuisance for geomagnetic observations of telluric currents associated with return currents from dc electrified railways has perhaps been the most widely investigated effect (La Cour and Hoge, 1937; Rössiger, 1942; Yanagihara and Oshima, 1953; Mikerina, 1962; Yanagihara and Yokouchi, 1965; Yanagihara, 1977). The spatial extent within the ground of telluric currents from railway operations has been investigated by Kovalevskiy *et al.* (1961) in the southern Urals. They detected telluric current pulses with periods between a few seconds and 20 minutes and amplitudes of about 0.5 to 3 V/km. They found the effects to drop off rapidly within 10 to 15 km from the railway, although still being dominant over natural telluric currents at 30 km, and still detectable at 60 km (where the measurements stopped). Meunier (1969), following a previous investigation by Dupouy (1950), detected telluric current effects related to a specific operation (lowering and raising the pantograph) of the Paris-Toulouse railway at 115 km distance from the railroad. This effect, in fact, can sometimes be detected on the magnetograms from the Chambon-la-Forêt observatory. An example of the effect is shown in Figure 16.12. Jones and Kelly (1966) detected Earth currents in Montreal, clearly correlated with a dc powered railway some 20 km distant.

F. Molina (private communication, Osservatorio

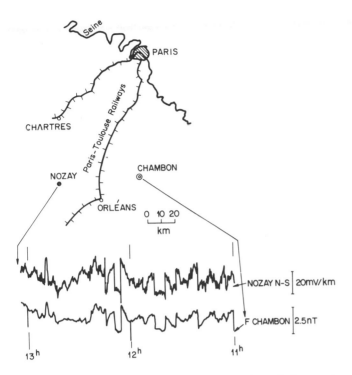

FIGURE 16.12 Artificial Earth currents (north-south direction) measured (Fournier and Rossignol, 1974) at Nozay (France) and concurrent fluctuations in the total magnetic field measured at Chambon-la-Forêt, on opposite sides of the Paris-Toulouse electrified railway line.

Geofisico Monte Parzio, Italy), in close examination of standard magnetogram records from the observatory at L'Aquila, has found a difference in the width of the trace depending on whether the Italian railways are on strike or not. In the former case, the Z-component trace is about a factor of 2 less thick than during normal operations. The noise introduced appears to be about 0.5 nT. The closest railway is about 30 km away.

A most impressive telluric current effect (Fraser-Smith and Coates, 1978; Fraser-Smith, 1981) in the San Francisco Bay area has been produced by BART (the San Francisco Bay Area Rapid Transit system). ULF waves (frequency less than 5 Hz) are observed, having energy at a frequency predominantly below about 0.3 Hz. Their amplitudes are at least ten times greater than the natural background environment, i.e., they are comparable with the levels reached during great geomagnetic storms. The effect originated by BART appears to occur over an area of about 100 km^2.

A similar effect has been detected by Lowes (1982) in Newcastle upon Tyne (U.K.), produced by the dc rapid transit underground railway system. F. J. Lowes (University of Newcastle upon Tyne, private communication, 1985) also noted that when the system starts up in the morning he can follow individual train movements over about 12 km of track before there is too much superimposition of the signals.

Corrosion

Corrosion in buried metal structures (in addition to pipelines) is significantly enhanced by the occurrence of telluric currents, presumably via electrolytic processes. This is a well-known phenomenon to people routinely working on repairs of telephone cables or of pipes (for water or otherwise). Severe damage comes mainly from man-made telluric currents when the conductors are buried close to dc electrified railways or tramways. A simple insulating coating, provided that it has no holes, appears to be the best protection. The problem is discussed to some extent by Peabody (1979). A much older reference (given by Kovalevskiy et al., 1961) is Tsikerman (1960). The problem can exist also for buried powerlines that have, unlike aerial power lines, some relevant problems of heat flow (Salvage, 1975).

APPLICATIONS OF TELLURIC CURRENT MEASUREMENTS

Listing all possible applications of telluric current measurements is presumptuous and almost impossible. A tentative scheme is given here, which, perhaps, can provide a first approach to such a complex topic.

Detection of Electromagnetic Signals from Space

The Earth (including natural conductivity structures and man-made systems, such as communication cables and powerlines) can be treated as a receiving antenna, useful for monitoring the external origin electromagnetic fields. Communication cables of varying length can provide information on the spatial scale as a function of frequency of the external signal. See Meloni et al. (1983) for additional discussions of this point.

Prospecting of Underground Structures

Prospecting of underground structures is the most developed application of telluric currents. The methodology is quite extensive. There are two principal approaches, viz., Magneto-Tellurics (MT), which uses measurements of the two horizontal components of both the geomagnetic and the geoelectric fields, and Geomagnetic Depth Sounding (GDS), which uses measurements of all three components of the geomagnetic field. "Active" methods (see, e.g., Keller, 1976; Ward, 1983) make use of man-made electromagnetic fields. Such methods are well suited for shallow prospecting but can hardly be applied to deeper layers, because of the skin-depth phenomenon. Such required long-period electromagnetic waves cannot be generated practically. Long powerlines have been used for generating electromagnetic induction fields for prospecting purposes (e.g., Gill and MacDonald, 1967).

A distinction should be made between prospecting techniques that use the static geomagnetic field of the Earth and techniques that use electromagnetic induction effects. In handling the standard aeromagnetic and oceanographic magnetic surveys, whose purposes are to understand the static fields, the time-varying fields recorded at a ground-based site "close" to the area of the survey must be subtracted from the air or the ocean signal. This introduces some errors, whose actual values are often difficult to estimate (e.g., Reford, 1979). The time-varying component from such surveys can be used for geomagnetic depth-sounding studies (Gregori and Lanzerotti, 1979a).

Deep-Earth Studies

Telluric currents are likely eventually to be important tools for prospecting the deep structure of the Earth, thus providing valuable complementary information to that provided by seismic waves. A great advantage of GDS methods is that, while many studies concentrate on magnetic storm events, the studies can also be carried out using inducing signals during more quiet times, signals that are always in existence. The use of induced currents from natural electromagnetic waves for deep-Earth research is being pursued actively in a number of countries, particularly the Soviet Union. Recent reviews include books by Rokityansky (1982), Patra and Mallick (1980), Parkinson (1982), and Berdichevsky and Zhdanov (1984).

Tidal Phenomena and Water Flows

There are three types of tides: atmospheric, oceanic, and solid Earth. Atmospheric tides generate a large part of the external-origin inducing field of long period; oceanic tides produce a time-varying geomagnetic field associated with water flows; and solid-Earth tides can similarly produce a geomagnetic field because they can produce an eventual water flow that will produce a magnetic field. Theoretical and observational aspects of the phenomena have been discussed by Meloni *et al.* (1983).

Within the past decade extensive use of shorter undersea cables (such as those across the Dover Strait; e.g., Prandle, 1978) has been made for studies of tidal oscilla-

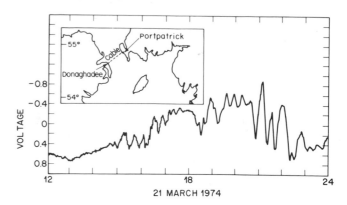

FIGURE 16.13   Cable voltage on the Donaghadee-Port Patrick cable on a geomagnetically disturbed day (from Prandle and Harrison, 1975).

tions and water flow. Cables across the Irish Sea have been used for such studies (for example, Prandle and Harrison, 1975; Prandle, 1979). Geomagnetic disturbances can affect the measurement capabilities and, hence, results of such a cable-monitoring system. The data presented in Figure 16.13 are from chart recordings of the cable voltage on the Donoghadee-Port Patrick cable on a day of geomagnetic disturbances (Prandle and Harrison, 1975). The low-frequency variation in the voltage, spanning the record, is produced by tidal flow. The higher-frequency variations, produced by geomagnetic storm induction of currents in the cable, obscure the variations in such a manner that the data cannot be used reliably for water-flow information on such a day.

Earlier, Wertheim (1954), in studying water flow across the Florida straits using the Key West-Havana cable, found occasional rapid variations in the cable voltage. He attributed these to geomagnetic effects and tried to model them using magnetometer data from the San Juan Observatory. Recent work in studies of the Florida current were reported by Larsen and Sanford (1985).

Earth's Astronomical Motion

The variation in the length of the day and the displacements of the positions of the geographic poles are among the most precise and fascinating topics in geophysics and have now become a vast discipline. The problem, however, of a possible role of telluric currents in producing a braking or an acceleration in the Earth's rotation or in displacing the Earth's poles appears still basically unsolved and/or is not considered important by many. This is discussed in some detail by Meloni *et al.* (1983) (and references therein), where also the possible use of a transatlantic communication cable is discussed as a possible experimental device to detect such an effect.

Earthquakes, Volcanoes, and Geodynamics

Since telluric currents are excellently suited for deep-Earth investigations, they are in principle also suitable for monitoring long-scale time variations as well. The most investigated aspect from this viewpoint is concerned with earthquake precursors (e.g., review by Honkura, 1981, and references therein). A clear distinction should be made among three different, possible types of phenomena: (1) geomagnetic effects that can presumably be said to be "very shallow" and are likely related to piezomagnetism, following changes in local stresses in the upper crust, which is an effect strictly local and can completely change in a distance of a few kilometers or less; (2) "shallow" effects that can be detected by ground resistivity changes or by suitably short-period MT or GDS investigations; and (3) "deep" or "very deep" effects that can be most suitably detected by means of long-period GDS investigations. This latter category of effects is more strictly related to telluric currents than are the first two types of effect.

Additional possible applications in this area include (1) slowly varying effects correlated with geodynamic and tectonic features, (2) shallow effects related to magma migration in volcanic areas, and (3) the monitoring of temporal variations in underground structures as related to fluid extraction (or reinjection). The shallow effects could, however, possibly be better detected by use of man-made electromagnetic fields than by means of natural fields. Telluric currents are also well suited for investigating ocean-bottom and geothermal areas (see Law, 1983; Berktold, 1983).

Communications

Within the last 10 to 15 years suggestions have been made that a natural waveguide in the Earth's crust, composed of the insulating layer of dry rocks sandwiched between the upper hydrated conducting and the underlying conducting hot layer, could be used for communication purposes. This suggestion, however, does not seem to have been followed by any known application. Existing literature is referenced in Gregori and Lanzerotti (1982). A practical problem is certainly concerned with the spatial nonuniformity of such a waveguide, the width and depth of which is undoubtedly widely varying (e.g, compare a cratonic area with a mid-oceanic ridge area) and is essentially unknown in many regions.

In a similar fashion, magma chambers in mid-oceanic ridges can be considered the natural "equivalent" of man-made submarine communication cables. The conductivities are such that the mid-Atlantic ridge is equivalent to about 1000 such cables in parallel (see Gregori and Lanzerotti, 1982).

An interesting communication experiment related to artificial telluric currents was reported by Fraser-Smith et al. (1977). They operated, as a transmitting antenna (using a simple car battery), a circuit loop composed of seawater encircling a small peninsula in a nearly enclosed area.

Biological Effects

The response of living species to electromagnetic fields (such fields being either responsible for, or a consequence of, telluric currents) is a difficult but important problem. Several examples discussed in the literature include the induced currents in a tree produced by geomagnetic fluctuations (Fraser-Smith, 1978) and the use of magnetic fields for orientation by aquatic bacteria (e.g., Blakemore, 1975) and by migrating birds (e.g., Moore, 1977; Larkin and Sutherland, 1977; Alerstam and Högstedt, 1983; Beason and Nichols, 1984). Telluric currents could play a role in some control of fish (e.g., Leggett, 1977; Kalmijn, 1978; Brown et al., 1979; Fainberg, 1980; Fonarev, 1982). Magnetite crystals have been reported as isolated from a sinus in the yellowfin tuna (Walker et al., 1984). Enhanced DNA synthesis has been reported for human fibroblasts exposed to magnetic-field fluctuations with frequencies and amplitudes similar to many geomagnetic occurrences (Liboff et al., 1984). The entire area is fraught with controversy, particularly that related to magnetic effects, and has been reviewed by Parkinson (1982) and commented on by Thomson (1983).

CONCLUSIONS

Historically, telluric currents were intensively investigated in the second half of the nineteenth century, particularly because of their influences on long telegraph conductors (see Appendix). The intrinsic difficulties encountered in obtaining fundamental understanding, basically related to the several causes that can be coresponsible for the observed effects, discouraged geophysicists from pursuing such investigations vigorously. However, with respect to a century ago, large improvements have been made in a number of areas, including recording techniques, density of available observations, international data exchanges, computational facilities, mathematical methodologies, and general geophysical understanding. Hence, research in this area would appear to be poised for achieving significant new understanding.

Investigations using, and studying, the telluric currents suffer from three principal drawbacks: (a) the spatial coverage by the recording equipment is often too sparse compared to the extent and spatial gradients of the phenomena under investigation; (b) there is not general agreement on the experiment and analysis methodologies, often leading to difficulties when comparing the results from different investigations; and (c) the use, within the general understanding of deep geophysical structures, of the information provided by electromagnetic techniques is often neglected. It is hoped that these deficiences will be ameliorated in future telluric current work.

Telluric currents are a relevant part of our electromagnetic environment, both as a consequence of the time-varying natural electromagnetic field and as a consequence of moving seawater. Human interactions with telluric currents and their related effects are definitely important, even though such interactions are only incompletely understood. A considerable effort on geomagnetic source characteristics and Earth conductivity characteristics are required before a satisfactory and comprehensive understanding can be achieved. The critical role of telluric currents within the geophysical environment, encompassing such areas as the Earth's astronomical motion, to current channeling, to eventual implications for ocean-water motion, still appears as a challenging frontier. As well, further assessments must be made on the actual role of telluric currents, as well as of the geomagnetic field *in toto*, on biological systems. The field is multidisciplinary and fascinating. Substantial achievements can be expected in the near future.

ACKNOWLEDGMENTS

We thank the referees and F. J. Lowes for helpful comments and references.

REFERENCES

Achace, J., J. L. LeMouel, and V. Courtillot (1981). Long-period geomagnetic variations and mantle conductivity: An inversion using Bailey's method, *Geophys. J. R. Astron. Soc. 65*, 579-601.

ACRES Consulting Services Limited (1975). Study of the disruption of electric power systems by magnetic storms, *Earth Phys. Branch Open File 77-19*, Niagara Falls, Ontario, Department of Energy, Mines and Resources, Ottawa.

Airy, G. B. (1868). Comparison of magnetic disturbances recorded by self-registering magnetometers at the Royal Observatory, Greenwich, with magnetic disturbances described from the corresponding terrestrial galvanic currents recorded by the self-registering galvanometers of the Royal Observatory, *Phil. Trans. R. Soc. 158A*, 465-472.

Akasofu, S.-I., and J. D. Aspnes (1982). Auroral effects on power transmission line systems, *Nature 295*, 136-137.

Akasofu, S.-I., and R. P. Merritt (1979). Electric currents in power transmission line induced by auroral activity, *Nature 279*, 308-310.

Akasofu, S.-I., S. Chapman, and A. B. Meinel (1966). The aurora, *Handb. Phys. 49*(1), 1-158.

Alabi, A. D. (1983). Magnetometer array studies, *Geophys. Surv. 6*, 153-172.

Albertson, V. D., and J. A. Van Baelen (1970). Electric and magnetic fields at the Earth's surface due to auroral currents, *IEEE Trans. Power Appar. Syst. PAS-89*(4), 578-584.

Albertson, V. D., S. C. Tripathy, and R. E. Clayton (1970). Electric power systems and geomagnetic disturbances, in *Proceedings of Midwest Power Symposium*, October 29-30.

Albertson, V. D., J. M. Thorson, Jr., R. E. Clayton, and S. C. Tripathy (1973). Solar-induced currents in power systems: Causes and effects, *IEEE Trans. Power Appar. Syst. PAS-92*, 471-477.

Albertson, V. D., J. M. Thorson, Jr., and S. A. Miske, Jr. (1974). The effects of geomagnetic storms on electrical power systems, *IEEE Trans. Power Appar. Syst. PAS-93*, 1031-1044.

Alerstam, T., and G. Högstedt (1983). The role of the geomagnetic field in the development of birds' compass sense, *Nature 306*, 463-465.

Anderson, C. W. (1979). Magnetic storms and cable communications, in *Solar System Plasma Physics*, L. J. Lanzerotti, C. F. Kennel, and E. N. Parker, eds., North-Holland, Amsterdam, pp. 323-327.

Anderson, C. W., L. J. Lanzerotti, and C. G. Maclennan (1974). Outage of the L-4 system and the geomagnetic disturbances of August 4, 1972, *Bell Syst. Tech. J. 53*, 1817.

Anderson, D. L. (1984). The Earth as a planet: Paradigms and paradoxes, *Science 223*, 347-355.

Ashour, A. A. (1950). The induction of electric currents in a uniform circular disk, *Q. J. Mech. Appl. Math. 3*, 119-128.

Aspnes, J. D., R. D. Merritt, and S. I. Akasofu (1981). The effects of geomagnetically induced current on electric power systems, *Northern Eng. 13*(3), 34.

Axe, G. A. (1968). The effects of Earth's magnetism on submarine cables, *Electr. Eng. J. 61*(1), 37-43.

Babour, K., and J. Mosnier (1980). Direct determination of the characteristics of the currents responsible for the magnetic anomaly of the Rhinegraben, *Geophys. J. R. Astron. Soc. 60*, 327-331.

Barber, N., and M. S. Longuet-Higgins (1948). Water movements and Earth currents: Electrical and magnetic effects, *Nature 161*, 192-193.

Barlow, W. H. (1849). On the spontaneous electrical currents observed in the wires of the electric telegraph, *Phil. Trans. R. Soc. 61A*, 61-72.

Battelli, A. (1888). Sulle correnti telluriche, *Atti R. Accad. Naz. Lincei 4*, 25.

Beamish, D. (1982). Anomalous geomagnetic variations on the island of South Georgia, *J. Geomagn. Geoelectr. 34*, 479-496.

Beamish, D., R. C. Hewson-Browne, P. C. Kendall, S. R. C. Malin, and D. A. Quinney (1980). Induction in arbitrarily shaped oceans IV, Sq for a simple case, *Geophys. J. R. Astron. Soc. 60*, 435-443.

Beason, R. C., and J. E. Nichols (1984). Magnetic orientation and magnetically sensitive material in a transequatorial migratory bird, *Nature 309*, 151-153.

Berdichevsky, M. N., and E. B. Fainberg (1972). Possibility of experimental separation of the variable geomagnetic field into poloidal and toroidal parts, *Geomagn. Aeron. 12*, 826-830.

Berdichevsky, M. N., and E. B. Fainberg (1974). Separation of the field of Sq variations into poloidal and toroidal parts, *Geomagn. Aeron. 14*, 315-318.

Berdichevsky, M. N., and M. S. Zhdanov (1984). *Advanced Theory of Deep Geomagnetic Sounding*, Elsevier, Amsterdam, 408 pp.

Berktold, A. (1983). Electronmagnetic studies in geothermal regions, *Geophys. Surv. 6*, 173-200.

Blackwell, O. B. (1928). Transatlantic telephony—the technical problem, *Bell System Tech. J. 7*, 168-186.

Blakemore, R. P. (1975). Magnetoactive bacteria, *Science 190*, 377-379.

Blavier, E. E. (1884). Sur les courants telluriques, *C. R. Hebd. Séana Acad. Sci. 98*, 1043-1045.

Boerner, W.-M., J. B. Cole, W. R. Goddard, M. Z. Tarnawecky, L. Shafai, and D. H. Hall (1983). Impacts of solar and auroral storms on power line systems, *Space Sci. Rev. 35*, 195-205.

Booker, J. R., and G. Hensel (1982). Nature's hidden power line, *Sci. Digest 90*, 18.

Bown, R. (1930). Transoceanic telephone service—Short wave transmission, *Bell Syst. Tech. J. 9*, 258-269.

Bown, R. (1937). Transoceanic radiotelephone development, *Proc. IRE 25*, 1124-1135.

Brooks, J. (1959). A reporter at large: A subtle storm, *New Yorker* (Feb. 7), 39-69.

Brown, H. R., O. B. Ilyinsky, V. M., Muravejko, E. S. Corshkov, and G. A. Fonarev (1979). Evidence that geomagnetic variations can be detected by Lorenzian ampullae, *Nature 277*, 648-649.

Burbank, J. E. (1905). Earth-currents and a proposed method for their investigation, *Terr. Magn. Atmos. Electr. 10*, 23-49.

Camfield, P. A. (1981). Magnetometer array study in a tectonically active region of Quebec, Canada, *Geophys. J. R. Astron. Soc. 65*, 553-570.

Campbell, W. H. (1976). Spatial distribution of the geomagnetic spectral composition for disturbed days, *J. Geomagn. Geoelectr. 28*, 481-496.

Campbell, W. H. (1978). Induction of auroral zone electric currents within the Alaska pipeline, *Pure Appl. Geophys. 116*, 1143-1173.

Campbell, W. H. (1979). The Alaskan pipeline, in *Solar System Plasma Physics*, L. J. Lanzerotti, C. F. Kennel, and E. N. Parker, eds., North-Holland, Amsterdam, pp. 352-356.

Campbell, W. H. (1980). Observation of electric currents in the Alaska oil pipeline resulting from auroral electrojet current sources, *Geophys. J. R. Astron. Soc. 61*, 437-499.

Campbell, W. H., and J. E. Zimmerman (1980). Induced electric currents in the Alaskan oil pipeline measured by gradient fluxgate and Squid magnetometers, *IEEE Trans. Geosci. Electron. GE-18*, 244-250.

Cermak, V. (1982). A geothermal model of the lithosphere and a map of the thickness of the lithosphere on the territory of USSR, *Izv. Akad. Nauk SSR Ser. Phys. Solid Earth 18*, 18-27.

Chan, G. H., H. W. Dosso, and L. K. Law (1981a). Electromagnetic induction in the San Juan Bay region of Vancouver island, *Phys. Earth Planet. Inter. 27*, 114-121.

Chan, G. H., H. W. Dosso, and W. Nienaber (1981b). An analogue model study of electromagnetic induction in the Queen Charlotte Island region, *J. Geomagn. Geoelectr. 33*, 587-606.

Chapman, D. S., and H. N. Pollack (1975). Global heat flow: A new look, *Earth Planet. Sci. Lett. 28*, 23-22.

Chapman, D. S., and H. N. Pollack (1977). Regional isotherms and lithospheric thickness, *Geology 5*, 265-268.

Chapman, S., and J. Bartels (1940). *Geomagnetism* (2 volumes), Clarendon Press, Oxford, 1049 pp.

Clement, K. T. (1860). *Das grosse Nordlicht in der Nacht zum 29 August 1859 und die Telegraphenverwirrung in Nord-Amerika und Europa*, Hamburg, pp. 1-121.

Davidson, M. J., and J. R. Heirtzler (1968). Spatial coherence of geomagnetic rapid variations, *J. Geophys. Res. 73*, 2143-2162.

Davidson, W. F. (1940). The magnetic storm of March 24, 1940—effects in the power system, *Bull. Edison Electr. Inst. 365*.

DeBeer, J. H., R. M. J. Huyssen, S. J. Joubert, and J. S. V. Van Zijl (1982). Magnetometer array studies and deep Schlumberger sounding in the Damara orogenic belt, South West Africa, *Geophys. J. R. Astron. Soc. 70*, 11-23.

De Laurier, J. M., E. R. Niblett, F. Plet, and P. A. Camfield (1980). Geomagnetic depth sounding over the central Arctic islands, Canada, *Can. J. Earth Sci. 17*, 1642-1652.

Dupouy, G. (1950). Perturbation de champ magnétique terestre et des courants telluriques par les chemins de fer électrofiés, *Ann. Geophys. 6*, 18-50.

Fainberg, E. B. (1980). Electromagnetic induction in the world ocean, *Geophys. Surv. 4*, 157-171.

Fainberg, E. B. (1983). Global geomagnetic sounding, Preprint No. 50a, pp. 1-66, Izmiran, Moscow.

Fainberg, E. B., and W. A. Sidorov (1978). Integrated conductivity of the sedimentary cover and water shell of the Earth [in Russian], Izmiran, with 5 maps, Nauka, Moscow, pp. 1-11.

Fischer, G. (1979). Electromagnetic induction effects at an ocean coast, *Proc. IEEE 67*, 1050-1060.

Fonarev, G. A. (1982). Electromagnetic research in the ocean, *Geophys. Surv. 4*, 501-508.

Fournier, H., and J. C. Rossignol (1974). A magnetotelluric experiment in Nozay-en-Dunois (Eure-et-Loir, France): Results, interpretation and critical study, *Phys. Earth Planet. Inter. 8*, 13-18.

Fraser, D. C. (1965). Magnetic fields of ocean waves, *Nature 206*, 605-606.

Fraser-Smith, A. C. (1978). ULF tree potentials and geomagnetic pulsations, *Nature 271*, 641-642.

Fraser-Smith, A. C. (1981). *Adv. Space Res. 1*, 455.

Fraser-Smith, A. C., and D. B. Coates (1978). Large-amplitude ULF electromagnetic fields from BART, *Radio Sci. 13*, 661-668.

Fraser-Smith, A. C., O. G. Villard, Jr., and D. M. Dubenick (1977). A study of the "Peninsula method" for the controlled artificial generation of ULF waves in the ionosphere and magnetosphere, Technical Report No. 4207-7-SEL-77-041, Radioscience Laboratory, Stanford Electronics Labs., Stanford Univ., pp. 1-35.

Germaine, L. W. (1942). The magnetic storm of March 24, 1940—effects in the communication system, *Bull. Edison Electr. Inst.*, 367-368.

Gill, P. J., and W. J. P. MacDonald (1967). Large scale earth resistivity experiment in New Zealand, *Nature 216*, 1195-1197.

Gish, O. H. (1936a). Electrical messages from the Earth, their reception and interpretation, *J. Wash. Acad. Sci. 26*, 267-289.

Gish, O. H. (1936b). Electrical messages from the Earth, their reception and interpretation, *Sci. Mon. N.Y. 43*, 47-57.

Gorely, K. I., and O. I. Uvarov (1981). Currents of geomagnetic disturbances in power lines of the Norilsk district [in Russian], *Issled. Geomagn. Aeron. Fiz. Solntsa 53*, 220-223.

Gough, D. I., and M. R. Ingham (1983). Interpretation methods for magnetometer arrays, *Rev. Geophys. Space Phys. 21*, 805-827.

Graves, J. (1873). On Earth currents, *J. Telegr. Eng. 2*, 102-123.

Green, A. R. (1977). The evolution of the Earth's crust and sedimentary basin development, in *The Earth's Crust*, J. G. Heacock, ed., Geophysical Monograph 20, Am. Geophysical Union, Washington, pp. 1-17.

Greener, J. G., and D. M. Schlapp (1979). A study of day-to-day variability of Sq over Europe, *J. Atmos. Terr. Phys. 41*, 217-223.

Gregori, G. P., and L. J. Lanzerotti (1979a). Geomagnetic depth sounding by means of oceanographic and aeromagnetic surveys, *Proc. IEEE 67*, 1029-1034.

Gregori, G. P., and L. J. Lanzerotti (1979b). The effect of coast lines on geomagnetic measurements—a commentary, *Riv. Ital. Geofis. Sci. Affini. 5*, 81-86.

Gregori, G. P., and L. J. Lanzerotti (1980). Geomagnetic depth sounding by induction arrow representation: A review, *Rev. Geophys. Space Phys. 18*, 203-209.

Gregori, G. P., and L. J. Lanzerotti (1982). The electrical conductivity structure in the lower crust *Geophys. Surv.*, 467-499.

Hansteen, C. (1860). The great auroral expedition of August 28, to September 4, 1859, 4th article, Observations at Christiania, *Am. J. Sci. Arts 29*, 386-399.

Harang, L. (1951). *The Aurorae*, Chapman and Hall, Ltd., London, 163 pp.

Hermance, J. F. (1983). Electromagnetic induction studies, *Rev. Geophys. Space Phys. 21*, 652-665.

Hessler, V. P. (1974). Causes, recording techniques, and characteristics of telluric currents, *Corrosion 56*, 1-18.

Hewson-Browne, R. C. (1981). The numerical solution of oceanic electromagnetic induction problems, *Geophys. J. R. Astron. Soc. 67*, 235-238.

Hewson-Browne, R. C., and P. C. Kendall (1981). Electromagnetic induction in the Earth in electrical contact with the oceans, *Geophys. J. R. Astron. Soc. 66*, 333-343.

Hobbs, B. A., and A. M. M. Brignall (1976). A method for solving general problems of electromagnetic induction in the oceans, *Geophys. J. R. Astron. Soc. 45*, 527-542.

Hobbs, B. A., and G. J. K. Dawes (1980). The effect of a simple model of the Pacific ocean on Sq variations, *J. Geomagn. Geoelectr. 32*, Suppl. I, SI59-SI66

Hohmann, G. W. (1983). Three-dimensional EM modeling, *Geophys. Surv. 6*, 27-53.

Honkura, Y. (1974). Electrical conductivity anomalies beneath the Japan Arc, *J. Geomagn. Geoelectr. 26*, 147-171.

Honkura, Y. (1981). Electric and magnetic approach to earthquake prediction, in *Current Research in Earthquake Prediction*, T. Rikitake, ed., pp. 301-383.

Honkura, Y., N. Isezaki, and K. Yaskawa (1981). Electrical conductivity structure beneath the northwestern Philippine sea as inferred from the island effect on Mimami-Daito island, *J. Geomagn. Geoelectr. 33*, 365-377.

Hutton, V. R. S. (1976). The electrical conductivity of the earth and planets, *Rep. Prog. Phys. 39*, 487-572.

Jones, A. G. (1983). The problem of current channeling: A critical review, *Geophys. Surv. 6*, 79-122.

Jones, F. W., and A. M. Kelly (1966). Man-made telluric micropulsations, *Can. J. Phys. 44*, 3025-3031.

Kalmijn, A. J. (1978). Electric and magnetic sensory world of sharks, skates and rays, in *Sensory Biology of Sharks: Skates and Rays*, H. S. Hodgson and R. F. Matthewson, eds., U.S. Government Printing Office, Washington, D.C.

Keller, G. V. (1966). Electrical properties of rocks and minerals, in *Handbook of Physical Constants*, S. P. Clark, Jr., ed., Memoir 97, Geological Society of America, Boulder, Colo., pp. 553-557.

Keller, G. V. (1976). Foreword, *IEEE Trans. Geosci. Electron. GE-14*, 218-220.

Kendall, P. C., and D. A. Quinney (1983). Induction in the oceans, *Geophys. J. R. Astron. Soc 74*, 239-255.

Kirkwood, S. C., V. R. S. Hutton, and J. Sik (1981). A geomagnetic study of the Great Glen Fault, *Geophys. J. R. Astron. Soc. 66*, 481-490.

Klein, M., P. Louvet, and P. Morat (1975). Measurement of electromagnetic effects generated by swell, *Phys. Earth Planet. Inter. 10*, 49-54.

Kovalevskiy, I. V., N. V. Mikerina, V. V. Novysh, and O. P. Gorodnicheva (1961). Distribution of the Earth currents from an electrified railroad in the Southern Urals, *Geomag. Aeron. 1*, 723-726.

Kurtz, R. D., E. R. Niblett, M. Chouteau, W. J. Scott, and L. R. Newitt (1981). An anomalous electrical resistivity zone near Ste-Mathilde, Quebec, Contr. 922, Earth Phys. Branch, Department of Energy, Mines, and Resources, Ottawa, pp. 56-67.

La Cour, D., and E. Hoge (1937). Note on the effects of electrification of railways passing in the area of the Copenhagen Observatory, *Bull. Int. Assoc. Terr. Magn. Atmos. Electr. 10*, 302-306.

Langel, R. A. (1982). Results from the MAGSAT mission, *Johns Hopkins APL Tech. Dig. 3*, 307-324.

Lanzerotti, L. J. (1979a). Impacts of ionospheric/magnetospheric processes on terrestrial science and technology, in *Solar System Plasma Physics* (3), L. J. Lanzerotti, C. F. Kennel, and E. N. Parker, eds., North-Holland, Amsterdam, pp. 317-363.

Lanzerotti, L. J. (1979b). Geomagnetic influences on man-made systems, *J. Atmos. Terr. Phys. 41*, 787-796.

Lanzerotti, L. J. (1979c). Geomagnetic disturbances and technological systems, *National Comm. Confer.* (Nov.), 7.1.1-7.1.4.

Lanzerotti, L. J. (1983). Geomagnetic induction effects in ground-based systems, *Space Sci. Rev. 34*, 347-356.

Lanzerotti, L. J., C. G. Maclennan, and C. Evans (1977). Magnetic fluctuations and ionosphere conductivity changes measured at Siple Station, *Antarct. J. 12*, 186-189.

Larkin, R. P., and P. J. Sutherland (1977). Migrating birds respond to Project Seafarer's electromagnetic field, *Science*, 777-779.

Larsen, J. C., and T. B. Sanford (1985). Florida current volume transports from voltage measurements, *Science 227*, 302-304.

Law, L. K. (1983). Marine electromagnetic research, *Geophys. Surv. 6*, 123-135.

Leggett, W. C. (1977). The ecology of fish migration, *Ann. Rev. Ecol. Syst. 8*.

Le Mouël, J. L., and M. Menvielle (1982). Geomagnetic variation anomalies and deflection of telluric currents, *Geophys. J. R. Astron. Soc. 68*, 575-587.

Liboff, A. R., T. Williams, Jr., D. M. Strong, and R. Wistar, Jr. (1984). Time-varying magnetic fields: Effect on DNA synthesis, *Science 223*, 818-820.

Lilley, F. E. M., and D. V. Woods (1978). The channeling of natural electric currents by orebodies, *Bull. Austr. Soc. Explor. Geophys. 9*, 62-63.

Lowes, F. J. (1982). On magnetic observations of electric trains, *Observatory 102*(1047), 44.

Malin, S. R. C. (1970). Separation of lunar daily geomagnetic variations into parts of ionospheric and oceanic origin, *Geophys. J. R. Astron. Soc. 21*, 447-455.

Malin, S. R. C. (1973). World-wide distribution of geomagnetic tides, *Phil. Trans. R. Soc. London 274A*, 551-593.

Matsushita, S. (1967). Solar quiet and lunar daily variation fields, in *Physics of Geomagnetic Phenomena*, S. Matsushita and W. H. Campbell, eds., Academic Press, New York, pp. 301-424.

Matsushita, S., and H. Maeda (1965a). On the geomagnetic solar quiet daily variation field during the IGY, *J. Geophys. Res. 70.*, 2535-2558.

Matsushita, S., and H. Maeda (1965b). On the geomagnetic lunar daily variation field, *J. Geophys. Res. 70*, 2559-2578.

Maxwell, J. C. (1984). What is the lithosphere? *EOS 65*(17), 321-325.

McKinnon, J. (1972). The August 1972 solar activity and related geophysical effects, NOAA Space Environment Laboratory Report, December.

Meloni, A., L. J. Lanzerotti, and G. P. Gregori (1983). Induction of currents in long submarine cables by natural phenomena, *Rev. Geophys. Space Phys. 21*, 795-803.

Meunier, J. (1969). Sondage électrique de le croûte utilisant des courants de retour industriels, *C. R. Acad. Sci. Ser. B 268*, 514-517.

Mikerina, N. V. (1962). The study of interference at the Voyeykovo magnetic observatory, *Geomagn. Aeron. 2*, 941-944.

Miller, S. L. (1974). The nature and occurrence of clathrate hydrates, in *Natural Gases in Marine Sediments*, I. R. Kaplan, ed., Plenum Press, New York, pp. 151-177.

Miyakoshi, J. (1980). Electrical conductivity structure beneath the Japan island arc by geomagnetic induction study, *Adv. Earth Planet. Sci. 8*, 153-161.

Moore, F. R. (1977). Geomagnetic disturbance and the orientation of nocturnally migrating birds, *Science 196*, 682-684.

Moureaux, T. H. (1896). L'installation d'une station d'étude des courants telluriques a l'Observatoire du Parc Saint Maur, *Soc. Met. France Ann. 4*, 25-38.

Nienaber, W., R. D. Hibbs, H. W. Dosso, and L. K. Law (1982). An estimate of the conductivity structure for the Vancouver Island region from geomagnetic results, *Phys. Earth Planet. Inter. 27*, 300-305.

Oxburgh, E. R. (1981). Heat flow and differences in lithosphere thickness, *Phil. Trans. R. Soc. London A301*, 337-346.

Parker, R. L. (1983). The magnetotelluric inverse problem, *Geophys. Surv. 6*, 5-25.

Parkinson, W. D. (1975). The computation of electric currents induced in the oceans, *J. Geomagn. Geoelectr. 27*, 33-46.

Parkinson, W. D. (1982). *Introduction to Geomagnetism*, Elsevier, Amsterdam, 433 pp.

Parkinson, W. D., and F. W. Jones (1979). The geomagnetic coast effect, *Rev. Geophys. Space Phys. 17*, 1999-2015.

Patra, H. P., and K. Mallick (1980). Geosounding principles, 2, in *Time Varying Geoelectric Soundings*, Elsevier, Amsterdam, pp. 1-419.

Paulikas, G. A., and L. J. Lanzerotti (1982). Impact of geospace on terrestrial technology, *Astronaut. Aeronaut.*, 42-47.

Peabody, A. W. (1979). Considerations of telluric current effects on pipelines, in *Solar System Plasma Physics*, L. J. Lanzerotti, C. F. Kennel, and E. N. Parker, eds., North-Holland, Amsterdam, pp. 349-352.

Peckover, R. S. (1973). Oceanic electric currents induced by fluid convection, *Phys. Earth Planet. Inter. 7*, 137-142.

Pirjola, R. (1983). Induction in power transmission lines during geomagnetic disturbances, *Space Sci. Rev. 85*, 185-193.

Porstendorfer, G. (1976). Electromagnetic field components and magnetotelluric sounding curves above horizontally inhomogeneous or anisotropic model media (catalogue of models), in *Geoelectric and Geothermal Studies (East-Central Europe, Soviet Asia)*, KAPG Monograph, A. Adam, ed., Akademiai Kiado, Budapest, pp. 152-164.

Prandle, D. (1978). Monthly-mean residual flows through the Dover straits, 1949-1972, *J. Mar. Biol. Assoc. U.K. 58*, 965-973.

Prandle, D. (1979). Recordings of potential difference across the Port Patrick-Donaghadee submarine cable (1977/78), *Rep. Inst. Oceanog. Sci. Bidston Obs. 83*.

Prandle, D., and A. J. Harrison (1975). Recordings of the potential difference across the Port Patrick-Donaghedee submarine cable, *Rep. Inst. Oceanog. Sci. Bidston Obs. 21*.

Prescott, G. B. (1860). The great auroral exhibition of August 28 to September 4, 1859 (2nd article), *Am. J. Sci. Arts 29*, II ser. (85), 92-95.

Prescott, G. B. (1866). *History, Theory and Practice of the Electric Telegraph*, IV ed., Ticknor and Fields, Boston.

Price, A. J. (1967). Electromagnetic induction within the earth, in *Physics of Geomagnetic Phenomena 1*, S. Matsushita, and W. H. Campbell, eds., Academic Press, New York, pp. 235-298.

Reford, M. S. (1979). Problems of magnetic fluctuations in geophysical exploration, in *Solar System Plasma Physics*, L. J. Lanzerotti, C. F. Kennel, and E. N. Parker, eds., North-Holland, Amsterdam, pp. 356-360.

Reiff, P. H. (1983). Polar and auroral phenomena, *Rev. Geophys. Space Phys. 21*, 418-433.

Riddihough, R. P. (1969). A geographical pattern of daily magnetic variation over Northwest Europe, *Ann. Geophys. 25*, 739-745.

Rikitake, T. (1966). *Electromagnetic Induction and the Earth's Interior*, Elsevier, Amsterdam, 308 pp.

Rokityansky, I. I. (1982). *Geoelectromagnetic Investigation of the Earth's Crust and Mantle*, Springer-Verlag, Berlin, 381 pp.

Root, H. G. (1979). Earth-current effects on communication-cable power subsystems, *IEEE Trans. Electromagn. Compat. EMC-21*, 87-922.

Rössiger, M. (1942). Die Entstörung magnetischer Beobachtungsräume und erdmagnetischer Observatorium von Gleichstrom-Magnetfeldern der elektrischen Bahnen, *Naturwissenschaften 50/51*, 753-755.

Salvage, B. (1975). Overhead lines or underground cables: The problem of electrical power transmission, *Endeavour 34*, 3-8.

Sanders, R. (1961). Effect of terrestrial electromagnetic storms on wireless communications, *IRE Trans. Commun. Syst.* (December), 367-377.

Sato, T. (1965). Long-period geomagnetic oscillations in southern high latitudes, in *Geomagnetism and Aeronomy*, A. H. Waynick, ed., American Geophys. Union, Washington, D.C., pp. 173-188.

Saunders, H. (1880). Earth-currents, *Electrician 167* (August 21).

Saunders, H. C. (1881). Discussion of the paper by A. J. S. Adams "Earth Currents" (2nd paper), *J. Soc. Telegr. Eng. Electricians 10*, 46-48.

Schelleng, J. C. (1930). Some problems in short-wave telephone transmission, *Proc. IRE 18*, 913-938.

Schlapp, D. M. (1968). World-wide morphology of day-to-day variability of Sq, *J. Atmos. Terr. Phys. 30*, 1761-1776.

Schmucker, U. (1970). Anomalies of geomagnetic variations in the Southwestern United States, *Bull. Scripps Inst. Oceanog. 13*, 1-165.

Sebesta, D. (1979). Impact of magnetic storms probed, *Electr. World* (March 1), 52-53.

Semevskiy, R. B., K. G. Stavrov, and B. N. Demin (1978). Possibilities of investigating the magnetic fields of ocean waves by towed magnetometers, *Geomagn. Aeron. 18*, 345-347.

Sik, J. M., V. R. S. Hutton, G. J. K. Dawes, and S. C. Kirkwood (1981). A geomagnetic variation study of Scotland, *Geophys. J. R. Astron. Soc. 66*, 491-512.

Soller, D. R., R. D. Day, and R. D. Brown (1982). A new global crustal thickness map, *Tectonics 1*, 125-149.

Southwood, D. J., and W. J. Hughes (1978). Source induced vertical components in geomagnetic pulsation signals, *Planet. Space Sci. 26*, 715-720.

Srivastava, B. J. (1981). Geomagnetic results from Sabhawalla, Yangi-Bazar and Alma-Ata in relation to the asthenosphere beneath the Pamir-Himalaya, *Phys. Earth Planet. Inter. 25*, 210-218.

Srivastava, B. J., and H. Abbas (1980). An interpretation of the induction arrows at Indian stations, *J. Geomagn. Geoelectr. 32* (Suppl. II), SI187-SI196.

Störmer, C. (1955). *The Polar Aurora*, Clarendon Press, Oxford, 403 pp.

Summers, D. M. (1982). On the frequency response of induction anomalies, *Geophys. J. R. Astron. Soc. 70*, 487-503.

Thakur, N. K., M. V. Mahashabde, B. R. Arora, and B. P. Singh (1981). Anomalies in geomagnetic variations on peninsular India near Palk Straight, *Geophys. Res. Lett. 8*, 947-950.

Thomson, K. S. (1983). The sense of discovery and vice-versa, *Am. Sci. 71*, 522-524.

Tsikerman, L. Ya. (1960). Insulation of buried metal pipes to prevent corrosion, Gosstroiizdat, Moscow.

U.S. Geodynamics Committee (1983). *The Lithosphere—Report of a Workshop*, National Academy Press, Washington, D.C., 84 pp.

Varentsov, I. M. (1983). Modern trends in the solution of forward and inverse 3D electromagnetic induction problems, *Geophys. Surv. 6*, 55-78.

Varley, C. F. (1873). Discussion of a few papers on Earth currents, *J. Soc. Telegr. Eng. 2*, 111-114.

Walker, M. M., J. L. Kirschvink, S.-B. R. Chang, and A. E. Dizon (1984). A candidate magnetic sense organ in the yellowfin tuna, *Thunnus albacares*, *Science 224*, 751-753.

Ward, S. H. (1983). Controlled source electrical methods for deep exploration, *Geophys. Surv. 6*, 137-152.

Wertheim, G. K. (1954). Studies of electrical potential between Key West, Florida, and Havana, Cuba, *Trans. AGU 35*, 872.

Williams, D. J. (1979). Magnetosphere impacts on ground-based power systems, in *Solar System Plasma Physics*, L. J. Lanzerotti, C. F. Kennel, and E. N. Parker, eds., North-Holland, Amsterdam, pp. 327-330.

Winch, D. E. (1981). Spherical harmonic analysis of geomagnetic tides, 1964-1965, *Phil. Trans. R. Soc. London A303*, 1-104.

Winckler, J. R., L. Peterson, R. Hoffman, and R. Arnoldy (1959). Auroral x rays, cosmic rays, and related phenomena during the storm of February 10-11, 1958, *J. Geophys. Res. 64*, 597-610.

Wollaston, C. (1881). Discussion of the paper by A. J. S. Adams "Earth currents" (2nd paper), *J. Soc. Telegr. Eng. Electricians 10*, 50-51; 85-87.

Woods, D. V., and F. E. M. Lilley (1980). Anomalous geomagnetic variations and the concentration of telluric currents in south-west Queensland, Australia, *Geophys. J. R. Astron. Soc. 62*, 675-689.

Yanagihara, K. (1977). Magnetic field disturbance produced by electric railway, *Mem. Kakioka Magn. Obs.* (Suppl. 7), 17-35.

Yanagihara, K., and H. Oshima (1953). On the Earth-current disturbances at Haranomachi caused by the leakage current from the electric railway Fukushima-Yonezawa [in Japanese], *Mem. Kakioka Magn. Obs. 6*, 119-134.

Yanagihara, K., and T. Yokouchi (1965). Local anomaly of Earth-currents and Earth-resistivity [in Japanese], *Mem. Kakioka Magn. Obs. 12*, 105-113.

Yoshimatsu, T. (1964). Results of geomagnetic routine observations and earthquakes; locality of time changes of short period variations [in Japanese], *Mem. Kakioka Magn. Obs. 11*, 55-68.

APPENDIX: HISTORICAL DEVELOPMENT

A selective sketch of the historical development of the understanding of telluric currents follows. It is essentially impossible for the present authors to attempt to give full justice to all authors of the most recent investigations. It is particularly difficult to evaluate these recent works in a historical context. For more extensive general aspects of the subject and for recent literature references, the interested reader should refer to Dosso and Weaver (1983). A recent excellent review of primarily American work is contained in Hermance (1983), while Rokityansky (1982) and Berdichevsky and Zhdanov (1984) contain many references to Eastern literature.

1540 — First reported measurement of geomagnetic declination and dip in London (as discussed, for example, in Malin and Bullard, 1981; Barraclough, 1982). For the early history of geomagnetism, including the works of Gilbert and Gauss, refer also to Mitchell (1932a, 1932b; 1937), Chapman (1963), Mattis (1965, Chap. 1), Parkinson (1982, Chap. 6), and Merrill and McElhinny (1983).

1600 — First modeling of the geomagnetic field by Gilbert's (1600) terrella (Malin, 1983).

1821 — Davy (1821) suggested the existence of Earth currents that, he argued, could be responsible for variations in the geomagnetic declination (Burbank, 1905).

1832 — Faraday (1832) envisaged for the first time the existence of induced currents in water, related to water flows and tides. He also attempted, without success, to detect, from the Waterloo Bridge, such currents flowing within the Thames. Gauss (1833) reported the first measurements, on May 21, 1832, of the absolute value of the geomagnetic field (Malin, 1982).

1846-1847 — Barlow (1849) made the first observations, in England, "on the spontaneous electric currents observed in the wires of the electric telegraph."

1848 — Matteucci detected induced currents in the telegraph wire between Florence and Pisa, while Highton observed the same effect in England (see section on Communication Cables).

1850 — Similar effects were reported in the United States.

1859 — A telegraph line in the United States was reported operated by means of the natural induced currents during geomagnetic disturbances on September 2.

1862 — Lamont (1862) reported one of the first experiments to specifically address Earth currents (carried out in the Munich Alps).

1865 — Experiment by Airy (1868) on two wires of 13 and 16 km from Greenwich.

1867 — Secchi (1867) reported measurements on two almost orthogonal telegraph lines of lengths 58 km (Rome-Arsoli) and 52 km (Rome-Anzio).

1881 — The Electrical Congress, meeting in Paris, recommended that certain short lines be set apart in each country for the study of Earth current phenomena and that longer lines be used as frequently as possible (Burbank, 1905).

1884-1887 — Four complete years of records on two telegraph wires in Germany (262 and 120 km) investigated by Weinstein (1902) and Steiner (1908).

1883-1884 — Blavier (1884) recorded, for 9 months, Earth potentials on five long telegraph lines extending from Paris, ranging in length from 200 to 390 km. See also Counil *et al.* (1983).

1886 — Shyda (1886) reported an Earth current study on the

1889   land line plus ocean cable route from Nagasaki, Japan, to Fusan, Korea.

1889   Schuster (1883, 1908) performed the first investigations on the diurnal variation of the geomagnetic field. He concluded that the origin is external, that the Earth must have an upper layer less conducting than that deep in the interior, and he proposed the "suggestive cause" of tidal motion in the atmosphere for the origin of the observed diurnal variation.

1892-   Two orthogonal Earth current lines, ~ 15 km each,
1985   were established at Saint-Maur-des-Fossés Observatory southwest of Paris (Moureaux, 1895, 1896; Bossler, 1912; Rougerie, 1940; Counil *et al.*, 1983).

1893   Moureaux (1893) found that the east-west Earth currents in the Paris basin were "exactly" correlated with the H-component of the geomagnetic field (i.e., the horizontal, north-south component), while this did not appear to be true for the north-south Earth current and the declination (east-west horizontal) geomagnetic field. This was the first reported detection of what is now interpreted in terms of telluric currents channeled east-west in the Seine basin from the Atlantic Ocean.

1905   Burbank (1905) provided a comprehensive bibliography on Earth currents.

1908   Van Bemmelen (1908) found that geomagnetic storm sudden commencements (ssc's) have opposite signs at Kew (close to London) and at St. Maur (close to Paris). He correctly explained this in terms of electric currents flowing in the English Channel.

1909   Schmidt (1909) investigated geomagnetic storms at Potsdam and at the Hilf Observatory (13 km south of Potsdam).

1912-   Van Bemmelen (1912, 1913) investigated the lunar
1913   period magnetic variation at 15 observatories.

1917-   Terada (1917) and Dechévren (1918a, 1918b) inves-
1918   tigated Earth currents in Japan and in England (Jersey), respectively.

1918   The British Admiralty succeeded for the first time to detect electro-magnetic disturbances related to seawater flows (Young *et al.*, 1920; figure reported in Chapman and Bartels, 1940).

1919   Chapman (1919) performed a systematic (and still quite valuable) analysis on the diurnal magnetic variation at 21 observatories, based on records collected in 1905.

1922   Bauer (1922) reviewed the status of Earth current studies.

Some historical points of interest in the past 60 years include the following:

1923   Chapman and Whitehead (1923) appear to have been the first investigators to be concerned with induction effects associated with the auroral electrojet (a localized current system). They erroneously concluded that geomagnetic storm effects at low latitudes are produced by Earth currents induced by the auroral electrojet.

1927-   Baird (1927) and Skey (1928) detected for the first
1928   time (at Watheroo in Australia and at Amberley and Christchurch in New Zealand, respectively) the intersection of what is now called the Parkinson plane (see, e.g., Gregori and Lanzerotti, 1980) with the DZ plane (i.e., the vertical, east-west oriented plane).

1930   Chapman and Price (1930) reconsidered the Chapman and Whitehead (1923) analysis and clearly stated that "the storm-time variations of the geomagnetic field in low latitudes cannot be due to currents, induced either the Earth or in a conducting layer of the atmosphere, by varying primary currents in the auroral zones."

1931   Cooperative project between the U.S. Coast and Geodetic Survey, the Carnegie Institution of Washington, and the American Telephone and Telegraph Company initiated at Tucson magnetic observatory to study Earth currents.

1936   Bossolasco (1936) detected for the first time (from measurements performed at Mogadiscio, Somalia, during the second International Polar Year, 1932-33) what is now called the Parkinson plane.

1949   De Wet (1949) attempted a numerical computation of the induction effects in oceans taking into account the coastal shapes.

1950   Ashour (1950) estimated the decay time of induced telluric currents within oceans. Constantinescu (1950) discovered what is now called the Parkinson plane and draw a plot, which is quite similar to a Wiese plot (see, e.g., Gregori and Lanzerotti, 1980).

1953   Rikitake and Yokoyama (1953) clearly stated the existence of the Parkinson plane. Banno (1953) detected for the first time the coast effect on Earth currents at Memambetsu (Hokkaido).

1954   Fleischer (1954a, 1954b, 1954c) hypothesized an east-west electric conductor 70 to 100 km deep beneath Bremen. Kertz (1954) stated that it cannot be lower than 80 km. Bartels (1957) estimated a depth of 50 to 100 km. Schmucker (1959) estimated a cylinder 63 km in radius, 100 km deep. Porstendorfer (1966) estimated high conductivity (0.2-0.5 mho/m) down to 10 km depth, an insulator (0.0001 mho/m) down to 100 km, a conductor (0.1 mho/m) between 100 and 130 km, an insulator (0.0001 mho/m) between 130 and 400 km, and 0.1 mho/m underneath. Vozoff and Swift (1968) reported a sedimentary layer (1.0 mho/m) 6 km deep in North Germany (8 sites from Braunschweig to Luebeck). The North German conductivity anomaly is now believed to be principally produced by surface-hydrated sedimentary layers that channel electric currents from the North Sea eastward to Poland. This is a classic example of how difficult the inversion (interpretation) problem is for geomagnetic measurements.

1955   Rikitake and Yokoyama (1955) appear to be the first authors to use the term "coast effect." In theoretically calculating a model of electromagnetic induction in a hemispherical ocean, they noted an enhanced magnetic field close to the coasts.

1958   Mansurov (1958) used the term "coastal effect" in analyzing geomagnetic measurements made at Mirny Station, Antarctica.

1959   Parkinson (1959, 1962a, 1962b, 1964), in a series of classic papers, analyzed in detail what is now called the Parkinson plane for geomagnetic measurements.

APPENDIX REFERENCES

Airy, G. B. (1868). Comparison of magnetic disturbances recorded by self-registering magnetometers at the Royal Observatory, Greenwich, with magnetic disturbances described from the corresponding terrestrial galvanic currents recorded by the self-registering galvanometers of the Royal Observatory, *Phil. Trans. R. Soc. 158A*, 465-472.

Ashour, A. A. (1950). The induction of electric currents in a uniform circular disk, *Q. J. Mech. Appl. Math. 3*, 119-128.

Baird, H. F. (1927). A preliminary investigation of some features of four magnetic storms recorded at seven magnetic observatories in the Pacific Ocean region during 1924, Thesis, Univ. of New Zealand.

Banno, N. (1953). On the Earth-current potentials at the Memambetsu magnetic observatory [in Japanese], *Mem. Kakioka Magn. Obs. 6*, 114-118.

Barlow, W. H. (1849). On the spontaneous electrical currents observed in the wires of the electric telegraph, *Phil. Trans. R. Soc. 61A*, 61-72.

Barraclough, D. R. (1982). Historical observations of the geomagnetic field, *Phil. Trans. R. Soc. A306*, 71-78.

Bartels, J. (1957). Erdmagnetische Tiefen-sondierungen, *Geol. Rdsch. 46*, 99-101.

Bauer, L. A. (1922). Some results of recent Earth-current observations and relations with solar activity, terrestrial magnetism and atmospheric electricity, *Terr. Magn. Atmos. Electr. 27*, 1-30.

Berdichevsky, M. N., and M. S. Zhdanov (1984). *Advanced Theory of Deep Geomagnetic Sounding*, Elsevier, 408 pp.

Blavier, E. E. (1884). Sur les courants telluriques, *C. R. Hebd. Séanc Acad. Sci. 98*, 1043-1045.

Bossler, J. (1912). Sur les relations des orages magnétiques et des phénomènes solaries, Thèse de specialité, Paris.

Bossolasco, M. (1936). Sur la nature des perturbations magnétiques, *C. R. Hebd. Séanc. Acad. Sci. 203*, 676-678.

Burbank, J. E. (1905). Earth-currents and a proposed method for their investigation, *Terr. Magn. Atmos. Electr. 10*, 23-49.

Chapman, S. (1919). The solar and lunar diurnal variations of terrestrial magnetism, *Phil. Trans. R. Soc. A218*, 1-118.

Chapman, S. (1963). Solar plasma, geomagnetism and aurora, in *Geophysics, The Earth's Environment*, C. DeWitt, J. Hieblot, and A. Lebaeu, eds., Gordon and Breach, pp. 371-502.

Chapman, S., and J. Bartels (1940). *Geomagnetism* (2 volumes), Clarendon Press, Oxford, 1049 pp.

Chapman, S., and A. Price (1930). The electric and magnetic state of the interior of the Earth as inferred form terrestrial magnetic variations, *Phil. Trans. R. Soc. A229*, 427-460.

Chapman, S., and T. T. Whitehead (1923). The influence of electrically conducting material within the Earth on various phenomena of terrestrial magnetism, *Trans. Camb. Phil. Soc. 22*, 463-482.

Constantinescu, L. (1950). Sudden commencements of magnetic storms in the years 1944-1949 [In Rumanian], *Lucr. Ses. Gen. Stiint. Acad. Rep. Pop. Rom.*

Counil, J. L., J. L. LeMouel, and M. Menvielle (1983). A study of the diurnal variation of the electromagnetic field in northern France using ancient recordings, *Geophys. J. R. Astron. Soc. 78*, 831-845.

Davy, H. (1821). On magnetic phenomena produced by electricity, *Phil. Trans. R. Soc.*, 7-19.

Dechévren, M. (1918a). Results of observations of Earth-currents made at Jersey, England, 1916-1917 (abstract) *Terr. Magn. Atmos. Electr. 23*, 37-39.

Dechévren, M. (1918b). Additional results of Earth-current observations at Jersey, England, *Terr. Magn. Atmos. Electr. 23*, 145-147.

De Wet, J. M. (1949). Numerical methods of solving electromagnetic induction problems, with applications of the theory of electric currents induced in the oceans, Thesis, Univ. of London.

Dosso, H. W., and J. T. Weaver, eds. (1983). Electromagnetic induction in the Earth and Moon: Invited review papers presented at the 6th Induction Workshop, *Geophys. Surv. 6*, 1-216.

Faraday, M. (1832). Experimental researches in electricity, *Phil. Trans. R. Soc.*, 163-194.

Fleischer, U. (1954a). Charakteristische erdmagnetische Baystörungen in Mitteleuropa and ihr innerer Anteil, *Z. Geophys. 20*, 120-136.

Fleischer, U. (1954b). Ein Erdstrom in tieferen untergrund Norddeutschlands waerend erdmagnetischer Baystörungen, *Naturwissenschaften 41*, 114-115.

Fleischer, U. (1954c). Ein Erdstrom in tiefereu untergrund Norddentschlands und sein anteil in dem magnetischen Baysörungen, Diss. Göttingen.

Gauss, C. F. (1833). Intensity of the terrestrial magnetic force, reduced to absolute units [in Latin], Göttingen.

Gilbert, W. (1600). *De Magnete* [in Latin], P. Short, London, English translation—1900, Chiswick Press, London.

Gregori, G. P., and L. J. Lanzerotti (1980). Geomagnetic depth sounding by induction arrow representation: A review, *Rev. Geophys. Space. Phys. 18*, 203-209.

Hermance, J. F. (1983). Electromagnetic induction studies, *Rev. Geophys. Space Phys. 21*, 652-665.

Kertz, W. (1954). Modelle für erdmagnetsche induzierte elektrische störm ein Untergrund, *Nachr. Akad. Wiss. Göttingen, Math. Phys. Kl. 2A*, 101-110.

Lamont, J. V. (1862). Die Erdstrom und der zusammenhang Desselben mit dem Magnetismus der Erde, Leopold-Voss-Verlag, Leipzig und Muenchen.

Malin, S. R. C. (1982). Sesquicentenary of Gauss' first measurement of the absolute value of magnetic intensity, *Phil. Trans. R. Soc. A306*, 5-8.

Malin, S. R. C. (1983). Modelling the geomagnetic field, *Geophys. J. R. Astron. Soc. 74*, 147-157.

Malin, S. R. C., and E. Bullard (1981). The direction of the Earth's magnetic field at London, 1570-1975, *Phil. Trans. R. Soc. 299*, 357-423.

Mansurov, S. M. (1958). Causes of local geomagnetic variation in the Mirny region, *Inf. Byull. Sov. Antarkt. Eksped. 2*, 37-41, 1958; English translation, in *Sov. Antarctic Expedition, Inf. Bull. 1*, 82-85.

Mattis, D. C. (1965). *The Theory of Magnetism, An Introduction to the Study of Cooperative Phenomena*, Harper and Row, New York, pp. 1-303.

Merrill, R. T., and M. W. McElhinny (1983). *The Earth's Magnetic Field—Its History, Origin and Planetary Perspective*, Academic Press, 401 pp.

Mitchell, C. A. (1932a). On the directive property of a magnet in the Earth's field and the origin of the nautical compass, *Terr. Magn. Atmos. Electr. 37*, 105-146.

Mitchell, C. A. (1932b). The discovery of magnetic declination, *Terr. Magn. Atmos. Electr. 37*, 105-146.

Mitchell, C. A. (1937). The discovery of the magnetic inclination, *Terr. Magn. Atmos. Electr. 42*, 241-280.

Moureaux, Th. (1893). *Soc. Met. France, Annu. 1*, 825-828.

Moureaux, Th. (1895). Notice sur l'installation d'une station d'étude des courants telluriques, *Soc. Met. France, Annu. 3*.

Moureaux, Th. (1896). L'installation d'une station d'étude des courants telluriques a l'Observatoire du Parc Saint Maur, *Soc. Met. France, Annu. 4*, 25-38.

Parkinson, W. D. (1959). Direction of rapid geomagnetic fluctuations, *Geophys. J. R. Astron. Soc. 2*, 1-14.

Parkinson, W. D. (1962a). The influence of continents and oceans on geomagnetic variations, *Geophys. J. R. Astron. Soc. 6*, 441-449.

Parkinson, W. D. (1962b). Magnetic variations over the oceans, in *Geomagnetica* (publicacao comemorativa de 50-mo aniversario do Observatorio Magnetico de S. Mignel, Azores—Servico Meteorologico Nacional, Lisbona), 97-108.

Parkinson, W. D. (1964). Conductivity anomalies in Australia und the ocean effect, *J. Geomagn. Geoelectr. 15*, 222-226.

Parkinson, W. D. (1982). *Introduction to Geomagnetism*, Elsevier, 433 pp.

Porstendorfer, G. (1966). Die raumlische verteilung tellurischer Ströme mit Perioden von 30 Min in Mitteleuropa, *Bergakademie 8*, 461-463.

Rikitake, T., and I. Yokoyama (1953). Anomalous relations between H and Z components of transient geomagnetic variations, *J. Geomagn. Geoelectr. 5*, 59-65.

Rikitake, T., and I. Yokoyama (1955). *Bull. Earthquake Res. Inst.* (Tokyo Univ.) *33*, 297.

Rokityansky, I. I. (1982). *Geoelectromagnetic Investigation of the Earth's Crust and Mantle*, Springer-Verlag, Berlin, 381 pp.

Rougerie, P. (1940). Contribution a l'étude des courants telluriques, Thèse de specialité, Paris.

Schmidt, A. (1909). Die magnetische störung am 25 September 1909 zu Potsdam und Seddin, *Met. Z. 26*, 509-511.

Schmucker, H. (1959). Erdmagnetische Tiefensondierung in Deutschland 1957-1959. Magnetogramme und erste Auswertung, *Abh. Akad. Wiss. Göttingen (Beitr. IGJ) 5*, 1-51.

Schuster, A. (1883). The diurnal variation of terrestrial magnetism, *Phil. Trans. R. Soc. A180*, 467-518.

Schuster, A. (1908). The diurnal variation of terrestrial magnetism, *Phil. Trans. R. Soc. A208*, 163-204.

Secchi, P. (1867). Earth currents on telegraph lines in Italy [in Italian], *C. R. Hebd. Séanc Acad. Sci. 58*, 1181.

Shyda, R. (1886). Earth-currents, *Trans. Seismol. Soc. Japan 9, Part I*.

Skey, H. F. (1928). Christchurch magnetic observatory—Report for the year 1926-27—Records of the survey of New Zealand 4, 42.

Steiner, L. (1908). On Earth-currents and magnetic variations, *Terr. Magn. Atmos. Elect. 13*, 57-62.

Terada, T. (1917). On rapid periodic variations of terrestrial magnetism, *J. Coll. Sci. Imp. Univ. Tokyo 37*, 56-84.

Van Bemmelen, W. (1908). The starting impulse of geomagnetic disturbances [in Dutch], *Versl. Gewone Vergad. Wis-en Natuurk, Afd. K. Akad. Wet.*, English edition, *Proc. Sec. Sci. K. Ned. Akad. Wet. 16*, 728-737.

Van Bemmelen, W. (1912). Die lunare variation des Erdmagnetismus, *Met. Z. 29*, 218-225.

Van Bemmelen, W. (1913). Berichtigung, *Met. Z. 30*, 589.

Vozoff, K., and C. M. Swift (1968). Magneto-telluric measurements in the North German basin, *Geophys. Prospect. 16*, 454-473.

Weinstein, B. (1902). Die Erdströme im Deutschen Reichstelegraphengebeit, und ihr Zusammenhamp mit den erdmagnetischen Erscheinungen, mit einen Atlas, Friedrich Vieweg und Shon, Braunschweig, 88 pages plus 19 plates, 1900. (A review by W. G. Cady is given in *Terr. Magn. Atmos. Electr. 7*, 149-152.

Young, F. B., H. Gerrard, and W. Jevons (1920). On electrical disturbances due to tides and waves, *Phil. Mag. 40*, 149-159.

Index